CAMBRIDGE MONOGRAPHS ON APPLIED AND COMPUTATIONAL MATHEMATICS

Series Editors
P. G. CIARLET, A. ISERLES, R. V. KOHN, M. H. WRIGHT

12   Radial Basis Functions:
Theory and Implementations

The *Cambridge Monographs on Applied and Computational Mathematics* reflects the crucial role of mathematical and computational techniques in contemporary science. The series publishes expositions on all aspects of applicable and numerical mathematics, with an emphasis on new developments in this fast-moving area of research.

State-of-the-art methods and algorithms as well as modern mathematical descriptions of physical and mechanical ideas are presented in a manner suited to graduate research students and professionals alike. Sound pedagogical presentation is a prerequisite. It is intended that books in the series will serve to inform a new generation of researchers.

Radial Basis Functions:
Theory and Implementations

M. D. BUHMANN
University of Giessen

CAMBRIDGE
UNIVERSITY PRESS

CAMBRIDGE UNIVERSITY PRESS
Cambridge, New York, Melbourne, Madrid, Cape Town, Singapore, São Paulo, Delhi

Cambridge University Press
The Edinburgh Building, Cambridge CB2 8RU, UK

Published in the United States of America by Cambridge University Press, New York

www.cambridge.org
Information on this title: www.cambridge.org/9780521633383

First published 2003
This digitally printed version 2008

A catalogue record for this publication is available from the British Library

Library of Congress Cataloguing in Publication data
Buhmann, M. D. (Martin Dietrich), 1963–
Radial basis functions : theory and implementations / Martin Buhmann.
p.   cm. – (Cambridge monographs on applied and computational mathematics; 12)
Includes bibliographical references and index.
ISBN 0 521 63338 9
1. Radial basis functions.   I. Title.   II. Series.
QA223 .B84   2003
511′.42 – dc21   2002034983

ISBN 978-0-521-63338-3 hardback
ISBN 978-0-521-10133-2 paperback

Für Victoria und Naomi.

Contents

Preface

The radial basis function method for multivariate approximation is one of the most often applied approaches in modern approximation theory when the task is to approximate scattered data in several dimensions. Its development has lasted for about 25 years now and has accelerated fast during the last 10 years or so. It is now in order to step back and summarise the basic results comprehensively, so as to make them accessible to general audiences of mathematicians, engineers and scientists alike.

This is the main purpose of this book which aims to have included all necessary material to give a complete introduction into the theory and applications of radial basis functions and also has several of the more recent results included. Therefore it should also be suitable as a reference book to more experienced approximation theorists, although no specialised knowledge of the field is required. A basic mathematical education, preferably with a slight slant towards analysis in multiple dimensions, and an interest in multivariate approximation methods will be suitable for reading and hopefully enjoying this book.

Any monograph of this type should be self-contained and motivated and need not much further advance explanations, and this one is no exception to this rule. Nonetheless we mention here that for illustration and motivation, we have included in this book several examples of practical applications of the methods at various stages, especially of course in the Introduction, to demonstrate how very useful this new method is and where it has already attracted attention in real life applications. Apart from such instances, the personal interests of the author mean that the text is dominated by theoretical analysis. Nonetheless, the importance of applications and practical methods is underlined by the aforementioned examples and by the chapter on implementations. Since the methods are usually applied in more than two or three dimensions, pictures will unfortunately not help us here very much which explains their absence.

After an introduction and a summary in Chapter 2 of the types and results of analysis that are used for the radial basis functions, the order of the remaining chapters essentially follows the history of the development: the convergence analysis was first completed in the setting of gridded data, after some initial and seminal papers by Duchon, and then further results on scattered data and their convergence orders were found; subsequently, radial basis functions on compact support were studied, then (and now) efficient implementations and finally wavelets using radial basis functions are the most topical themes.

Few can complete a piece of work of this kind without helping hands from various people. In my case, I would like to thank first and foremost my teacher Professor Michael Powell who introduced me into radial basis function research at Cambridge some 17 years ago and has been the most valuable teacher, friend and colleague to me ever since. Dr David Tranah of Cambridge University Press suggested once that I should write this book and Dr Alan Harvey as an editor kept me on the straight and narrow thereafter. Dr Oleg Davydov, Mr Simon Hubbert, Dr Ulrike Maier, Professor Tomas Sauer and Professor Robert Schaback looked at various parts of the manuscript and gave many helpful suggestions. Finally, I would like to thank Mrs Marianne Pfister of ETH who has most expertly typed an early version of the manuscript and thereby helped to start this project.

At the time of proofreading this book, the author learnt about the death of Professor Will Light of Leicester University. Will's totally unexpected death is an irreplaceable less to approximation theory and much of what is being said in this book would have been unthinkable without his many contributions to and insights into the mathematics of radial basis functions.

1

Introduction

In the present age, when computers are applied almost anywhere in science, engineering and, indeed, all around us in day-to-day life, it becomes more and more important to implement mathematical functions for efficient evaluation in computer programs. It is usually necessary for this purpose to use all kinds of 'approximations' of functions rather than their exact mathematical form. There are various reasons why this is so. A simple one is that in many instances it is not possible to implement the functions exactly, because, for instance, they are only represented by an infinite expansion. Furthermore, the function we want to use may not be completely known to us, or may be too expensive or demanding of computer time and memory to compute in advance, which is another typical, important reason why approximations are required. This is true even in the face of ever increasing speed and computer memory availability, given that additional memory and speed will always increase the demand of the users and the size of the problems which are to be solved. Finally, the data that define the function may have to be computed interactively or by a step-by-step approach which again makes it suitable to compute approximations. With those we can then pursue further computations, for instance, or further evaluations that are required by the user, or display data or functions on a screen. Such cases are absolutely standard in mathematical methods for modelling and analysing functions; in this context, analysis can mean, e.g., looking for their stationary points with standard optimisation codes such as quasi-Newton methods.

As we can see, the applications of general purpose methods for functional approximations are manifold and important. One such class of methods will be introduced and is the subject area of this book, and we are particularly interested when the functions to be approximated (the approximands)

(a) depend on many variables or parameters,
(b) are defined by possibly very many data,

1

(c) and the data are 'scattered' in their domain.

The 'radial basis function approach' is especially well suited for those cases.

1.1 Radial basis functions

Radial basis function methods are the means to approximate the multivariate functions we wish to study in this book. That is, in concrete terms, given data in n dimensions that consist of data sites $\xi \in \mathbb{R}^n$ and 'function values' $f_\xi = f(\xi) \in \mathbb{R}$ (or $\mathbb{C}$ but we usually take $\mathbb{R}$), we seek an approximant $s: \mathbb{R}^n \to \mathbb{R}$ to the function $f: \mathbb{R}^n \to \mathbb{R}$ from which the data are assumed to stem. Here $n > 0$ is the dimension of underlying space and, incidentally, one often speaks – not quite correctly – of 'data' when referring just to the ξ. They can also be restricted to a domain $D \subset \mathbb{R}^n$ and if this D is prescribed, one seeks an approximation $s: D \to \mathbb{R}$ only. In the general context described in the introduction to this chapter, we consider $f(\xi)$ as the explicit function values we know of our f, which itself is unknown or at least unavailable for arbitrarily large numbers of evaluations. It could represent magnetic potentials over the earth's surface or temperature measurements over an area or depth measurements over part of an ocean.

While the function f is usually not known in practice, for the purpose of (e.g. convergence) analysis, one has to postulate the existence of f, so that s and f can be compared and the quality of the approximation estimated. Moreover, some smoothness of f normally has to be required for the typical error estimates.

Now, given a linear space S of approximants, usually finite-dimensional, there are various ways to find approximants $s \in S$ to approximate the approximand (namely, the object of the approximation) f. In this book, the approximation will normally take place by way of interpolation, i.e. we explicitly require $s|_\Xi = f|_\Xi$, where $\Xi \subset \mathbb{R}^n$ is the discrete set of data sites we have mentioned above. Putting it another way, our goal is to interpolate the function between the data sites. It is desirable to be able to perform the interpolation – or indeed any approximation – without any further assumptions on the shape of Ξ, so that the data points can be 'scattered'. But sometimes we assume $\Xi = (h\mathbb{Z})^n$, h a positive step size, $\mathbb{Z}$ the integers, for example, in order that the properties of the approximation method can more easily be analysed. We call this type of data distribution a square (cardinal) grid of step size h. This is only a technique for analysis and means no restriction for application of the methods to scattered ξ. Interpolants probably being the most frequent choice of approximant, other choices are nonetheless possible and used in practice, and they can indeed be very desirable such as least squares approximations or 'quasi-interpolation', a variant of interpolation, where s still depends in a simple way on f_ξ, $\xi \in \Xi$, while not necessarily matching each f_ξ exactly. We will come back to this type of approximation at many places in this book. We remark that if we know how to approximate a function $f\colon \mathbb{R}^n \to \mathbb{R}$ we can always approximate a vector-valued approximand, call it $F\colon \mathbb{R}^n \to \mathbb{R}^m$, $m > 1$, componentwise.

From these general considerations, we now come back to our specific concepts for the subject area of this monograph, namely, for radial basis function approximations the approximants s are usually finite linear combinations of translates of a radially symmetric basis function, say $\phi(\| \cdot \|)$, where $\| \cdot \|$ is the Euclidean norm. Radial symmetry means that the value of the function only depends on the Euclidean distance of the argument from the origin, and any rotations thereof make no difference to the function value.

The translates are along the points $\xi \in \Xi$, whence we consider linear combinations of $\phi(\| \cdot -\xi \|)$. So the data sites enter already at two places here, namely as the points where we wish to match interpolant s and approximand f, and as the vectors by which we translate our radial basis function. Those are called the centres, and we observe that their choice makes the space S dependent on the set Ξ. There are good reasons for formulating the approximants in this fashion used in this monograph.

Indeed, it is a well-known fact that interpolation to arbitrary data in more than one dimension can easily become a singular problem unless the linear space S from which s stems depends on the set of points Ξ – or the Ξ have only very restricted shapes. For any fixed, centre-independent space, there are some data point distributions that cause singularity.

In fact, polynomial interpolation is the standard example where this problem occurs and we will explain that in detail in Chapter 3. This is why radial basis functions always define a space $S \subset C(\mathbb{R}^n)$ which depends on Ξ. The simplest example is, for a finite set of centres Ξ in $\mathbb{R}^n$,

$$(1.1) \qquad S = \left\{ \sum_{\xi \in \Xi} \lambda_\xi \| \cdot -\xi \| \ \middle| \ \lambda_\xi \in \mathbb{R} \right\}.$$

Here the 'radial basis function' is simply $\phi(r) = r$, the radial symmetry stemming from the Euclidean norm $\| \cdot \|$, and we are shifting this norm in (1.1) by the centres ξ.

More generally, radial basis function spaces are spanned by translates

$$\phi(\| \cdot -\xi \|), \qquad \xi \in \Xi,$$

where $\phi\colon \mathbb{R}_+ \to \mathbb{R}$ are given, continuous functions, called radial basis functions. Therefore the approximants have the general form

$$s(x) = \sum_{\xi \in \Xi} \lambda_\xi \phi(\| \cdot - \xi \|), \qquad x \in \mathbb{R}^n,$$

with real coefficients λ_ξ.

Other examples that we will encounter very often from now on are $\phi(r) = r^2 \log r$ ('thin-plate splines'), $\phi(r) = \sqrt{r^2 + c^2}$ (c a positive parameter, 'multiquadrics'), $\phi(r) = e^{-\alpha r^2}$ (α a positive parameter, 'Gaussian'). As the later analysis will show, radial symmetry is not the most important property that makes these functions such suitable choices for approximating smooth functions as they are, but rather their smoothness and certain properties of their Fourier transform. Nonetheless we bow to convention and speak of radial basis functions even when we occasionally consider general n-variate $\phi\colon \mathbb{R}^n \to \mathbb{R}$ and their translates $\phi(\cdot - \xi)$ for the purpose of approximation. And, at any rate, most of these basis functions that we encounter in theory and practice *are* radial. This is because it helps in applications to consider genuinely radial ones, as the composition with the Euclidean norm makes the approach technically in many respects a univariate one; we will see more of this especially in Chapter 4. Moreover, we shall at all places make a clear distinction between considering general n-variate $\phi\colon \mathbb{R}^n \to \mathbb{R}$ and radially symmetric $\phi(\| \cdot \|)$ and carefully state whether we use one or the other in the following chapters.

Unlike high degree spline approximation with scattered data in more than one dimension, and unlike the polynomial interpolation already mentioned, the interpolation problem from the space (1.1) is *always* uniquely solvable for sets of distinct data sites ξ, and this is also so for multiquadrics and Gaussians. For multivariate polynomial spline spaces on nongridded data it is up to now not even possible in general to find the exact dimension of the spline space! Thus we may very well be unable to interpolate uniquely from that spline space. Only several upper and lower bounds on the spatial dimension are available. There exist radial basis functions ϕ of compact support, where there *are* some restrictions so that the interpolation problem is nonsingular, but they are only simple bounds on the dimension n of $\mathbb{R}^n$ from where the data sites come. We will discuss those radial basis functions of compact support in Chapter 6 of this book.

Further remarkable properties of radial basis functions that render them highly efficient in practice are their easily adjustable smoothness and their powerful convergence properties. To demonstrate both, consider the ubiquitous multiquadric function which is infinitely often continuously differentiable for $c > 0$ and only continuous for $c = 0$, since in the latter case $\phi(r) = r$ and

$\phi(\| \cdot \|)$ is the Euclidean norm as considered in (1.1) which has a derivative discontinuity at zero. Other useful radial basis functions of any given smoothness are readily available, even of compact support, as we have just mentioned. Moreover, as will be seen in Chapter 4, on $\Xi = (h\mathbb{Z})^n$, e.g. an approximation rate of $O(h^{n+1})$ is obtained with multiquadrics to suitably smooth f. This is particularly remarkable because the convergence rate increases linearly with dimension, and, at any rate, it is very fast convergence indeed. Of course, the amount of work needed (e.g. the number of centres involved) for performing the approximation also increases at the same rate. Sometimes, even exponential convergence orders are possible with multiquadric interpolation and related radial basis functions.

1.2 Applications

Consequently, it is no longer a surprise that in many applications, radial basis functions have been shown to be most useful. Purposes and applications of such approximations and in particular of interpolation are manifold. As we have already remarked, there are many applications especially in the sciences and in mathematics. They include, for example, mappings of two- or three-dimensional images such as portraits or underwater sonar scans into other images for comparison. In this important application, interpolation comes into play because some special features of an image may have to be preserved while others need not be mapped exactly, thus enabling a comparison of some features that may differ while at the same time retaining others. Such so-called 'markers' can be, for example, certain points of the skeleton in an X-ray which has to be compared with another one of the same person, taken at another time. The same structure appears if we wish to compare sonar scans of a harbour at different times, the rocks being suitable as markers this time. Thin-plate splines turned out to be excellent for such very practical applications (Barrodale and Zala, 1999).

Measurements of potential or temperature on the earth's surface at 'scattered' meteorological stations or measurements on other multidimensional objects may give rise to interpolation problems that require the aforementioned scattered data. Multiquadric approximations are performing well for this type of use (Hardy, 1990).

Further, the so-called track data are data sites which are very close together on nearly parallel lines, such as can occur, e.g., in measurements of sea temperature with a boat that runs along lines parallel to the coast. So the step size of the measurements is very small along the lines, but the lines may have a distance of 100 times that step size or more. Many interpolation algorithms fail on such awkward distributions of data points, not so radial basis function (here multiquadric) methods (Carlson and Foley, 1992).

The approximation to so-called learning situations by neural networks usually leads to very high-dimensional interpolation problems with scattered data. Girosi (1992) mentions radial basis functions as a very suitable approach to this, partly because of their availability in arbitrary dimensions and of their smoothness.

A typical application is in fire detectors. An advanced type of fire detector has to look at several measured parameters such as colour, spectrum, intensity, movement of an observed object from which it must decide whether it is looking at a fire in the room or not. There is a learning procedure before the implementation of the device, where several prescribed situations (these are the data) are tested and the values zero (no fire) and one (fire) are interpolated, so that the device can 'learn' to interpolate between these standard situations for general situations later when it is used in real life.

In another learning application, the data come from the raster of a screen which shows the reading of a camera that serves as the eye of a robot. In this application, it is immediately clear why we have a high-dimensional problem, because each point on the square raster represents one parameter, which gives a million points even on a relatively low resolution of 1000 by 1000. The data come from showing objects to the robot which it should recognise as, for instance, a wall it should not run into, or a robot friend, or its human master or whatever. Each of these situations should be interpolated and from the interpolant the robot should then be able to recognise other, similar situations as well. Invariances such as those objects which should be recognised independently of angle etc. are also important in measurements of neural activity in the brain, where researchers aim to recognise those activities of the nerve cells that appear when someone is looking at an object and which are invariant of the angle under which the object is looked at. This is currently an important research area in neuro-physics (Eckhorn, 1999, Kremper, Schanze and Eckhorn 2002) where radial basis functions appear often in the associated physics literature. See the above paper by Eckhorn for a partial list.

The numerical solution of partial differential equations also enters into the long list of mathematical applications of radial basis function approximation. In the event, Pollandt (1997) used them to perform approximations needed in a multidimensional boundary element method to solve nonlinear elliptic PDEs on a domain Ω, such as $\Delta u_\ell = p_\ell(u(x), x)$, $x \in \Omega \subset \mathbb{R}^n$, $\ell = 1, \ldots, N$, with Dirichlet boundary conditions $u_\ell|_{\partial\Omega} = q_\ell$, when $u = (u_1, \ldots, u_N)^T$ are suitably smooth functions and p_ℓ are multivariate polynomials. Here, Δ denotes the Laplace operator. The advantage of radial basis functions in this application is the already mentioned convergence power, in tandem with easy formulation of interpolants and quasi-interpolants and their introduction into the PDE. Moreover, especially for boundary element methods, it is relevant that several radial basis functions are Green's functions of elliptic differential operators, i.e. the elliptic differential operator applied to them including the composition with the ubiquitous Euclidean norm yields the Dirac δ-operator.

The same reasons led Sonar (1996) to use radial basis functions for the local reconstruction of solutions within algorithms which solve numerically hyperbolic conservation laws. It was usual to employ low order polynomial approximation (mostly linear) for this purpose so far, but it turned out that radial basis functions, especially thin-plate splines, help to improve the accuracy of the finite volume methods notably to solve the hyperbolic equations, because of their ability to approximate locally ('recover' in the language of hyperbolic conservation laws) highly accurately.

They appear to be remarkably resilient against irregular data distributions, for not only track data but also those that occur, for instance, when local models are made for functions whose stationary points (or extrema) are sought (Powell, 1987). This is problematic because algorithms that seek such points will naturally accumulate data points densely near the stationary point, where now an approximation is made, based on those accumulated points, to continue with the approximant instead of the original function (which is expensive to evaluate). Furthermore, it turned out to be especially advantageous for their use that radial basis functions have a variation-diminishing property which is explained in Chapter 5. Thin-plate splines provide the most easily understood variant of that property and thus they were used for the first successful experiments with radial basis functions for optimisation algorithms. Not only do the variation-diminishing properties guarantee a certain smoothness of the approximants, but they are tremendously helpful for the analysis because many concepts of orthogonal projections and norm-minimising approximants can be used in the analysis. We shall do so often in this book.

In summary, our methods are known from practice to be good and general purpose approximation and interpolation techniques that can be used in many instances, where other methods are unlikely to deliver useful results or fail completely, due to singular interpolation matrices or too high dimensionality. The methods are being applied widely, and important theoretical results have been found that support the experimental and practical observations, many of which will enter into this book. Among them are the exceptional accuracy that can be obtained, when interpolating smooth functions.

Thus the purpose of this book is to demonstrate how well radial basis function techniques work and why, and to summarise and explain efficient implementations and applications. Moreover, the analysis presented in this work will allow a user to choose which radial basis functions to use for his application based on its individual merits. This is important because the theory itself, while mathematically stimulating, stands alone if it does not yield itself to support practical use.

1.3 Contents of the book

We now outline the contents and intents of the following chapters. The next chapter gives a brief summary of the schemes including precise mathematical details of some specific methods and aspects, so that the reader can get sufficient insight into the radial basis function approach and its mathematical underpinning to understand (i) the specifics of the methods and (ii) the kind of mathematical analysis typically needed. This may also be the point to decide whether this approach is suitable for his needs and interests and whether he or she wants to read on to get the full details in the rest of the book or, e.g., just wants to go directly to the chapter about implementations.

Chapter 3 puts radial basis functions in the necessary, more general context of multivariate interpolation and approximation methods, so that the reader can compare and see the 'environment' of the book. Especially splines, Shepard's method, and several other widely used (mostly interpolation) approaches will be reviewed briefly in that chapter. There is, however, little on practice and implementations in Chapter 3. It is really only a short summary and not comprehensive.

Chapter 4 introduces the reader to the very important special case of $\Xi = (h\mathbb{Z})^n, h > 0$, i.e. radial basis functions on regularly spaced (integer) grids. This was one of the first cases when their properties were explicitly and comprehensively analysed and documented, because the absence of boundaries and the periodicity of Ξ allow the application of powerful analysis tools such as Fourier transforms, Poisson summation formula etc. While the analysis is easier, it still gives much insight into the properties of the functions and the spaces generated by them, such as unique existence of interpolants, conditioning of interpolation matrices and exactness of approximations to polynomials of certain degrees, and, finally but most importantly, convergence theorems. Especially the latter will be highly relevant to later chapters of the book. Moreover, several of the results on gridded Ξ will be seen to carry through to scattered data, so that indeed the properties of the spaces generated by translates of radial basis functions were documented correctly. Many of the results are shown to be best possible, too, that is, they explicitly give the best possible convergence results.

The following chapter, 5, generalises the results of Chapter 4 to scattered data, confirming that surprisingly many *approximation order results* on gridded data carry through with almost no change, whereas, naturally, new and involved proofs are needed. One of the main differences is that there are usually finitely many data for this setting and, of course, there are boundaries of the domain wherein the $\xi \in \Xi$ reside, which have to be considered. It is usual that there are less striking convergence results in the presence of boundaries and this is what we will find there as well.

This Chapter 5, dealing with the many and deep theorems that have been established concerning the convergence rates of approximations, is the core of the book. This is because, aside from the existence and uniqueness theorems about interpolation, convergence of the methods is of utmost importance in applications. After all, the various rates of convergence that can be achieved are essential to the choice of a method and the interpretation of its results. Besides algebraic rates of convergence that are related to the polynomial exactness results already mentioned, the aforementioned spectral rates are discussed.

In Chapter 6, radial basis functions with compact support are constructed. They are useful especially when the number of data or evaluations of the interpolant is massive so that any basis functions of global support incur prohibitive costs for evaluation. Many of those radial basis functions are piecewise polynomials, and all of them have similar nonsingularity properties for the interpolation problem to the ones we have mentioned before. Moreover, radial basis functions with compact support are suitable for, and are now actually used in, solving linear partial differential equations by Galerkin methods. There, they provide a suitable replacement for the standard piecewise polynomial finite elements. It turns out that they can be just as good as means for approximation, while not requiring any triangulation or mesh, so they allow meshless approximation which is easier when the amount of data has to be continuously enlarged or made smaller. That is often the case when partial differential equations are solved numerically. By contrast, finite elements can be difficult to compute in three or more dimensions for scattered data due to the necessity of triangulating the domain before using finite elements and due to the complicated spaces of piecewise polynomials in more than two dimensions.

While many such powerful theoretical results exist for radial basis functions, the implementation of the methods is nontrivial and requires careful attention. Thus Chapter 7 describes several modern techniques that have been developed to implement the approach, evaluate and compute interpolants fast and efficiently, so that real-time rendering of surfaces that interpolate the data is possible now, for example. The methods we describe are iterative and they include so-called particle methods, efficient preconditioners and the Beatson–Faul–Goodsell–Powell (BFGP) algorithm of local Lagrange bases.

As outlined above, the principal application of radial basis functions is clearly with interpolation. This notwithstanding, least squares methods are frequently asked for, especially because often in applications, data are inaccurate, too many, and/or need smoothing. Hence Chapter 8 is devoted to least squares approaches both using the standard Euclidean least squares setting and with the so-called Sobolev inner products. Existence and convergence questions will be considered as well as, briefly, implementation.

Closely related to the least squares problem, whose solution is facilitated by computing orthogonal or orthonormal bases of the radial basis function spaces in advance, are 'wavelet expansions' by radial basis functions. In these important wavelet expansions, the goal is to decompose a given function simultaneously into its local parts in space *and* in frequency. The purpose of this can be the analysis, approximation, reconstruction, compression, or filtering of functions and signals. In comparison with the well-known Fourier analysis we can, e.g., tell from a wavelet expansion *when* a frequency appears in a melody, say, and not just that it appears and with what amplitude. This is what we call *localness*, not only in frequency for the wavelet expansions. Radial basis functions are bound to be useful for this because of their approximational efficacy. After all, the better we can approximate from a space, the fewer coefficients are needed in expansions of functions using bases of that space. All this is detailed, together with several examples of radial basis (especially multiquadric) wavelets, in Chapter 9.

Chapter 10 concerns the most recent and topical results in review form and an outlook and ideas towards further, future research. Many aspects of these tools are studied in research articles right now and in Chapter 10 we attempt to catch up with the newest work. Of course this can only be discussed very briefly. Several important questions are still wide open and we will outline some of those.

We conclude with an extensive bibliography, our principal aim being to provide a good account of the state of the art in radial basis function research. Of course not all aspects of current or past interest in radial basis functions can be covered within the scope of a book of this size but the aim is at least to provide up to date references to those areas that are not covered. We also give a commentary on the bibliography to point the reader to other interesting results that are not otherwise in this book on one hand, and to comment on generalisations, other points of view etc. on those results that are.

Now, in the following chapter, we give the already mentioned summary of some aspects of radial basis functions in detail in order to exemplify the others.

2

Summary of Methods and Applications

We have seen in the introduction what a radial basis function is and what the general purposes of multivariate interpolation are, including several examples. The aim of this chapter is more specifically oriented to the mathematical analysis of radial basis functions and their properties in examples.

That is, in this chapter, we will demonstrate in what way radial basis function interpolation works and give several detailed examples of its mathematical, i.e. approximation, properties. In large parts of this chapter, we will concentrate on one particular example of a radial basis function, namely the multiquadric function, but discuss this example in much detail. In fact, many of the very typical properties of radial basis functions are already contained in this example which is indeed a nontrivial one, and therefore quite representative. We deliberately accept the risk of being somewhat repetitive here because several of the multivariate general techniques especially of Chapter 4 are similar, albeit more involved, to the ones used now. What is perhaps most important to us in this chapter, among all current radial basis functions, the multiquadric is the best-known one and best understood, and very often used. One reason for this is its versatility due to an adjustable parameter c which may sometimes be used to improve accuracy or stability of approximations with multiquadric functions.

2.1 Invertibility of interpolation matrices

The goal is, as before, to provide interpolants to data $(\xi, f_\xi) \in \mathbb{R}^n \times \mathbb{R}$ which are arbitrary (but the ξ are distinct and there are, at this point, just finitely many of them). The data sites ξ are from the set Ξ, which is still a finite subset of $\mathbb{R}^n$ with more than one element, and our interpolants are required to have the form

$$s(x) = \sum_{\xi \in \Xi} \lambda_\xi \, \phi(\|x - \xi\|), \quad x \in \mathbb{R}^n.$$

11

The λ_ξ are real coefficients. They are required to satisfy the interpolation conditions $s|_\Xi = f|_\Xi$, that is $f(\xi) = s(\xi)$ for all ξ from Ξ, where we think of the f_ξ as $f(\xi)$ for an $f: \mathbb{R}^n \to \mathbb{R}$, as before, which usually needs to have a certain minimal smoothness. One of the central results of this chapter shows that the matrix that determines the λ_ξ using the f_ξ, namely the so-called interpolation matrix

$$(2.1) \qquad A = \{\phi(\|\zeta - \xi\|)\}_{\zeta,\xi \in \Xi},$$

is always nonsingular for the multiquadric function $\phi(r) = \sqrt{r^2 + c^2}$. Indeed, the entries λ_ξ of the vector $\boldsymbol{\lambda} = \{\lambda_\xi\}_{\xi \in \Xi}$ are found by premultiplying $\{f_\xi\}_{\xi \in \Xi}$ by A^{-1}, as the linear system we solve is

$$A\boldsymbol{\lambda} = \{f_\xi\}_{\xi \in \Xi} = \mathbf{f}.$$

In practical computations, however, one uses a solver for the above linear system and does not invert the matrix. We have much more on this in Chapter 7. As it turns out, for our approach the interpolation matrix is sometimes (for some radial basis functions, especially for the compactly supported ones) even positive definite, i.e. for all vectors $\boldsymbol{\lambda} = \{\lambda_\xi\}_{\xi \in \Xi}$ that are not identically zero, the quadratic form

$$\boldsymbol{\lambda}^T A \boldsymbol{\lambda}$$

is positive at $\boldsymbol{\lambda}$.

Hence using the fact that A is nonsingular or even positive definite, we can conclude that the λ_ξ and s exist uniquely, for all data and for all n. There are no restrictions for any of those parameters. After demonstrating this remarkable observation in detail, we shall describe some further important properties of this interpolation process.

We give a definition and several results that explain the nonsingularity properties of (2.1) for multiquadrics now.

The principal concept that serves to show nonsingularity of the interpolation matrix is complete monotonicity. We will define this concept and show its usefulness in the next few results. In its definition we use the standard notation C^∞ for the set of infinitely continuously differentiable functions on a set stated in parentheses thereafter. Of course the analogous notation C^ℓ, say, stands for ℓ times continuously differentiable functions.

Definition 2.1. *A function* $g \in C^\infty(\mathbb{R}_{>0})$ *is completely monotonic if and only if, for* $\ell = 0, 1, 2, \ldots,$ $(-1)^\ell g^{(\ell)}(t) \geq 0$ *for all positive* t.

The prototype of a completely monotonic function is the exponential function $g(t) = e^{-\alpha t}$ for any nonnegative α. This is a prototype because in some sense all completely monotonic functions are generated by integration of a measure with g as a kernel. Soon we will see in what way this is meant.

We will encounter many other simple examples of completely monotonic functions later, but we note at once that $g(t) = (t + c^2)^{-1/2}$ is an example for a continuous completely monotonic function for all c. Although this g is not the multiquadric function, it serves well to show the usefulness of the following proposition to which we will hark back very soon in connection with the actual multiquadric function. The following result was shown first by Schoenberg (1938), but he had quite different applications in mind from ours.

Proposition 2.1. *Let $g\colon \mathbb{R}_+ \to \mathbb{R}$ be a continuous completely monotonic function. Then, for all finite $\Xi \subset \mathbb{R}^n$ of distinct points and all n, the matrix $\mathbf{A}$ in (2.1) is positive definite for $\phi(r) = g(r^2)$, unless g is constant. In particular, the matrix $\mathbf{A}$ in (2.1) is nonsingular.*

As an example, we note that the above proposition immediately shows that the Gaussian kernel function $\phi(r) = e^{-r^2}$ gives rise to an invertible interpolation matrix $\mathbf{A}$. In fact, Schoenberg established a characterisation of positive semi-definite gs in his original theorem, but we only require the single implication stated in the proposition above.

The proof of Proposition 2.1 requires a lemma, which characterises completely monotonic functions. It is called the Bernstein–Widder representation, often just referred to as the Bernstein representation theorem (Widder, 1946).

Bernstein–Widder representation. *A function g is completely monotonic if and only if it is the Laplace transform*

$$g(t) = \int_0^\infty e^{-t\alpha}\, d\mu(\alpha), \quad t > 0,$$

of a nondecreasing measure μ that is bounded below, so that, in particular, $d\mu \geq 0$.

Trivial cases, i.e. constant completely monotonic functions, can be excluded by not letting μ be a point measure. Incidentally, the function g is also continuous at the origin if the measure remains finite.

Proof of Proposition 2.1: Let $\boldsymbol{\lambda} \in \mathbb{R}^\Xi \backslash \{0\}$, i.e. $\boldsymbol{\lambda}$ is a vector with components λ_ξ where ξ ranges over Ξ. Then,

$$(2.2) \qquad \boldsymbol{\lambda}^T \mathbf{A} \boldsymbol{\lambda} = \sum_{\zeta \in \Xi} \sum_{\xi \in \Xi} \lambda_\zeta \lambda_\xi\, g(\|\zeta - \xi\|^2),$$

where $\mathbf{A}$ still denotes the matrix (2.1).

This is, according to the Bernstein–Widder representation, by exchanging sums and integration (the sums are finite, so it is permitted to interchange integration and summation), the same as the quadratic form

$$\int_0^\infty \sum_{\zeta \in \Xi} \sum_{\xi \in \Xi} \lambda_\zeta \lambda_\xi \, e^{-\alpha \|\zeta - \xi\|^2} d\mu(\alpha),$$

now inside the integral. Here μ is not a point measure because g is not a constant, and moreover $d\mu \geq 0$ and $d\mu \neq 0$. This quadratic form and, *a fortiori*, (2.2) are positive, as required, by virtue of the positive definiteness of the matrix

$$\{e^{-\alpha \|\zeta - \xi\|^2}\}_{\xi, \zeta \in \Xi}$$

for all positive α.

Indeed, there are many straightforward ways to show positive definiteness of the matrix in the above display for all positive α; see, for example, Stein and Weiss (1971). We demonstrate a standard approach here. That is, one can use for instance the fact that the Fourier transform of $e^{-\alpha r^2}$ is

$$\pi^{n/2} \cdot e^{-r^2/(4\alpha)} \cdot \alpha^{-n/2}, \qquad r \geq 0,$$

to deduce for all nonzero vectors $\lambda \in \mathbb{R}^\Xi$ and positive α

$$\sum_{\xi, \zeta \in \Xi} \lambda_\zeta \lambda_\xi e^{-\alpha \|\zeta - \xi\|^2} = \frac{\pi^{n/2}}{(2\pi)^n} \int_{\mathbb{R}^n} \left| \sum_{\xi \in \Xi} \lambda_\xi e^{-i\xi \cdot y} \right|^2 e^{-\|y\|^2/(4\alpha)} \cdot \alpha^{-n/2} \, dy$$

which is positive, as required, because of the linear independence of exponential functions with distinct (imaginary) exponents. $\square$

In summary, we have by now made the observation that all **A** for the inverse multiquadric function $\phi(r) = (r^2 + c^2)^{-1/2}$ are positive definite, hence nonsingular. This is an application of the above proposition. It turns out to be a fairly simple consequence of this analysis that **A** is also nonsingular for $\phi(r) = \sqrt{r^2 + c^2}$.

Theorem 2.2. *Let $g \in C^\infty[0, \infty)$ be such that g' is completely monotonic but not constant. Suppose further that $g(0) \geq 0$. Then **A** is nonsingular for $\phi(r) = g(r^2)$.*

The result is due to Micchelli (1986). Before we embark on its proof, note that this theorem does apply to the multiquadric, namely $g(t) = \sqrt{t + c^2}$ and $\phi(r) = \sqrt{r^2 + c^2}$, and it gives the desired nonsingularity result.

Proof of Theorem 2.2: As $g \in C^\infty[0, \infty)$, we can rewrite it as

$$g(t) = g(0) + \int_0^t g'(x)dx.$$

We now insert the Bernstein–Widder representation of $g'(x)$ and exchange integrals. This is admissible because of Fubini's theorem. We get

$$g(t) = g(0) + \int_0^t \int_0^\infty e^{-\alpha x} \, d\mu(\alpha)\, dx.$$

Let $\boldsymbol{\lambda} \in \mathbb{R}^\Xi$ be such that its components sum to zero, i.e. $\sum_{\xi \in \Xi} \lambda_\xi = 0$. Thus, because of the Bernstein–Widder representation, used for $g'(x)$, we get

$$\boldsymbol{\lambda}^T \mathbf{A} \boldsymbol{\lambda} = -\int_0^\infty \sum_{\xi \in \Xi} \sum_{\zeta \in \Xi} \lambda_\zeta \lambda_\xi \, \alpha^{-1} e^{-\alpha \| \zeta - \xi \|^2} d\mu(\alpha).$$

We are using here that $\int_0^t e^{-\alpha x} \, dx = -\alpha^{-1} e^{-\alpha t} + \alpha^{-1}$ and that the sum over $\boldsymbol{\lambda}$'s components cancels the α^{-1} term which is constant with respect to x. Therefore $\boldsymbol{\lambda}^T \mathbf{A} \boldsymbol{\lambda} < 0$ for all such $\boldsymbol{\lambda}$ unless $\boldsymbol{\lambda} = \mathbf{0}$. Hence all but one of $\mathbf{A}$'s eigenvalues are negative. Otherwise, we could take γ and δ as two nonnegative ones and let z and t their associated orthonormal eigenvectors; they exist because $\mathbf{A}$ is symmetric. Thus there would be a nonzero vector $\boldsymbol{\lambda}$ whose components sum to zero and that has the representation $\boldsymbol{\lambda} = az + bt$. It fulfils

$$0 > \boldsymbol{\lambda}^T \mathbf{A} \boldsymbol{\lambda} = a^2\gamma + b^2\delta \geq 0,$$

and this is a contradiction. Thus, indeed, all but one of the matrix's eigenvalues are negative.

On the other hand, as $\mathbf{A}$'s trace is nonnegative, the remaining eigenvalue must be positive, since it is a well-known fact that the trace of the matrix is the sum of its eigenvalues. Hence $\det \mathbf{A} \neq 0$. Indeed, the sign of its determinant is $(-1)^{|\Xi|-1}$. □

Thus we have now established the important result that multiquadric interpolation is always nonsingular, i.e. uniquely possible. Note that this also applies to the special case $c = 0$, which is the case $\phi(r) = r$. The sign of the matrix determinant is always $(-1)^{|\Xi|-1}$, that is the same as multiquadrics also for this choice of the parameter c.

The fact that Euclidean norms are used here, incidentally, is of the essence. There are examples for ℓ^1 and ℓ^∞ norms and choices of Ξ, where $\{\| \zeta - \xi \|_p\}_{\zeta, \xi \in \Xi}$ is singular for $p = 1, \infty$. More precisely, the matrix is always nonsingular (for all n and Ξ, that is) for $p \in (1, 2]$. It always *can* be singular (for some n and/or Ξ) otherwise. The case $p = 1$ was studied in particular by Dyn, Light and

Cheney (1989) who give general theorems that characterise point constellations leading to singularity. The positive and negative results on p-norms, $p \in (1, 2]$ and $2 < p < \infty$, respectively, are due to Baxter (1991).

Many other radial basis functions exist for which either Proposition 2.1 or Theorem 2.2 applies: for example $\phi(r) = e^{-\alpha r^2}$, $\phi(r) = e^{-\alpha r}$, $\phi(r) = (r^2 + c^2)^{-1}$, and others. As has also been proved, the condition of complete monotonicity can be relaxed to λ-fold (finite or 'incomplete') monotonicity (Micchelli, 1986, Buhmann, 1989), which is closely related to certain radial basis functions of compact support. We will return to those later on in Chapter 6 and explain λ times monotonic functions with examples there.

2.2 Convergence analysis

After showing the unique existence of multiquadric interpolants, the next important question is that of their usefulness for approximation. Only if they turn out to be able to approximate smooth functions well (depending on step size of the centres and the actual smoothness of the approximand) will they be suitable for applications. This question is, within the scope of this chapter, best discussed for $n = 1$ and the infinite set of equally spaced centres $\Xi = h\mathbb{Z}$, where h is a positive step size. Note that this means that the approximand $f : \mathbb{R} \to \mathbb{R}$ must be defined everywhere and at least continuous on the real axis and that, in particular, no boundary conditions need be considered. In fact we give up the interpolation conditions altogether for the first part of the convergence analysis because the goodness of approximation can very well be discussed within the context of 'quasi-interpolation' which we introduce now as

$$s_h(x) = \sum_{j \in \mathbb{Z}} f(jh)\, \psi\left(\frac{x}{h} - j\right), \quad x \in \mathbb{R},$$

where f is the approximand and ψ is a finite linear combination of multiquadric functions

$$\psi(x) = \sum_{|k| \le N} \lambda_k\, \phi(|x - k|), \quad x \in \mathbb{R}.$$

It is important to notice that we are using here the so-called stationary case, where we are using a fixed ψ which is scaled by the reciprocal of h in its argument. It is also possible to study the nonstationary case, where ψ depends on h in another way and where $\psi = \psi_h$ is evaluated at $x - jh$ without scaling. We shall, however, use only stationary analysis in this chapter and address the nonstationary case briefly in Chapters 4 and 9.

The N is a positive integer in the expression for ψ. It is understood that s_h depends on h as well as on f. The function s_h need not satisfy any interpolatory properties but it should be such that $s_h \approx f$ is a good uniform approximation on the real line by virtue of properties of the function ψ. It is reasonable to address this form of approximants, as the quality of the approximation depends more on the *space* of approximants than on any particular choice of methods such as interpolation. The above form is especially helpful for our consideration because of its particularly simple form: f enters directly into the expression without any preprocessing and ψ is completely independent of f.

We want to explain this further and give an example. Note first that we indeed remain in the space spanned by translates of a radial basis function, in spite of the fact that we formulate the approximant as a linear combination of ψs. This is because we are using *finite* linear combinations. Thus, later on, we will be able to deduce properties of interpolation with multiquadrics on the equally spaced points from the analysis of the present situation.

The idea is to find λ_k such that ψ is local, e.g. by requiring the absolute sum

$$(2.3) \qquad \sum_{j \in \mathbb{Z}} |\psi(x - j)|$$

to be uniformly bounded for all $x \in \mathbb{R}$ and by demanding that the absolutely convergent series gives

$$\sum_{j \in \mathbb{Z}} \psi(x - j) \equiv 1,$$

so that $s = f$ at a minimum for constant f. Here, we abbreviate s_1 by s. Both conditions are eminently reasonable for approximation and in spite of their relative weakness they will provide good methods. However, all this should happen irrespective of the asymptotic linear growth of the multiquadric function! We will show now that this is possible and that furthermore $s = f$ for all linear f. This provides essentially second order convergence of s to f when f is smooth enough and $h \to 0$. Indeed, let ψ be a second divided difference of ϕ, i.e.

$$\psi(x) = \frac{1}{2}\phi(|x - 1|) - \phi(|x|) + \frac{1}{2}\phi(|x + 1|).$$

Then we can show that s is at least well-defined for at most linearly growing f and that in particular the boundedness condition that (2.3) be finite holds for all x. We let without loss of generality $c > 0$, because otherwise our quasi-interpolation is the usual piecewise interpolation and then the following statements are trivial, ψ being then the hat-function or equivalently the linear B-spline B_1, see Powell (1981) for example, for a comprehensive treatment of

B-splines. For us here, it suffices to remark that B_1 is the continuous piecewise linear function which is one at zero, zero at all nonvanishing integers and whose breakpoints are the integers.

The next proposition shows the boundedness of (2.3).

Proposition 2.3. *The above second divided difference ψ satisfies, for $|x|$ greater than 1,*

$$|\psi(x)| \leq Cc^2|x|^{-3},$$

where C is a generic positive constant, independent of x and c. We note in particular the trivial estimate $|\psi(x)| \leq C(1 + c^2)$ for all arguments.

We remark that this proposition shows that, although ϕ is an unbounded function, the linear combination ψ of ϕ's translates decays cubically as $x \rightarrow \pm\infty$. Uniform boundedness of (2.3) is thus a trivial consequence of absolute convergence of the series, because ψ is continuous and bounded anyway.

Proof of Proposition 2.3:   According to the Peano kernel theorem (Powell, 1981, p. 270)

$$\psi(x) = \frac{1}{2} \int_{-\infty}^{\infty} B_1(x - t)\, \phi''(t)dt,$$

where B_1 is the linear B-spline with knots at 0, ±1. Because $\phi''(t) = c^2(t^2 + c^2)^{-3/2}$, the proof now follows from the compact support of B_1, thus from the finiteness of the integral.   □

This proof will, incidentally, also apply to general second divided differences of ϕ with respect to *nonequidistant* ξ, as does the following result on linear polynomial reproduction.

Theorem 2.4. *The second divided difference ψ satisfies also the polynomial reproduction property*

$$\sum_{j \in \mathbb{Z}} (a + bj)\, \psi(x - j) = a + bx, \qquad x, a, b \in \mathbb{R}.$$

Note that Proposition 2.3 implies the series in Theorem 2.4 above converges uniformly and absolutely. Note also that Theorem 2.4 means, in particular, $s = f$ for constant and for linear approximands f.

Proof of Theorem 2.4:   As in the previous proof, we express ψ by the Peano kernel theorem and exchange integrals:

$$
\begin{aligned}
\sum_{j\in\mathbb{Z}} (a+bj)\,\psi(x-j) &= \frac{1}{2} \int_{-\infty}^{\infty} \sum_{j\in\mathbb{Z}} (a+bj)\,B_1(x-j-t)\,\phi''(t)dt \\
&= \frac{1}{2} \int_{-\infty}^{\infty} (a+b(x-t))\,\phi''(t)dt \\
&= (a+bx)\,\frac{1}{2} \int_{-\infty}^{\infty} \phi''(t)dt \\
&= a+bx,
\end{aligned}
$$

where we have used that sums over linear B-splines recover linear polynomials. We have also used that the integral over ϕ'' is 2 and the integral over $t\phi''(t)$ vanishes. Here, a and b are arbitrary reals. The proof is complete. $\square$

We observe that this result gives the polynomial recovery indicated above. This, in tandem with the localisation result Proposition 2.3 opens the door to a uniform convergence result by suitable Taylor series arguments when twice differentiable functions are approximated. Moreover, these two results above exemplify very well indeed how we are going to approach the convergence questions elsewhere in the book, namely almost always via polynomial recovery and locality estimates, that is asymptotic decay estimates. In most instances, the difficulty in the proofs for several dimensions lies much more in establishing the decay of the basis function, that is its locality, than in the polynomial recovery which is relatively standard and straightforward, once we know the principles behind polynomial exactness. More precisely, the same requirements on the aforementioned coefficients of ψ which lead to a suitable decay behaviour also imply polynomial exactness with no further work. The convergence estimates, however, need a more difficult analysis than the familiar ones from spline theory for instance, because compact support of the basis functions makes the proof techniques much simpler.

We point out once more that the above results are not just confined to integer data. Indeed, as Powell (1991) has shown, it suffices to have a strictly increasing sequence of centres ξ on the real axis that have $\pm\infty$ as upper and lower limits, respectively, to achieve the same results.

Convergence results are obtained in various guises. They always use the asymptotic decay at an algebraic rate of the basis function: not the radial basis functions but linear combinations of its translates; it is important to distinguish carefully between those. The basis function we use in the approximation is our ψ. They use this decay in tandem with polynomial recovery of a nontrivial order to show that smooth functions can be locally approximated by Taylor polynomials which are recovered by the approximant. An easy but quite representative convergence theorem is the following theorem. In it $\| \cdot \|_\infty$ denotes the uniform or Chebyshev norm on the whole axis $\mathbb{R}$, as is usual.

Theorem 2.5. *Let f be twice differentiable and such that $\|f'\|_\infty$ and $\|f''\|_\infty$ are finite. Then for any nonnegative c*

$$\|f - s_h\|_\infty = O(h^2 + c^2 h^2 |\log h|), \quad h \to 0.$$

Proof: Let $x \in \mathbb{R}$ be arbitrary. Let $p(y) := f(x) + (y - x) f'(x)$ be a local Taylor polynomial. Thus p is x-dependent, but recall that we fix x. We have therefore

$$|f(y) - p(y)| \le 2\|f'\|_\infty \|x - y\|_\infty$$

and

$$|f(y) - p(y)| \le \frac{1}{2}\|f''\|_\infty \|x - y\|_\infty^2.$$

Both estimates follow from Taylor expansions, with two and three terms, respectively, and with the respective remainder terms estimated by their maximum. We note that by the assumptions $|f'(x)|$ is bounded and f is therefore of at most linear growth. Thus, by Theorem 2.4 and the definition of the Taylor polynomial

$$|f(x) - s_h(x)| = \left| \sum_{j \in \mathbb{Z}} \left(f(x) + (jh - x)f'(x) - f(jh) \right) \psi \left(\frac{x - jh}{h} \right) \right|$$

$$= \left| \sum_{j \in \mathbb{Z}} \left(p(jh) - f(jh) \right) \psi \left(\frac{x - jh}{h} \right) \right|.$$

Using the bound in Proposition 2.3 several times, we get, for a generic (i.e. independent of x and h) positive constant C which may even change its associated value from line to line, the required estimates by dividing the sums up into three parts, as follows.

$$|f(x) - s_h(x)| \le \sum_{|x - jh| < 2h} |p(jh) - f(jh)| \left| \psi \left(\frac{x - jh}{h} \right) \right|$$

$$+ \sum_{2h \le |x - jh| \le 1} |p(jh) - f(jh)| \left| \psi \left(\frac{x - jh}{h} \right) \right|$$

$$+ \sum_{|x - jh| > 1} |p(jh) - f(jh)| \left| \psi \left(\frac{x - jh}{h} \right) \right|.$$

These are at most a fixed positive multiple of

$$\|f''\|_\infty h^2 \max_{|x-jh|<2h} \left| \psi\left(\frac{x-jh}{h}\right) \right|$$

$$+ C\|f''\|_\infty c^2 h^2 \sum_{2h \le |x-jh| \le 1} \frac{h}{|x-jh|}$$

$$+ C\|f'\|_\infty c^2 h^2 \sum_{|x-jh|>1} \frac{h}{|x-jh|^2}.$$

This is the same as a fixed multiple of

$$\|f''\|_\infty (h^2 + c^2 h^2) + C\|f''\|_\infty c^2 h^2 \int_h^1 y^{-1} dy$$

$$+ C\|f'\|_\infty c^2 h^2 \int_{1-h}^\infty y^{-2} dy.$$

We can summarise these expressions by an order term in c and h

$$O(h^2 + c^2 h^2 \,|\log h|),$$

thus finishing the proof. $\qquad\qquad\qquad\qquad\qquad\qquad\qquad\square$

We note that for the first derivatives also, a similar convergence statement can be made. Indeed, with the same assumptions as in Theorem 2.5 in place, the derivatives of f and s_h have the property that

$$\|f' - s_h'\|_\infty = O(h + c^2/h), \qquad h \to 0.$$

Of course, in order for the above to tend to zero, it is necessary that c tends to zero as well. With the above proof of Theorem 2.5 in mind, establishing this second estimate is routine work.

Thus we have now made a first important step towards a quantitative analysis of the radial basis function method, specifically about convergence: Theorem 2.5 gives, in particular, uniform, essentially quadratic convergence of approximants from multiquadrics to sufficiently smooth approximands which clearly shows the usefulness of multiquadric approximation. It is the most important positive result in this chapter. On the other hand, we may ask whether such constructions are always possible with the well-known examples of radial basis functions, such as all the ones mentioned already in this book, multiquadrics, inverse multiquadrics, thin-plate splines etc. It turns out that they are not. That is, there is a further result relevant in this context, but it is a negative one, namely,

Theorem 2.6. *Let ϕ be the inverse multiquadric $\phi(r) = (r^2 + c^2)^{-1/2}$. Then, for all finite linear combinations ψ of translates of ϕ, absolute integrability can*

only be achieved for a basis function ψ of zero means, that is

$$\int_{-\infty}^{\infty} |\psi(x)|dx < \infty \implies \int_{-\infty}^{\infty} \psi(x)dx = 0.$$

Before we embark on the proof, we note that such a result is useful for radial basis function approximations in the context we are considering; that is, the theorem gives us the interesting information that no satisfactory approximations can be obtained from linear combinations of translates of inverse multiquadrics. This is because all such linear combinations will have zero means, whence even approximations to simple data such as constants f will be bad unless we use approximations in $L^2(\mathbb{R})$ such as will be discussed in Chapter 9 of this book. Therefore Theorem 2.6 shows an instance where, surprisingly, linear combinations of the increasing multiquadric give much better approximations than the decaying inverse multiquadrics. This was so far unknown to many users of the radial basis function method, to whom the inverse multiquadric appeared falsely to be more useful because of its decay – which is albeit too slow to give any actual advantage. Later in this book, there will be many further such results that classify radial basis functions according to their ability to approximate.

Now to the proof of Theorem 2.6.

Proof of Theorem 2.6:   Let $\varepsilon > 0$ be arbitrary and N the largest modulus $|i|$ of a translate $\phi(\cdot - i)$ of ϕ incorporated in the definition of ψ. Let M be such that $|\phi(x)| \leq \varepsilon \ \forall |x| \geq M - N$ and

$$\int_{\mathbb{R}\setminus[-M,M]} |\psi(x)|dx < \varepsilon.$$

It is possible to find such M and ε because ψ is absolutely integrable, and because ϕ decays linearly for large argument.

For any $m \geq M$,

$$\left| \int_{-\infty}^{\infty} \psi(x)dx \right| \leq \left| \int_{-m}^{m} \psi(x)dx \right| + \varepsilon$$

$$= \left| \int_{-m}^{m} \sum_{|i|\leq N} \lambda_i \, \phi(x-i)dx \right| + \varepsilon.$$

This is at most

$$\left| \sum_{|i|\leq N} \lambda_i \int_{-m+i}^{m+i} \phi(x-i)dx \right| + \varepsilon + 2\varepsilon N \sum_{|i|\leq N} |\lambda_i|$$

$$\leq \left| \sum_{|i|\leq N} \lambda_i \right| \left| \int_{-m}^{m} \phi(x)dx \right| + \varepsilon \left(1 + 2N \sum_{|i|\leq N} |\lambda_i| \right).$$

Now, the first term must be zero which can be seen by changing $+\varepsilon$ to $-\varepsilon$ and reversing the inequality signs in the above display, recalling $\phi \geq 0$ and $\phi \notin L^1(\mathbb{R})$, and finally letting $m \to \infty$. Since ε was chosen arbitrarily, the theorem is proved. $\qquad\square$

2.3 Interpolation and convergence

2.3.1 Central results about existence

Our Theorem 2.5 in the last section, providing upper bounds for the error of best approximation in the aforementioned stationary case, gives insight into the approximational accuracy of multiquadric quasi-interpolation. We shall see later that very similar bounds hold for both gridded and scattered data sites alike, also in higher dimensions.

Furthermore, for interpolation, the results can sometimes be significantly improved, both in one and in more dimensions, albeit for very special classes of functions f only. Although already our quasi-interpolation results give highly relevant information about radial basis function spaces and their efficiency as spaces for approximants, we are still keenly interested in interpolation.

Indeed, it is not hard to show that these convergence results for equally spaced data (and f with certain differentiability properties like those in Theorem 2.5) hold for interpolation as well. Therefore we now look at multiquadric interpolation in one dimension and still with centres $\Xi = h\mathbb{Z}$. We claim that we can find a 'cardinal function' (also known as a Lagrange function) denoted by L which is here a linear combination of shifts of ψ, namely

$$L(x) = \sum_{j \in \mathbb{Z}} c_j \, \psi(x - j), \quad x \in \mathbb{R},$$

where ψ is still the same as before. The existence of Lagrange functions is fundamental in interpolation theory, because, if they exist, we know the generic interpolation problem is well-posed, the interpolant being expressible as a simple linear combination of Lagrange functions multiplied by the function values. Furthermore, it is much simpler to work with ψ instead of ϕ in both theory and practice because of ψ's decay properties, which are in stark contrast with the unboundedness of most radial basis functions.

As a cardinal function, L is required to satisfy the 'cardinality conditions'

$$L(j) = \delta_{0j}, \qquad j \in \mathbb{Z}.$$

Here δ denotes the Kronecker δ, i.e. $\delta_{0j} = \delta(-j)$ with $\delta(0) = 1$, $\delta|_{\mathbb{Z}\setminus\{0\}} \equiv 0$. If that is so, we can build interpolants extremely easily in Lagrange form even on scaled lattices as

$$s_h(x) = \sum_{j\in\mathbb{Z}} f(jh)\, L\!\left(\frac{x}{h} - j\right), \quad x \in \mathbb{R},$$

which fulfil the interpolation conditions $s|_{h\mathbb{Z}} = f|_{h\mathbb{Z}}$. We are still using the stationary setting as before and want to study the error $f - s$ as $h \to 0$ in this set-up. For this, it is helpful to know more about L, and, indeed, we have not even proved L's existence yet. (Note that we are dealing with infinitely many data at present, so that our earlier results do not apply. This fact is related especially with the question of convergence of the infinite series that occur in the Lagrange formulation of interpolants.) To this end, consider the equation that has to be fulfiled by L's coefficients for every ℓ, namely

$$\sum_{j\in\mathbb{Z}} c_j\, \psi(\ell - j) = \delta_{0\ell}, \quad \ell \in \mathbb{Z}.$$

At first purely formally, ignoring the question of convergence of the following series, we can form the so-called discrete Fourier transform of these conditions by multiplying by exponentials to the ℓth imaginary power and summing over ℓ. By recalling that such a sum over exponentials, multiplied by convolutions as above, can be decomposed into a product, we get the product of Fourier series

$$\sum_{j\in\mathbb{Z}} c_j\, e^{-ij\vartheta} \sum_{\ell\in\mathbb{Z}} \psi(\ell)e^{-i\ell\vartheta} = 1, \quad \vartheta \in \mathbb{T},$$

through use of the Cauchy formula for products of infinite sums. This is equivalent to the previous display, denoting the convolution of the Fourier coefficients of the two Fourier series of the last display. Here $\mathbb{T} = [-\pi, \pi]$.

Hence, according to a theorem of Wiener, which is traditionally called Wiener's lemma (Wiener, 1933, Rudin, 1991), the desired $\{c_j\}_{j\in\mathbb{Z}}$ exist *uniquely* as an absolutely summable sequence if and only if the so-called symbol

$$\sigma(\vartheta) = \sum_{\ell\in\mathbb{Z}} \psi(\ell)\, e^{-i\ell\vartheta}, \quad \vartheta \in \mathbb{T},$$

which is an infinite, absolutely convergent series, has no zero. For later reference, we state the lemma already for several unknowns but restrict its application in this section to one variable only.

Wiener's lemma. *If the Fourier series*

$$\sum_{j\in\mathbb{Z}^n} d_j e^{-i\vartheta \cdot j}, \quad \vartheta \in \mathbb{T}^n,$$

is absolutely convergent and has no zero, its reciprocal can also be expanded in an absolutely convergent Fourier series with coefficients c_j:

$$\frac{1}{\sum_{j\in\mathbb{Z}^n} d_j e^{-i\vartheta\cdot j}} = \sum_{j\in\mathbb{Z}^n} c_j e^{-i\vartheta\cdot j}, \qquad \vartheta \in \mathbb{T}^n.$$

The statement of Wiener's lemma can also be phrased as a statement about the ring of absolutely convergent Fourier series, namely, the ring of absolutely convergent Fourier series is an algebra, the so-called Wiener algebra.

Now, in this event, the c_j of the above are the Fourier coefficients of the 2π-periodic continuous reciprocal $\sigma(\vartheta)^{-1}$ and we have to look for its positivity. Because ψ is such that $|\psi(\ell)|$ decays cubically, the series above converges absolutely and we may apply the Poisson summation formula (Stein and Weiss, 1971) to obtain the alternative formulation for the 'symbol' σ that will show us that the symbol has no zero and will be very important also later in this book. Following is the pertinent result, and we recall that the Fourier transform of an integrable function f is defined by

$$\hat{f}(x) = \int_{\mathbb{R}^n} e^{-ix\cdot t} f(t)\,dt, \qquad x \in \mathbb{R}^n,$$

cf. the Appendix. For the same reasons as above we define this and state the following result already in several dimensions; its usage in this chapter is only for $n = 1$.

Poisson summation formula. *Let $s \in L^1(\mathbb{R}^n)$ be such that its Fourier transform $\hat{s}$ is also absolutely integrable. Then we have the equality*

$$\sum_{j\in\mathbb{Z}^n} s(j)e^{-i\vartheta\cdot j} = \sum_{\ell\in\mathbb{Z}^n} \hat{s}(\vartheta + 2\pi\ell), \qquad \vartheta \in \mathbb{T}^n,$$

the convergence of the sums being in $L^1(\mathbb{T}^n)$. If s satisfies the two estimates $|s(x)| = O((1 + \|x\|)^{-n-\varepsilon})$ and $|\hat{s}(x)| = O((1 + \|x\|)^{-n-\varepsilon})$ for positive ε, then the two sums above are absolutely convergent and their limiting functions continuous. Therefore the above identity holds pointwise.

It follows for our purposes that the symbol is

$$\sigma(\vartheta) = \sum_{\ell\in\mathbb{Z}} \hat{\psi}(\vartheta + 2\pi\ell), \qquad \vartheta \in \mathbb{T}.$$

So $\hat{\psi} \geq 0$ or $\hat{\psi} \leq 0$ everywhere is sufficient for $\sigma(\vartheta) \neq 0$ for all ϑ, as long as $\hat{\psi}$ has no 2π-periodic zero. We will now check this condition as well as the absolute convergence of the series of the symbol in the new formulation. We commence with the latter which is the more difficult issue. The other condition will come out as a by-product.

To this end we note that $\hat{\psi}$ decays even exponentially, so that the series in the last display converges absolutely. Indeed, this can be verified by computing the Fourier transform of ψ explicitly. It is composed of the distributional Fourier transform of ϕ times a trigonometric polynomial coming from taking the second divided difference: that is,

$$\frac{1}{2}\,\phi(|x-1|) - \phi(|x|) + \frac{1}{2}\,\phi(|x+1|)$$

gives through Fourier transformation

$$\hat{\psi}(x) = \hat{\phi}(|x|)\left(\frac{1}{2}\,e^{-ix} - 1 + \frac{1}{2}\,e^{ix}\right).$$

The distributional or generalised Fourier transforms will be discussed in more detail in Section 4.1. Here it is only relevant that for our choice of ψ above,

$$\hat{\psi}(\vartheta) = (\cos\vartheta - 1)\,\hat{\phi}(|\vartheta|),$$

where $\hat{\phi}(|\vartheta|)$ is $-(2c/|\vartheta|)K_1(c|\vartheta|)$, the distributional Fourier transform in one dimension of

$$\phi(|x|) = \sqrt{x^2 + c^2}, \qquad x \in \mathbb{R}.$$

This K_1 is a modified Bessel function; the Fourier transform $\hat{\phi}$ is found in Jones (1982). The $\hat{\psi}$ satisfies our condition of nonnegativity because $K_1(z) > 0$ and $K_1(z) \sim z^{-1}$, $z \to 0_+$ (Abramowitz and Stegun, 1972). Moreover, it shows that $\hat{\psi}$ and $\hat{\phi}$ decay exponentially, because $K_1(z)$ does for growing argument $z \to \infty$. Finally, there is no 2π-periodic zero; in particular $\hat{\psi}(0) = \lim_{\vartheta \to 0} \hat{\psi}(\vartheta) \neq 0$.

Later on when we work in n dimensions instead of one, we use that the Fourier transform $\hat{\phi}$ for the multiquadric function is a constant multiple of $(c/|\vartheta|)^{-(n+1)/2} K_{(n+1)/2}(c|\vartheta|)$. We will also explain distributional Fourier transforms in more detail in the fourth chapter.

2.3.2 Properties of the Lagrange function

Hence we have shown the unique existence of a bounded cardinal function, where the uniqueness means 'the only one with absolutely summable coefficients' in the sense of Wiener's lemma. We claim that L does in fact decay cubically for large argument just as ψ does, and we will prove in Chapter 4, in a more general context, that its multidimensional analogue even decays for general n like $\|x\|^{-2n-1}$ as $\|x\| \to \infty$. For now, however, the following result suffices.

Theorem 2.7. *Let ϕ be the multiquadric function and ψ be the second divided difference mentioned above. Then there is a unique absolutely summable set of coefficients $\{c_j\}_{j \in \mathbb{Z}}$ such that L satisfies the cardinality conditions*

$$L(j) = \delta_{0j}, \qquad j \in \mathbb{Z},$$

and is bounded. Moreover, $|c_j| = O(|j|^{-3})$, and so in particular L decays cubically at infinity because ψ does as well.

Proof: The first part of the theorem has already been shown. We only have to show the cubic decay of the cardinal function's coefficients. That this implies that $|L(x)|$ also decays cubically is an easy consequence of the convolution form of L because ψ decays cubically as well (e.g. from Lemma 4.14 of Light and Cheney, 1992a). Indeed, it is straightforward to show that a convolution in one variable of two cubically decaying functions decays cubically.

Further, the coefficients are of the form of a discrete inverse Fourier transform as we have noted before, namely

$$c_j = \frac{1}{2\pi} \int_{-\pi}^{\pi} \frac{e^{ij\vartheta}}{\sigma(\vartheta)} \, d\vartheta.$$

This is well-defined because σ has no zero as we have seen already by using the Poisson summation formula. It is straightforward to verify that the c_j provide the desired cardinal function L once we have established their asymptotic decay; that follows from the fact that we can expand the reciprocal of the symbol in an absolutely convergent Fourier series whose coefficients are our desired coefficients. Assuming for the moment that we have already established their decay, we get by Cauchy's formula for the multiplication of infinite absolutely convergent series

$$\sum_{j, \ell \in \mathbb{Z}} c_j \psi(\ell - j) e^{-i\ell\vartheta} = \left(\sum_{\ell \in \mathbb{Z}} c_\ell e^{-i\ell\vartheta} \right) \left(\sum_{j \in \mathbb{Z}} \psi(j) e^{-ij\vartheta} \right) = \frac{\sigma(\vartheta)}{\sigma(\vartheta)} = 1,$$

as required. We have used here in particular that the c_ℓ are the Fourier coefficients of the reciprocal of the symbol and therefore the Fourier series with those coefficients reproduces $\sigma(\vartheta)^{-1}$. Thus we now only consider the coefficients' decay.

In order to establish the result, it suffices to prove that $\sigma(\vartheta)^{-1}$ is three times differentiable except perhaps at zero or indeed at any finite number of points, while all those derivatives are still integrable over $\mathbb{T}$. Then we can apply integration by parts to our above integral to show that $|c_j| \cdot |j|^3$ is uniformly bounded through the absolute integrability of the third derivative of the reciprocal of the symbol. For that, we integrate the exponential function and differentiate the reciprocal of the symbol:

$$\int_{-\pi}^{\pi} \frac{e^{ij\vartheta}}{\sigma(\vartheta)}\, d\vartheta = \frac{-1}{ji} \int_{-\pi}^{\pi} e^{ij\vartheta} \frac{d}{d\vartheta} \frac{1}{\sigma(\vartheta)}\, d\vartheta.$$

There are no boundary terms because of the periodicity of the integrand. Each time this integration by parts gives a factor of $\frac{-1}{ij}$ in front of the integral; performing it three times gives the desired result so long as the remaining integrand is still absolutely integrable. To this end, the symbol's reciprocal is further differentiated while the integration of the exponential function provides the required powers of j. Moreover, because $K_1 \in C^\infty(\mathbb{R}_{>0})$, and therefore the same is true for the whole Fourier transform $\hat{\phi}$, we only have to prove the integrability assertions in a neighbourhood of the origin.

Indeed, near zero, setting without loss of generality $c = 1$, and letting

$$\hat{c}_1,\, \hat{c}_2,\, \hat{c}_3,\, \hat{c}_4$$

be suitable real constants, we have the following short expansion of the reciprocal of $\sigma(\vartheta)$, where the expression $\sum'$ in the display denotes $\sum_{\ell \in \mathbb{Z}\setminus\{0\}}$, and where we have used the expression for $\hat{\psi}$ derived above:

$$\sigma(\vartheta)^{-1} = (1 - \cos\vartheta)^{-1} \left\{ \hat{\phi}(|\vartheta|) + \sum{}' \hat{\phi}(|\vartheta + 2\pi\ell|) \right\}^{-1}$$

$$= \left\{ \frac{1}{2}\vartheta^2 - \frac{1}{24}\vartheta^4 + O(\vartheta^6) \right\}^{-1}$$

$$\times \left\{ \hat{c}_1\, \vartheta^{-2} + \hat{c}_2 \log\vartheta + \hat{c}_3 + \hat{c}_4\vartheta + O(\vartheta^2 \log\vartheta) \right\}^{-1}, \quad \vartheta \to 0_+.$$

This is because we have in our case the particularly simple form

$$\hat{\phi}(r) = -(2/r)\, K_1(r), \qquad r > 0,$$

and because of the expansion of K_1 to be found in Abramowitz and Stegun (1972, p. 375). It is as follows:

$$K_1(z) = \frac{1}{z} + \frac{1}{2}z \log\left(\frac{1}{2}z\right) - \frac{1}{4}(1 - 2\gamma)z + O(z^3 \log z), \qquad z \to 0_+,$$

where γ denotes Euler's constant. The sum $\sum' \hat{\phi}(|\vartheta + 2\pi\ell|)$ is infinitely differentiable near zero. This reciprocal of the symbol given for the radial basis function in question is therefore

$$\sigma(\vartheta)^{-1} = \tilde{c}_1 + \tilde{c}_2\vartheta^2 \log\vartheta + \tilde{c}_3\vartheta^2 + \tilde{c}_4\, \vartheta^3 + O(\vartheta^4 \log\vartheta), \quad \vartheta \to 0_+.$$

Here, again, $\tilde{c}_1$, $\tilde{c}_2$, $\tilde{c}_3$, $\tilde{c}_4$ are suitable nonzero constants, whose particular values are unimportant to us.

The first four terms give no or an $O(|j|^{-3})$ contribution to the inverse generalised Fourier transform of $\sigma(\vartheta)^{-1}$, because the inverse Fourier transform of $\vartheta^2 \log \vartheta$ in one dimension is $O(\vartheta^{-3})$ and the Fourier transform of a polynomial is a linear combination of derivatives of the δ-distribution (Jones, 1982), which gives no contribution to the decay at infinity.

The remaining terms in the above short asymptotic expansion are all at least three times continuously differentiable, as is $\sigma(\vartheta)^{-1}$ everywhere else other than at zero. Therefore the c_j, which are computed by the inverse transform of the 2π-periodic function $\sigma(\vartheta)^{-1}$, are composed of terms which all decay at least as fast as a multiple of $|j|^{-3}$. Hence the theorem is true. $\qquad\square$

We note that a convergence theorem such as the one we have derived above for quasi-interpolation follows immediately for cardinal interpolation too. This is because, as we have just seen, L decays at least as fast as ψ does, and cardinal interpolation recovers linear polynomials as well, because it is a projection onto the space spanned by translates of ψ, by the uniqueness of interpolation. Therefore cardinal interpolation recovers all polynomials reproduced by quasi-interpolation, namely linear ones. The convergence proof, however, has made no use of any further properties of ψ.

All of this work on interpolation will be generalised considerably in Chapter 4. Most notably, it will apply to all n and to much larger classes of radial basis functions. We will also show strong decay results that lead to high approximation orders. Specifically, multiquadric interpolation on h-scaled integer grids in n dimensions incurs approximation errors of at most $O(h^{n+1})$ if f is sufficiently smooth. (Interestingly enough, the polynomial recovery properties are the same for interpolation and quasi-interpolation, so, e.g., linear polynomial recovery is the best we can do in one dimension with multiquadrics. This says also that it is not necessary to perform all the work which interpolation requires, as quasi-interpolation will do from the viewpoint of asymptotic error analysis.) Nonetheless, the work demonstrated so far gives insight into the achievable results and the proof techniques. We now give some concrete examples for mathematical applications.

2.4 Applications to PDEs

Perhaps the most important concrete example of applications is the use of radial basis functions for solving partial differential equations. These methods

are particularly interesting when nonlinear partial differential equations are
solved and/or nongrid approaches are used, e.g. because of nonsmooth domain boundaries, where nonuniform knot placement is important to modelling the solution to good accuracy. This is no contradiction to our analysis
above, where equal spacing was chosen merely for the purpose of theoretical analysis. As we shall see in the next chapter, there are, for more than
two or three-dimensions, not many alternative methods that allow nongrid
approaches.

Two ways to approach the numerical solution of elliptic boundary value problems are by collocation and by the dual reciprocity method. We begin with a
description of the collocation approach. This involves an important decision
whether to use the well-known, standard globally supported radial basis functions such as multiquadrics or the new compactly supported ones which are
described in Chapter 6 of this book. Since the approximation properties of the
latter are not as good as the former ones, unless multilevel methods (Floater
and Iske, 1996) are used, we have a trade-off between accuracy on one hand
and sparsity of the collocation matrix on the other hand. Compactly supported
ones give, if scaled suitably, banded collocation matrices while the globally
supported ones give dense matrices. When we use the compactly supported
radial basis functions we have, in fact, another trade-off, because even their
scaling pits accuracy against population of the matrix. We will come back to
those important questions later in the book.

One typical partial differential equation problem suitable for collocation
techniques reads

$$Lu(x) = f(x), \quad x \in \Omega \subset \mathbb{R}^n,$$
$$Bu|_{\partial\Omega} = q,$$

where Ω is a domain with suitably smooth, e.g. Lipschitz-continuous, boundary
$\partial\Omega$ and f, q are prescribed functions. The L is a linear differential operator and
B a boundary operator acting on functions defined on $\partial\Omega$. Often, B is just point
evaluation (this gives rise to the so-called Dirichlet problem) on the boundary or
taking normal derivatives (for Neumann problems). We will come to nonlinear
examples soon in the context of boundary element techniques.

For centres Ξ that are partitioned into two disjoint sets Ξ_1 and Ξ_2, the former
from the domain, the latter from its boundary, the usual approach to collocation
is to solve the so-called Hermite interpolation system

$$\Lambda_\xi u_h = f(\xi), \quad \xi \in \Xi_1,$$
$$\Lambda_\zeta u_h = q(\zeta), \quad \zeta \in \Xi_2,$$

which involves both derivatives of different degrees and function evaluations. The approximants u_h are defined by the sums

$$u_h(x) = \sum_{\xi \in \Xi_1} c_\xi \Lambda_\xi \phi(\|x - \xi\|) + \sum_{\zeta \in \Xi_2} d_\zeta \Lambda_\zeta \phi(\|x - \zeta\|).$$

The Λ_ξ and Λ_ζ are suitable functionals to describe our operators L and B on the discrete set of centres. This is usually done by discretisation, i.e. replacing derivatives by differences.

Thus we end up with a square symmetric system of linear equations whose collocation matrix is nonsingular if, for instance, the radial basis function is positive definite and the aforementioned linear functionals are linearly independent functionals in the dual space of the native space of the radial basis functions (see Chapter 5 for the details about 'native spaces' which is another name, commonly used in the literature, for the reproducing kernel semi-Hilbert spaces treated there).

An error estimate is given in Franke and Schaback (1998). For those error estimates, it has been noted that more smoothness of the radial basis function is required than for a comparable finite element setting, but clearly, the radial basis function setting has the distinct advantage of availability in any dimension and the absence of grids or triangulations which take much time to compute.

If a compactly supported radial basis function is used, it is possible to scale so that the matrix is a multiple of the identity matrix, but then the approximation quality will necessarily be bad. In fact, the *conditioning* of the collocation matrix is also affected which becomes worse the smaller the scaling η is with $\phi(\cdot/\eta)$ being used as scaled radial basis function. A Jacobi preconditioning by the diagonal values helps here, so the matrix $\mathbf{A}$ is replaced by $\mathbf{P}^{-1}\mathbf{A}\mathbf{P}^{-1}$ where $\mathbf{P} = \sqrt{\text{diag}(\mathbf{A})}$ (Fasshauer, 1999).

We now outline the second method, that is a boundary element method (BEM). The dual reciprocity method as in Pollandt (1997) uses the second Green formula and a fundamental solution $\phi(\| \cdot \|)$ of the Laplace operator

$$\Delta = \frac{\partial^2}{\partial x_1^2} + \cdots + \frac{\partial^2}{\partial x_n^2}$$

to reformulate a boundary value problem as a boundary integral problem over a space of one dimension lower. No sparsity occurs in the linear systems that are solved when BEM are used, but this we are used to when applying noncompactly supported radial basis functions (see Chapter 7).

The radial basis function that occurs in that context is this fundamental solution, and, naturally, it is highly relevant in this case that the Laplace operator is rotationally invariant. We wish to give a very concrete practical example from Pollandt (1997), namely, for a nonlinear problem on a domain $\Omega \subset \mathbb{R}^n$ with Dirichlet boundary conditions such as

$$\Delta u(x) = u^2(x), \quad x \in \Omega \subset \mathbb{R}^n,$$

$$u|_{\partial\Omega} = q,$$

one gets after two applications of Green's formula (Forster, 1984) the equation on the boundary (where g will be defined below)

$$(2.4) \qquad \frac{1}{2}\left(u(x) - g(x)\right) + \int_{\partial\Omega}\left(\phi(\|x - y\|)\,\frac{\partial}{\partial n_y}\left(u(y) - g(y)\right)\right.$$

$$\left. - \left(u(y) - g(y)\right)\frac{\partial}{\partial n_y}\,\phi(\|x - y\|)\right)d\,\Gamma_y = 0, \quad x \in \partial\Omega,$$

where $\frac{\partial}{\partial n_y}$ is the normal derivative with respect to y on $\Gamma_y = \Gamma = \partial\Omega$. We will later use (2.4) to approximate the boundary part of the solution, that is the part of the numerical solution which satisfies the boundary conditions. In order to define the function g which appears in (2.4), we have to *assume* that there are real coefficients λ_ξ such that the – usually infinite – expansion (which will be approximated by a finite series in an implementation)

$$(2.5) \qquad u^2(y) = \sum_{\xi \in \Xi} \lambda_\xi\,\widetilde{\phi}\,(\|y - \xi\|), \qquad y \in \Omega,$$

holds, and set

$$g(y) = \sum_{\xi \in \Xi} \lambda_\xi\,\widetilde{\Phi}\,(\|y - \xi\|), \qquad y \in \Omega,$$

so that $\Delta g = u^2$ everywhere with no boundary conditions. Here $\widetilde{\phi}$, $\widetilde{\Phi}$ are suitable radial basis functions with the property that $\Delta\widetilde{\Phi}\,(\|\cdot\|) = \widetilde{\phi}(\|\cdot\|)$ and the centres ξ are from Ω.

The next goal is to approximate the solution u of the PDE on the domain by g which is expanded in radial basis functions plus a boundary term $\tilde{r}$ that satisfies $\Delta\tilde{r} \equiv 0$ on Ω. To this end, we require that (2.4) holds at finitely many boundary points $x = \zeta_j \in \partial\Omega$, $j = 1, 2, \ldots, t$, only. Then we solve for the coefficients λ_ξ by requiring that (2.5) holds for all $y \in \Xi$. The points in $\Xi \subset \Omega$ must be chosen so that the interpolation problem is solvable.

Therefore we have fixed the λ_ξ by interpolation (collocation in the language of differential equations), whereas (2.4) determines the normal derivative $\frac{\partial}{\partial n_y}\,u(y)$

on Γ, where we are replacing $\frac{\partial}{\partial n_y} u(y)$ by another approximant, a spline, say, as in Chapter 3, call it $\tau(y)$. Thus the spline is found by requiring (2.4) for all $x = \zeta_j \in \Gamma$, $j = 1, 2, \ldots, t$, and choosing a suitable t. Finally, an approximation $\widetilde{u}(x)$ to $u(x)$ is determined on Ω by the identity

$$\widetilde{u}(x) := g(x) + \int_\Gamma (q(y) - g(y)) \frac{\partial}{\partial n_y} \phi(\|x - y\|) d\Gamma_y$$
$$- \int_\Gamma \phi(\|x - y\|) \left(\tau(y) - \frac{\partial g(y)}{\partial n_y} \right) d\Gamma_y, \quad x \in \Omega,$$

where $\widetilde{r}$ corresponds to the second and third terms on the right-hand side of the display (Pollandt, 1997).

Now, all expressions on the right-hand side are known. This is an outline of the approach but we have skipped several important details. Nonetheless, one can clearly see how radial basis functions appear in this algorithm; indeed, it is most natural to use them here, since many of them are fundamental solutions of the rotationally invariant Laplace operators in certain dimensions. In the above example and $n = 2$, $\phi(r) = \frac{1}{2\pi} \log r$, $\widetilde{\phi}(r) = r^2 \log r$ (thin-plate splines) and $\widetilde{\Phi}(r) = \frac{1}{16} r^4 \log r - \frac{1}{32} r^4$ are the correct choices. An undesirable feature of those functions for this application, however, is their unbounded support because it makes it harder to solve the linear systems for the λ_ξ etc., especially since in the approximative solution of partial differential equations usually very many collocation points are used to get sufficient accuracy.

One suitable approach to such problems that uses radial basis functions with compact support is with the 'piecewise thin-plate spline' that we shall describe now. With it, the general form of the thin-plate spline is retained as well as the nonsingularity of the interpolation matrix for nonuniform data. In fact, the interpolation matrix turns out to be positive definite. To describe our new radial basis functions, let ϕ be the radial basis function

$$(2.6) \qquad \phi(r) = \int_0^\infty (1 - r^2/\beta)_+^\lambda (1 - \beta^\mu)_+^\nu d\beta, \quad r \geq 0.$$

Here $(\cdot)_+^t$ is the so-called truncated power function which is zero for negative argument and $(\cdot)^t$ for positive argument. From this we see immediately that supp $\phi = [0, 1]$; it can be scaled for other support sizes. An example with $\mu = \frac{1}{2}$, $\nu = \lambda = 1$ is

$$(2.7) \qquad \phi(r) = \begin{cases} 2r^2 \log r - \frac{4}{3} r^3 + r^2 + \frac{1}{3}, & \text{if } r \leq 1, \\ 0 & \text{otherwise.} \end{cases}$$

which explains why we have called ϕ a 'piecewise thin-plate spline'. The positive definiteness of the interpolation matrix follows from a theorem which is stated and established in full generality in Chapter 6.

We now state a few more mathematical applications explicitly where the methods turned out to be good. Casdagli (1989) for instance used them to interpolate componentwise functions $F: \mathbb{R}^n \to \mathbb{R}^n$ that have to be iterated to simulate what is called a discrete dynamical system. In such experiments we especially seek the attractor of the discrete dynamical system that maps $F: \mathbb{R}^2 \to \mathbb{R}^2$. An example is the Hénon map

$$F(x, y) = (y, 1 + bx - ay^2)$$

(a and b being suitable parameters). Note that often in such mathematical applications, the dimension is much larger than two, so that radial basis functions are very suitable.

Since F often is far too expensive to be evaluated more than a few times, the idea is to interpolate F by s and then iterate with s instead. For instance, if F can reasonably be evaluated m times, beginning from a starting value $\xi_0 \in \mathbb{R}^n$, interpolation points

$$\xi_1 = F(\xi_0), \ \ \xi_2 = F(\xi_1), \ \ldots, \ \xi_m = F(\xi_{m-1})$$

are generated, and we let $\Xi = \{\xi_j\}_{j=0}^m$. Then we wish to interpolate F by s on the basis of that set Ξ. We note that thus the points in Ξ can be highly irregularly distributed, and at any rate their positions are not foreseeable. Moreover it is usual in this kind of application that n is large. Therefore both spline and polynomial interpolation are immediately ruled out, whereas Casdagli notes that, e.g., interpolation by multiquadrics is very suitable and gives good approximations to the short term and long term asymptotic behaviour of the dynamical system.

Hence radial basis functions are useful for such applications where interpolation is required to arbitrarily distributed data sites. There is, so far, no comprehensive theoretical explanation of this particular successful application, but the numerical results are striking as documented in Casdagli (1989).

In summary, this chapter has presented several concepts fundamental to radial basis functions and highly relevant to Chapters 4–10, namely complete monotonicity, positive definiteness, quasi- and cardinal interpolation, polynomial reproduction and convergence orders, localness of cardinal functions, radial basis functions of compact support. Three of the principal tools that we use here, namely the Bernstein representation theorem, Wiener's lemma, the

Poisson summation formula, are so central to our work that they will come up frequently in the later chapters as well.

In the following chapter we will show several other approaches to approximation and interpolation of functions with many variables. The main purpose of that chapter is to enable the reader to contrast our approach with other possible methods.

3

General Methods for Approximation and Interpolation

In this chapter we summarise very briefly some general methods other than radial basis functions for the approximation and especially interpolation of multivariate data. The goal of this summary is to put the radial basis function approach into the context of other methods for approximation and interpolation, whereby the advantages and some potential disadvantages are revealed. It is particularly important to compare them with spline methods because in one dimension, for example, the radial basis function approach with integral powers (i.e. $\phi(r) = r$ or $\phi(r) = r^3$ for instance) simplifies to nothing else than a polynomial spline method. This is why we will concentrate on polynomial and polynomial spline methods. They are the most important ones and related to radial basis functions, and we will only touch upon a few others which are non(-piecewise-)polynomial. For instance, we shall almost completely exclude the so-called local methods although they are quite popular. They are local in the sense that there is not one continuous function s defined over the whole domain, where the data are situated, through the method for approximating all data. Instead, there is, for every x in the domain, an approximation $s(x)$ sought which depends just on a few, nearby data. Thus, as x varies, this $s(x)$ may not even be continuous in x (it is in some constructions). Typical cases are 'natural neighbour' methods or methods that are not interpolating but compute local least-squares approximations.

Such methods are not to be confused with our global methods which usually should also depend locally on the data to approximate well; the difference is that our methods define just one continuous function normally over the whole of $\mathbb{R}^n$, or anyway over a very specific nontrivial range, a subset of $\mathbb{R}^n$.

We begin in the next section with polynomial methods, especially polynomial interpolation in more than one dimension, where the data Ξ are allowed to be scattered in $\mathbb{R}^n$. Then we will deal with piecewise polynomial methods, and we conclude the chapter with a few remarks about nonpolynomial methods.

36

3.1 Polynomial schemes

The most frequently employed techniques for multivariate approximation, other than radial basis functions, are straight polynomial interpolation, and piecewise polynomial splines. We begin with polynomial interpolation. There are various, highly specific techniques for forming polynomial interpolants. Very special considerations are needed indeed because as long as Ξ is a finite generic set of data sites from an open set in more than one dimension, and if we are interpolating from a polynomial space independent of Ξ, there can always be singularity of the interpolation problem. That is, we can always find a finite set of sites Ξ that causes the interpolation problem to be singular, whenever the dimension is greater than one and the data sites can be varied within an open subset of the underlying Euclidean space.

This is a standard result in multivariate interpolation theory and it can be shown as follows. Suppose that Ξ is such that the interpolation matrix for a fixed polynomial basis, call the matrix $\mathbf{A}$, is nonsingular. If Ξ stems from an open set in two or more dimensions, two of Ξ's points can be swapped, causing a sign change in det $\mathbf{A}$, where for the purpose of the swap the two points can be moved along paths that do not intersect. Hence there must be a constellation of points for which det $\mathbf{A}$ vanishes, det $\mathbf{A}$ being continuous in each $\xi \in \Xi$ due to the independence of the polynomial basis of the points in Ξ. So we have proved the result that singularity can always occur (see Mairhuber, 1956). Of course, this proof works for all continuous finite bases, but polynomials are the prime example for this case.

As a consequence of this observation, we need either to impose special requirements on the placement of Ξ – which is nontrivial and normally not very attractive in applications – or to make the space of polynomials dependent on Ξ, a more natural and better choice.

The easiest cases for multivariate polynomial interpolation with prescribed geometries of data points are the tensor-product approach (which is useless in most practical cases when the dimension is large because of the exponential increase of the required number of data and basis functions) and the interpolation e.g. on intersecting lines. Other approaches admitting m scattered data have been given by Goodman (1983), Kergin (1980), Cavaretta, Micchelli and Sharma (1980) and Hakopian (1982). All these approaches have in common that they yield unique polynomials in $\mathbb{P}_n^{m-1-\nu}$, i.e. polynomials of total degree $m-1-\nu$ in n unknowns, where $\nu < m$ varies according to the type of approach. They also have in common the use of ridge functions as a proof technique for establishing their properties, i.e. forming basis functions for the polynomial spaces which involve functions $g(\lambda \cdot x)$ where λ and x are from $\mathbb{R}^n$ so that this function is constant in directions orthogonal to λ and $g \in C^{m-1-\nu}(\mathbb{R})$. The approach by Goodman is the most general among these. The remarkable property of Kergin interpolation is that it simplifies to the standard Hermite, Lagrange or Taylor polynomials in one dimension, as the case may be. The work by Cavaretta *et al.* is particularly concerned with the question which types of Hermite data (i.e. data involving function evaluations and derivatives of varying degrees) may be generalised in this way.

A completely different approach for polynomial interpolation in several un-knowns is due to Sauer and Xu (1995) who use divided differences represented in terms of simplex splines and directional derivatives to express the polyno-mials. Computational aspects are treated in Sauer (1995), see also the survey paper, Gasca and Sauer (2000).

The representations of the approximants are usually ill-conditioned and there-fore not too useful in practical applications. Some convergence results for the approximation method are available in the literature (Bloom, 1981, Goodman and Sharma, 1984).

The interpolation of points on spheres by polynomials has been studied by Reimer (1990) including some important results about the interpolation oper-ator. The key issue is here to place the points at which we interpolate suitably on the sphere. 'Suitably' means on one hand that the interpolation problem is well-posed (uniquely solvable) and on the other hand that the norm of the interpolation operator does not grow too fast with increasing numbers of data points. The former problem is more easily dealt with than the latter. It is eas-ier to distribute the points so that the determinant of the interpolation matrix is maximised than to find point sets that give low bounds on operator norms. Surprisingly, the points that keep the operator norms small do not seem to be distributed very regularly, while we get a fairly uniform distribution if for instance the potentials in the three-dimensional setting

$$\sum_{\xi \neq \zeta} \frac{1}{\|\xi - \zeta\|}$$

are minimised with a suitable norm on the sphere. This work is so far only avail-able computationally for the two-dimensional sphere in $\mathbb{R}^3$, whereas theoretic analysis extends beyond $n = 3$.

Another new idea is that of de Boor and Ron which represents the interpo-lating polynomial spaces and is dependent on the given data points. In order to explain the various notions involved with this idea, we need to introduce some simple and useful new notations now. They include the so-called least term of an analytic function – it is usually an exponential – and the minimal totality of a set of functionals, which is related to our interpolation problem.

Definition 3.1. *We call the* least term $f_\downarrow$ *of a function f that is analytic at zero the homogeneous polynomial $f_\downarrow$ of largest degree j such that*

$$f(x) = f_\downarrow(x) + O(\|x\|^{j+1}), \qquad \|x\| \to 0.$$

Also, for any finite-dimensional space H of sufficiently smooth functions, the least of the space H is

$$H_\downarrow := \{ f_\downarrow \mid f \in H \}.$$

This is a space of polynomials.

Let P^* be a space of linear functionals on the continuous functions. We recall that such a space P^* of linear functionals is 'minimally total' for H if for any $h \in H$, $\lambda h = 0 \; \forall \lambda \in P^*$ implies $h = 0$, and if, additionally, P^* is the smallest such space. Using Definition 3.1 and the notion of minimal totality, de Boor and Ron (1990) prove the following important minimality property of $H_\downarrow$. Here, the overline means, as is usual, complex conjugation.

Proposition 3.1. *Among all spaces P of polynomials defined on $\mathbb{C}^n$ which have the property that P^* is minimally total for H, the least $\overline{H}_\downarrow$ is one of least total degree, that is, contains the polynomials of smallest degree.*

The reason why Proposition 3.1 helps us to find a suitable polynomial space for interpolation when the set of data Ξ is given is that we can reformulate the interpolation problem, which we wish to be nonsingular, in a more suitable form. That is, given that we wish to find a polynomial q from a polynomial space Q, say, so that function values on Ξ are met, we can represent the interpolation conditions alternatively in an inner product form as the requirement

$$f_\xi = q^* \exp\big(\xi \cdot (\cdot)\big), \qquad \xi \in \Xi.$$

Here, the first $\cdot$ in the exponential's argument denotes the standard Euclidean inner product, while the $\cdot$ in parentheses denotes the argument to which the functional q^* is applied. The latter is, in turn, defined by application to any sufficiently smooth p through the formula

$$q^* p = \sum_{\alpha \in \mathbb{Z}_+^n} \frac{1}{\alpha!} \big(D^\alpha q \big)(0) \cdot \big(D^\alpha p \big)(0),$$

using standard multiindex notation for partial derivatives

$$D^\alpha = \left(\frac{\partial^{\alpha_1}}{\partial x_1^{\alpha_1}}, \frac{\partial^{\alpha_2}}{\partial x_2^{\alpha_2}}, \ldots, \frac{\partial^{\alpha_n}}{\partial x_n^{\alpha_n}} \right)$$

and $\alpha! = \alpha_1! \cdot \alpha_2! \ldots \alpha_n!$, a notation that occurs often in the book. This functional is well-defined, whenever p is a function that is sufficiently smooth at the origin. It implies that our polynomial interpolation problem, as prescribed through Q and Ξ, is well-posed (i.e. uniquely solvable) if and only if the dual problem of interpolation from $H := \left\{ \sum_{\xi \in \Xi} a_\xi \exp\left(\xi \cdot (\cdot)\right) \,\middle|\, a_\xi \in \mathbb{C} \right\}$ with interpolation conditions defined through q^* is well-posed. Hence the minimal totality of the set $H_\downarrow$ can be used to prove the following important result.

Theorem 3.2.   *Given a finite set of data $\Xi \subset \mathbb{C}^n$, let H be as above. Then $\overline{H}_\downarrow$ is a polynomial space of least degree that admits unique interpolation to data defined on Ξ.*

The authors de Boor and Ron state this result more generally for Hermite interpolation, i.e. it involves interpolation of derivatives of various degrees and various centres.

There is also an algorithm for computing the least of a space that is a recursive method and is closely related to the Gram–Schmidt orthogonalisation procedure. We refer to the paper by de Boor and Ron for the details of this algorithm.

We give a few examples for the polynomials involved in two dimensions. If Ξ contains just one element, then

$$H = \left\{ a_\xi \exp\left(\xi \cdot (\cdot)\right) \right\},$$

with

$$\exp\left(\xi \cdot (\cdot)\right) = 1 + \xi \cdot (\cdot) + \frac{1}{2}\left(\xi \cdot (\cdot)\right)^2 + \cdots.$$

Thus $H_\downarrow = \mathrm{span}\{1\}$. Therefore our sought polynomial space, call it $\mathbb{P}$, is $\mathbb{P}_2^0$, i.e. constant polynomials in two variables. In general we let $\mathbb{P}_n^k$ be all polynomials in $\mathbb{P}_n$ of total degree at most k. If Ξ contains two elements ξ and τ, then

$$H = \{a_\xi \exp(\xi \cdot (\cdot)) + a_\tau \exp(\tau \cdot (\cdot))\},$$

hence $H_\downarrow = \mathrm{span}\{1, (\cdot) \cdot (\xi - \tau)\}$. Therefore $\mathbb{P} = \mathbb{P}_1^1 \circ (\boldsymbol{\lambda} \cdot)$, where $\circ$ denotes composition and where the vector $\boldsymbol{\lambda}$ is parallel to the affine hull of Ξ, a one-dimensional object. If $|\Xi| = 3$, then $\mathbb{P} = \mathbb{P}_1^2 \circ (\boldsymbol{\lambda} \cdot)$ or $\mathbb{P}_2^1$, depending on whether the convex hull of Ξ is a line parallel to the vector $\boldsymbol{\lambda}$ or not. Finally, if Ξ contains four elements and they are on a line, $\mathbb{P} = \mathbb{P}_1^3 \circ (\boldsymbol{\lambda} \cdot)$; otherwise $\mathbb{P}_2^1 \subset \mathbb{P} \subset \mathbb{P}_2^2$. E.g. if $\Xi = \{0, \xi, \tau, \xi + \tau\}$, ξ = first coordinate unit vector, τ = second coordinate unit vector, then $\mathbb{P}$ is the space of bi-linear polynomials, i.e. we have linear tensor-product interpolation.

3.2 Piecewise polynomials

Spline, i.e. piecewise polynomial, methods usually require a triangulation of the set Ξ in order to define the space from which we approximate, unless the data sites are in very special positions, e.g. gridded or otherwise highly regularly distributed. The reason for this is that it has to be decided where the pieces of the piecewise polynomials lie and where they are joined together. Moreover, it then has to be decided with what smoothness they are joined together at common vertices, edges etc. and how that is done. This is not at all trivial in more than one dimension and it is highly relevant in connection with the dimension of the space. Since triangulations or similar structures (such as quadrangulations) can be very difficult to provide in more than two dimensions, we concentrate now on two-dimensional problems – this in fact is one of the severest disadvantages of piecewise polynomial techniques and a good reason for using radial basis functions (in three or more dimensions) where no triangulations are required. Moreover, the quality of the spline approximation depends severely on the triangulation itself, long and thin triangles, for instance, often being responsible for the deterioration of the accuracy of approximation.

Let Ξ with elements ξ be the given data sites in $\mathbb{R}^2$. We describe the Delaunay triangulation which is a particular technique for triangulation, and give a standard example. We define the triangulation by finding first the so-called Voronoi tessellation which is in some sense a dual representation. Let, for ζ from Ξ, $T_\zeta = \{x \in \mathbb{R}^2 \mid \|x - \zeta\| = \min \|x - \xi\|, \xi \in \Xi\}$. These T_ζ are two-dimensional tiles surrounding the data sites. They form a Voronoi diagram and there are points where three of those tiles meet. These are the vertices of the tessellation. (In degenerate cases there could be points where more than three tiles meet.) Let t_ζ be any vertex of the tessellation; in order to keep the description simple, we assume that degeneracy does not take place. Let D_ζ be the set of those three ξ such that $\|t_\zeta - \xi\|$ is least. Then the set of triangles defined by the D_ζ is our triangulation, it is the aforementioned dual to the Voronoi diagram. This algorithm is a reliable method for triangulation with a well-developed theory, at least in two dimensions (cf., e.g., Braess, 1997, Brenner and Scott, 1994). In higher dimensions there can be problems with such triangulations, for instance it may be difficult to re-establish prescribed boundary faces when triangulations are updated for new sets of data, which is important for solving PDEs numerically with finite element methods.

Now we need to define interpolation by piecewise polynomials on such a triangulation. It is elementary how to do this with piecewise linears. However, often higher order piecewise polynomials and/or higher order smoothness of the interpolants are required, in particular if there is further processing of the approximants in applications to PDE solving etc. needed. For instance, piece-wise quadratics can be defined by interpolating at all vertices of the triangulation plus the midpoints of the edges, which gives the required six items of informa-tion per triangle. Six are needed because quadratics in two dimensions have six degrees of freedom. This provides an interpolant which is still only continuous. In order to get continuous differentiability, say, we may estimate the gradient of the proposed interpolant at the vertices, too. This can be done by taking suitable differences of the data, for example. In order to have sufficient freedom within each of the triangles, they have to be further subdivided. The subdivision into subtriangles requires additional, interior C^1 conditions.

Powell and Sabin (1977) divide the triangles into six subtriangles in such a way that the approximant has continuous first derivatives. To allow for this, the subdivision must be such that, if we extend any internal boundary from the common internal vertex to an edge, then the extension is an internal boundary of the adjacent element. Concretely, one takes the midpoint inside the big triangle to be the intersection of the normals at the midpoints of the edges. By this construction and by the internal C^1 requirement we get nine degrees of freedom for interpolating function values and gradients at the vertices, as required. Continuity of the first derivatives across internal edges of the triangulation is easy to show due to the interpolation conditions and linearity of the gradient.

Another case is the C^1-Clough–Tocher interpolant (Ciarlet, 1978). It is a particularly easy case where each triangle of the triangulation is divided into three smaller ones by joining the vertices of the big triangle to the centroid. If we wish to interpolate by these triangles over a given (or computed) triangulation, we require function and gradient values at each of the vertices of the big triangle plus the normal derivatives across its edges (this is a standard but not a necessary condition; any directional derivative not parallel to the edges will do). Therefore we get 12 data for each of the big triangles inside the triangulation, each of which is subdivided into three small triangles. On each of the small triangles, there is a cubic polynomial defined which provides 10 degrees of freedom each. The remaining degrees of freedom are taken up by the interior smoothness conditions inside the triangle.

In those cases where the points Ξ form a square or rectangular grid, be it finite or infinite, triangulations such as the above are not needed. In that event, tensor-product splines can be used or, more generally, the so-called box-splines that are comprehensively described in the book by de Boor, Höllig and Riemenschneider (1993). Tensor-product splines are the easiest multivariate splines, but here we start by introducing the more general notion of box-splines and then we will simplify again to tensor-product splines as particular examples. Box-splines are piecewise polynomial, compactly supported functions defined by so-called *direction sets* $X \subset \mathbb{Z}^n$ and the Fourier transform of the box-spline B,

$$(3.1) \qquad \hat{B}(t) = \prod_{x \in X} \frac{\sin \frac{1}{2} x \cdot t}{\frac{1}{2} x \cdot t}, \qquad t \in \mathbb{R}^n.$$

We recall the definition of the Fourier transform from the previous chapter. They can also be defined directly in the real domain without Fourier transforms, e.g. recursively, but for our short introduction here, the above is sufficient and indeed quite handy. In fact, many of the properties of box-splines are derived from their Fourier transform which has the above very simple form. Degree of the polynomial pieces, smoothness, polynomial recovery and linear independence are among the important properties of box-splines that can be identified from (3.1).

The direction sets X are fundamental to box-splines; they are responsible via the Fourier transform for not only degree and smoothness of the piecewise polynomial B but also its approximation properties and its support $X[0, 1]^{|\Xi|} \subset \mathbb{R}^n$. By the latter expression we mean all elements of the $|\Xi|$-dimensional unit cube, to which X seen as a *matrix* (and as a linear operator) is applied. Usually, X consists of *multiple entries* of vectors with components from $\{0, \pm 1\}$, but that is not a condition on X. Due to the possibly repeated entries, they are sometimes called multisets. The only condition is that always span $X = \mathbb{R}^n$. If in two dimensions, say, the vectors $\binom{1}{0}$, $\binom{0}{1}$, $\binom{1}{1}$ are used, one speaks of a three-directional box-spline, if $\binom{1}{-1}$ is added, a four-directional one, and any number of these vectors may be used. These two examples are the Courant finite element, and the Zwart–Powell element, respectively. If X contains just the two unit vectors in two dimensions, we get the characteristic function of the unit square.

In the simplest special case, X consists only of a collection of standard unit vectors of $\mathbb{R}^n$, where it is here particularly important that multiple entries in the set X are allowed. If that is so, B is a product $B(y) = B_{\ell_1}(y_1) \cdot B_{\ell_2}(y_2) \ldots B_{\ell_n}(y_n)$ of univariate B-splines, where $y = (y_1, y_2, \ldots, y_n)^T$ and the degrees $\ell_i - 1$ of the B-splines are defined through the multiplicity ℓ_i of the corresponding unit vector in X. When X has more complicated entries, other choices of box-splines B occur, i.e. not tensor-products, but they are still piecewise polynomials of which we have seen two examples in the paragraph above. In order to determine the accuracy that can be obtained from approximations by B and its translates along the grid (or the h-scaled grid) it is important to find out which polynomials lie in the span of those translates. This again depends on certain properties of X, as does the linear independence of the translates of the box-spline. The latter is relevant if we want to interpolate with linear combinations of the translates of the box-spline.

Linear independence, for instance, is guaranteed if X is 'unimodular', i.e. the determinants of each collection of n vectors from X are either 0 or ± 1 (Dahmen and Micchelli, 1983a), which is important, not only for interpolation from the space but also if we wish to create multiresolution analyses as defined in Chapter 9.

Out of the many results which are central to the theory and applications of box-splines we choose one that identifies the polynomials in the linear span of the box-splines. It is especially important to the approximational power of box-splines. Another one, which we do not prove here, is the fact that the multiinteger translates of a box-spline such as the above form a partition of unity.

Theorem 3.3. *Let S be the linear span of the box-spline B defined by the direction set $X \subset \mathbb{Z}^n$, span $X = \mathbb{R}^n$. Let $\mathbb{P}_n$ be the space of all n-variate polynomials. Then*

$$\mathbb{P}_n \cap S = \bigcap_{\{Z \subset X \mid \operatorname{span}(X \setminus Z) \neq \mathbb{R}^n\}} \ker \prod_{z \in Z} D_z,$$

where D_z, $z \in \mathbb{R}^n$, denotes in this theorem directional derivative in the direction of z.

For the proof and further discussion of this result, see de Boor, Höllig and Riemenschneider (1993). A corollary whose simple proof we present is

Theorem 3.4. *Let $\mathbb{P}_n^k$ be all polynomials in $\mathbb{P}_n$ of total degree at most k, let $d := \max\{r \mid \operatorname{span} X \setminus Z = \mathbb{R}^n, \forall Z \subset X$ with $|Z| = r\}$. Then $\mathbb{P}_n^k \subset S \Longleftrightarrow k \leq d$.*

Proof: '$\Longleftarrow$': Let Z be a subset of X. Since $\prod_{z \in Z} D_z$ reduces the degree of any polynomial by $|Z|$ or less ($|Z|$ being attained) and since by the definition of d

$$\min_{\{Z \subset X \mid \operatorname{span}(X \setminus Z) \neq \mathbb{R}^n\}} |Z| = d + 1,$$

it follows that $\mathbb{P}_n^d \subset \ker \prod D_z$, as required, for such Z.

There is, again for one such Z, $|Z| = d+1$ attained, whence $\prod_{z \in Z} D_z\, p \neq 0$ for some $p \in \mathbb{P}_n^{d+1}$. This proves the other implication. $\square$

We remark that the number d used in Theorem 3.4 is also related to the smoothness of the box-spline. The box-spline with direction set X and the quantity d as defined above is $d-1$ times continuously differentiable and its partial derivatives of the next order are bounded if possibly discontinuous.

In what way convergence rates are obtained from such results as the above Theorem 3.4 will be exemplified in the chapter about interpolation (albeit with radial basis functions, not with box-splines) but the approach there is similar to how we would go about it here (and as it is done, for instance, in the book by de Boor, Höllig and Riemenschneider). In fact, convergence proofs would be in principle much simpler here than they will be with our radial basis functions, because the latter are not of compact support which the box-splines are.

Additionally to the simple examples of tensor-product (univariate) splines that have been given above, we wish to give further examples of box-splines. Two well-known ones that are not tensor-product are the aforementioned Courant finite element or hat-function that occurs if

$$X = \left\{ \binom{1}{0}, \binom{0}{1}, \binom{1}{1} \right\}$$

and the Zwart–Powell element

$$X = \left\{ \binom{1}{0}, \binom{0}{1}, \binom{1}{1}, \binom{1}{-1} \right\}.$$

The Courant finite element is well-known to be continuous piecewise linear; the Zwart–Powell element is piecewise quadratic and $C^1(\mathbb{R}^2)$. Using certain recursions for derivatives of box-splines, it is in fact quite easy to establish that $B \in C^{d-1}(\mathbb{R}^2)$, where d is the same as in Theorem 3.4. We will not go into further details here because this is not our goal in this book on radial basis functions.

3.3 General nonpolynomial methods

Perhaps the best-known global, multivariate interpolation scheme for universal scattered distributions of data sites, which is not using polynomials, is Shepard's method. It is, however, not really the most successful one in the sense of accuracy of approximations, although it does give easy-to-define interpolants in any dimension which are not hard to evaluate either. These two facts give a clear advantage to the application of the method in practice. With finitely many data Ξ prescribed as in Chapter 1, a Shepard approximant is usually of the form

$$(3.2) \quad s(x) = \frac{\sum\limits_{\xi \in \Xi} f_\xi \omega(x - \xi)}{\sum\limits_{\xi \in \Xi} \omega(x - \xi)} = \frac{\sum\limits_{\xi \in \Xi} f_\xi \|x - \xi\|^{-\mu}}{\sum\limits_{\xi \in \Xi} \|x - \xi\|^{-\mu}}, \quad x \in \mathbb{R}^n,$$

where $\mu > 0$ and $\omega(x - \xi) = \|x - \xi\|^{-\mu}$ are the so-called weight functions; also weights ω other than $\omega(x) = \|x\|^{-\mu}$ are admitted, namely exponentially decaying ones for example. It is easy to see that s does indeed yield $s|_\Xi = f|_\Xi$, as is required for interpolation, due to the singularities of the weight functions at the data sites.

Unless μ is very small, s depends locally on the data, because the influence of f_ξ diminishes quickly at a rate of $-\mu$ when the argument of s moves away from ξ. Shepard's method's Lagrange functions are

$$L_\xi(x) = \frac{\|x - \xi\|^{-\mu}}{\sum_{\zeta \in \Xi} \|x - \zeta\|^{-\mu}}, \qquad x \in \mathbb{R}^n,$$

so that we can reformulate $s = \sum_{\xi \in \Xi} f_\xi L_\xi$. This L_ξ clearly decays when x moves away from ξ. If we wish to have completely local dependence of s on the f_ξ, we can arrange for that too: that is, even compactly supported weights which make s completely local, are possible. For example, a useful weight function ω is defined by

$$\omega(x) = \begin{cases} \dfrac{\exp\left(-\hat{r}^2/(\hat{r} - \|x\|)^2\right)}{\exp(\|x\|^2/h^2) - 1}, & \text{if } \|x\| \leq \hat{r}, \\ 0, & \text{otherwise}, \end{cases}$$

with positive radius $\hat{r}$ and scaling h parameters given by the user. Thus ω has compact support in $B_{\hat{r}}(0)$, namely the ball of radius $\hat{r}$ about the origin, a notation we shall use frequently in the book.

A severe disadvantage of Shepard's method (3.2) is that s has stationary points (vanishing gradients) at all data sites ξ if $\mu > 1$ which is a strange and undesirable property, as there is no reason to believe that all underlying functions f with $f_\xi = f(\xi)$ should have this feature. Several possible remedies for this unsatisfactory behaviour have been proposed, mostly in the shape of adding derivative information about the data. Such modifications are described in Powell (1996), e.g. one remedy is to modify s so as to satisfy $\nabla s(\xi) = g_\xi$ for some approximations g_ξ to the gradient of f at ξ, or indeed for the actual gradient of the approximand, where we are assuming that now a differentiable function f underlies the data in the aforementioned manner. The weight functions ω are further modified so as to give the limits $\lim_{\|x\| \to 0} \|x\| \, \omega(x) = 0$. Then the approximant s is defined afresh by the formula

$$(3.3) \qquad s(x) = \frac{\sum_{\xi \in \Xi} \left(f_\xi + (x - \xi) \cdot g_\xi \right) \omega(x - \xi)}{\sum_{\xi \in \Xi} \omega(x - \xi)}, \qquad x \in \mathbb{R}^n.$$

Thus s interpolates f_ξ and ∇s interpolates the prescribed approximative gradient g_ξ at each point ξ in the finite set Ξ. More generally, the term $f_\xi \omega(x - \xi)$ in (3.3) can be replaced by any 'good' approximation $h_\xi(x)$ to $f(x)$ for $x \rightarrow \xi$. For Shepard's method, there is also a convergence analysis in Farwig (1986) that considers it both with and without augmentation by derivatives. However, the convergence of Shepard's method is unsatisfactory unless it is augmented by derivative information about the approximand. This is usually an undesirable requirement as for most data sets, derivative information is not available, or too 'expensive' to obtain.

Other multivariate approximation schemes are so-called 'moving' least squares schemes, such as the one proposed by McLain (1974). These methods are local in the sense of the first paragraph in our chapter. The famous McLain scheme seeks, for each prescribed $x \in \mathbb{R}^n$, a multivariate function g from a function space X that is given in advance, and which minimises the expression

$$\sum_{\xi \in \Xi} \omega_\xi(x)(g(\xi) - f_\xi)^2, \quad g \in X.$$

We call the solution g of this requirement g^x because it depends on x, so that $s(x) := g^x(x)$ for all x. If we choose weight functions ω_ξ that are nonnegative and continuous, except that $\omega_\xi(x) \rightarrow +\infty$ for $x \rightarrow \xi$, then this clearly implies that interpolation in Ξ is achieved. Usually, the weight's supports are required to be small in practical algorithms and X is a space with dimension much smaller than $|\Xi|$, so that a least squares approach is suitable. In particular, some smoothing of rough or noisy data can be achieved in this way (cf. our Chapters 8 and 9). Thus the values of s depend indeed locally on x. Uniqueness is achieved by asking X to be such that, for each x, the only function g from that space which satisfies

$$g \in X \wedge g(\xi) = 0 \; \forall \xi \; \text{with} \; \omega_\xi(x) > 0$$

is the zero function. The main disadvantage of those methods is that they normally do not give an explicit analytic expression for one approximant for all the provided data at once.

In summary, we have seen that there are many approximation methods in several dimensions other than radial basis functions, the most attractive ones being probably the ones that generate piecewise polynomials. However, those require much set-up work especially in more than two dimensions and this, among others previously mentioned, is a strong argument in favour of radial basis functions.

4

Radial Basis Function Approximation on Infinite Grids

Inasmuch as radial basis function approximations are used mainly for the purpose of scattered data approximation in applications, the assumption that the data lie at the vertices of an infinite regular grid gives rise to an interesting special case. In this chapter, data at the vertices of a scaled grid will lead us to highly relevant theoretical results about their approximation power in this chapter. Such results are important in terms of understanding the usefulness of the approximation spaces, especially if the best approximation orders (so-called saturation orders) are known. The $L^p(\mathbb{R}^n)$-approximation order is at least μ for approximants from an h-dependent space $\mathcal{S} = \mathcal{S}_h$ of the approximants with centres $h\mathbb{Z}^n$ if

$$\text{dist}_{L^p(\mathbb{R}^n)}(f, \mathcal{S}) := \inf_{g \in \mathcal{S}} \| f - g \|_p = O(h^\mu)$$

for all f from the given space of approximands. (In some of our estimates an extra factor of $\log h$ occurs.) The approximation order is exactly μ if the O cannot be replaced by o on the right-hand side (as used elsewhere sometimes: if μ cannot be replaced by $\mu + \varepsilon$ for any positive ε). In this chapter, the h-dependence of $\mathcal{S}$ comes only from the fact that shifts of the radial basis function on a scaled integer grid $h\mathbb{Z}^n = \Xi$, which is our set of data sites, are employed.

The results in this chapter, regarding interpolation and quasi-interpolation, are extensions of the results in Chapter 2 in various directions. The remark about reasons for studying regular grids is, incidentally, true for the study of multivariate polynomial splines as well (e.g. de Boor, Höllig and Riemenschneider, 1993), whose approximation properties were intensely investigated on regular square grids. Although we do not normally intend to compute approximations in practice on such grids, statements about their approximation power – especially if they are optimal (best approximation orders, saturation orders) – are helpful for making choices of radial basis functions for approximation when they are used in applications. It must be kept in mind that for applications for data *on square grids* there are many other suitable methods, such as tensor-product methods or the box-splines mentioned in Chapter 3. Among those, tensor-product B-splines are most often used and best understood, because the B-splines have a particularly simple univariate form, are piecewise polynomials and are of small, compact support, while radial basis functions are usually not compactly supported.

Further, for the analysis of their properties in computer experiments it is usual to do computations on *finite* cardinal (i.e. regular, square, equally spaced) grids, in order to elucidate their approximational behaviour. Finally, approximation on so-called sparse grids, which are being studied at present and which are regular grids thinned out by removal of many points, are useful in applications for the numerical solution of partial differential equations. In short, radial basis functions were born to be applied to scattered data interpolation, but for the purpose of analysis, it makes perfect sense to study them on infinite grids, and we shall do so now.

4.1 Existence of interpolants

The interpolants we consider now are defined on equally spaced grids. A crucial property of grids with equal spacing, e.g.

$$\mathbb{Z}^n = \{(j_1, j_2, \ldots, j_n)^T \mid j_i \in \mathbb{Z}, \ i = 1, 2, \ldots, n\}$$

or $(h\mathbb{Z})^n$, is that they are periodic and boundary-free. This enables us to apply Fourier transform techniques during the analysis, since discrete Fourier transforms are defined on infinite equally spaced lattices. The spaces spanned by shifts of a basis function, call it ψ (not to be confused with our *radial* basis function), namely by

$$\psi(\cdot - j), \qquad j \in \mathbb{Z}^n,$$

are called *shift-invariant* because for any f in such a space, its shift $f(\cdot - k)$, $k \in \mathbb{Z}^n$, is an element of the same space. Shift-invariant spaces were studied extensively in the literature, see for instance the various papers by de Boor, DeVore and Ron mentioned in the bibliography. We are only using them implicitly at this stage, because their analysis in the literature is (mostly) restricted to square-integrable or even compactly supported basis functions which is not the case for our radial basis functions, although most of the known theory of shift-invariant spaces can be adapted to and used for radial basis functions (e.g. Halton and Light, 1993).

One of the main tools in the analysis of our approximants is the Poisson summation formula, which will be used often in this chapter. In fact, the Lagrange functions of our interpolants will be defined only by stating their form in the Fourier domain. Even if we actually wish to compute approximants on such grids, their periodicity can be used e.g. by the application of FFT techniques.

The goal of approximation on grids is to find approximants which in our study have the relatively simple form

$$\sum_{j \in \mathbb{Z}^n} f(j)\, \psi(x - j), \quad x \in \mathbb{R}^n,$$

where $f \colon \mathbb{R}^n \to \mathbb{R}$ is the function we wish to approximate (the approximand) and ψ has the following expansion which is a finite sum in many instances:

$$\psi(x) = \sum_{k \in \mathbb{Z}^n} c_k\, \phi(\|x - k\|), \quad x \in \mathbb{R}^n,$$

all sums being assumed at present to be absolutely convergent, ϕ being our radial function and $\{c_k\}_{k \in \mathbb{Z}^n}$ suitable, f-independent coefficients that may or may not be of compact support with respect to k. Of course, so far we have written an approximant based on $\mathbb{Z}^n$-translates of $\phi(\|\cdot\|)$ but to get convergence on the whole underlying space $\mathbb{R}^n$ of the approximant to f, it is necessary to base the approximant on $h\mathbb{Z}^n$, h being positive and becoming small. In fact, it is desirable in the latter case to remain with exactly the same ψ, but then we must scale the argument of the ψ as follows:

$$\sum_{j \in \mathbb{Z}^n} f(jh)\, \psi\!\left(\frac{x}{h} - j\right), \quad x \in \mathbb{R}^n.$$

In the language of shift-invariant spaces and multiresolution analysis which we shall use in Chapter 9, what we do here is a 'stationary' scaling as mentioned already in Chapter 2, since for all h the same function ψ is used which is scaled by h^{-1} inside its argument.

In contrast, a nonstationary approximation process would correspond to making h-dependent choices ψ_h of ψ, which would typically result in approximations

$$\sum_{j \in \mathbb{Z}^n} f(jh)\, \psi_h(x - hj), \quad x \in \mathbb{R}^n.$$

We shall encounter this in Chapter 9 and briefly in Chapter 5 but give no further attention to this more complicated nonstationary version at present.

In this chapter about radial basis functions on infinite lattices, we are particularly interested in the question of uniform approximation orders of the above to sufficiently differentiable approximands f when $h \to 0$. This order should only depend on ϕ and n (and in certain ways on the smoothness of the approximand f), ψ being assumed to be chosen in some way optimally for any given ϕ. This function may in fact not at all be unique, but 'optimally' is to be understood anyway only in a very weak sense: it means that the error for a class of sufficiently smooth approximands decreases for $h \to 0$ at the best possible asymptotic rate for the particular radial function under investigation. This does not exclude, for instance, that an error which is smaller by a constant factor cannot be achieved with another ψ. It also does not exclude a $-\log h$ term as a multiplier of the determining power of h since the logarithm grows so slowly that its presence will hardly show up in practical computations.

Incidentally, we only consider uniform approximations and convergence orders; Binev and Jetter (1992) in contrast offer a different approach to radial basis functions approximation altogether and in particular cardinal interpolation *only* by means of L^2-theory. This facilitates many things in the theory through the exclusive use of Hilbert space and Fourier transform techniques.

There is a variety of methods to achieve our goal outlined above. One way is via the so-called Strang and Fix theory. This is the method of choice especially when the functions ϕ are compactly supported, such as box-splines or other multivariate piecewise polynomials. Strang–Fix theory is, in particular, almost always used when finite element methods are discussed. Indeed, it is possible to apply this in the present context of our noncompactly supported radial basis functions and we shall discuss this briefly below. The main feature of this approach is that certain zero-conditions are imposed on the Fourier transform of ψ at 2π-multiples of multiintegers – which guarantee that the approximants reproduce polynomials of a certain degree that in turn is linked to the order of the imposed zero-conditions at 2π-multiples of multiintegers.

At this point, however, we shall address the problem from a slightly different point of view, namely that of the theory of interpolation by radial basis functions. Starting with interpolation, the question of approximation orders obtainable via the Strang and Fix conditions is answered almost automatically, as we shall see.

One way to choose ψ is by interpolation, and in that special case it is normally unique in the sense that the coefficients are unique within the class of all absolutely summable coefficient sequences. So in this instance we choose ψ to be a Lagrange function

$$L(x) = \sum_{k \in \mathbb{Z}^n} c_k\, \phi(\|x - k\|), \quad x \in \mathbb{R}^n,$$

that satisfies the Lagrange conditions which are also familiar, e.g. from univariate polynomial Lagrange interpolation and from Chapter 2 of this book,
namely

$$L(j) = \delta_{0j} = \begin{cases} 1 & \text{if } j = 0, \\ 0 & \text{if } j \in \mathbb{Z}^n \backslash \{0\}. \end{cases}$$

If L is so, then our above approximants, setting $\psi = L$, will automatically
interpolate f on $h\mathbb{Z}^n$ or $\mathbb{Z}^n$, depending on whether we scale or not.

We want to pursue this approach further and ask first, whether such L exist
and what their properties are. Much of the analysis of our radial basis function
methods on lattices boils down to exactly those questions. Many of the properties
of the approximants can be identified through relatively simple facts about
Lagrange functions. It is very helpful to the analysis at this time to fix the
coefficients of the functions through the Lagrange conditions, because this way
they are identified easily, and fortunately, the Lagrange functions have the best
possible properties, as we shall see. This means that they have the highest
polynomial exactness of the approximation and best local behaviour. This is
in spite of the fact that for the formulation of the approximants the Lagrange
conditions are not always necessary unless we really want to interpolate. We
shall see that this is true later in our convergence theorems where the Lagrange
conditions as such are not required.

Under the following weak conditions we can show that there are $\{c_k\}_{k\in\mathbb{Z}^n}$
which provide the property $L(j) = \delta_{0j}$. To begin our analysis, let us assume
that the radial basis function in use decays so fast that $\phi(\|\cdot\|) \in L^1(\mathbb{R}^n)$, an as-
sumption which we will have to drop soon in order to apply the
theory to our general class of radial basis functions, most of which are anyway not integrable or even decaying for large argument. At first, however, we
can prove the following simple result with fairly weak conditions that hold
for decaying and integrable radial basis functions such as the Gauss-kernel
$\phi(r) = e^{-\alpha r^2}$.

Proposition 4.1. *Let ϕ be such that $|\phi(r)| \leq C(1+r)^{-n-\varepsilon}$ for some fixed
positive constant C and $\varepsilon > 0$, so that in particular $\phi(\|\cdot\|) \in L^1(\mathbb{R}^n)$. Suppose
that*

$$\sigma(\vartheta) = \sum_{j\in\mathbb{Z}^n} \phi(\|j\|) e^{-i\vartheta\cdot j} \neq 0, \quad \forall\, \vartheta \in \mathbb{T}^n.$$

*Then there are unique absolutely summable coefficients $\{c_k\}_{k\in\mathbb{Z}^n}$ such that the
above cardinal function L satisfies $L(j) = \delta_{0j}$. Moreover, the conditions that
the Fourier transform of the radially symmetric $\phi(\|\cdot\|)$, namely $\hat{\phi}(\|\cdot\|)$, satisfy*

$|\hat{\phi}(r)| \le C(1+r)^{-n-\varepsilon}$ *and be positive, are sufficient for the symbol* $\sigma(\vartheta)$ *to be positive everywhere.*

Remark: We recall that the Fourier transform of a radially symmetric function is always radially symmetric, too. Specifically, if $f: \mathbb{R}^n \to \mathbb{R}$ is radially symmetric, $F: \mathbb{R}_+ \to \mathbb{R}$ being its radial part, its Fourier transform $\hat{f}$ is as well and the *radial part* of that Fourier transform, call it $\hat{F}$, is

$$\hat{F}(r) = (2\pi)^{n/2} r^{-n/2+1} \int_0^\infty F(s) s^{n/2} J_{n/2-1}(rs)\, ds, \qquad r \ge 0.$$

Here, J denotes the standard Bessel function (Abramowitz and Stegun, 1972, p. 358). The above expression comes from the fact that the Fourier transform of a radially symmetric function has the following shape that leads to Bessel functions:

$$\int_{\mathbb{R}^n} e^{-ix\cdot t} F(\|t\|)\, dt = \int_0^\infty F(s) \int_{\|t\|=s} e^{ix\cdot t}\, dt\, ds.$$

The Bessel function occurring in the previous display is a result of the second integral on the right-hand side in this display, because of the following identity:

$$\int_{\|t\|=s} e^{ix\cdot t} dt = s^n \int_{\|t\|=1} e^{isx\cdot t} dt$$
$$= (2\pi)^{n/2} \|x\|^{-n/2+1} s^{n/2} J_{n/2-1}\left(\|x\| s\right).$$

This fact about radially symmetric Fourier transforms remains true for the distributional or generalised Fourier transforms which we shall use below. We also remark that, instead of the Fourier transform $\hat{\phi}$ being required to be positive in the statement of Proposition 4.1, it suffices that it is nonnegative and has no 2π-periodic zero.

Proof of Proposition 4.1: The first statement of the proposition is an application of Wiener's lemma, performed exactly as in our work in the second chapter. The coefficients of the Lagrange function are the Fourier coefficients of the periodic reciprocal of the symbol exactly as before, even though we are now in more than one dimension.

The second assertion follows from an application of the Poisson summation formula:

$$\sigma(\vartheta) = \sum_{j \in \mathbb{Z}^n} \phi(\|j\|) e^{-i\vartheta \cdot j} = \sum_{j \in \mathbb{Z}^n} \hat{\phi}(\|\vartheta - 2\pi j\|)$$

which is positive whenever the radial function's Fourier transform is positive.

$\square$

An example of a suitable radial function for the above is $\phi(r) = e^{-\alpha r^2}$ for a positive parameter α. As we noted in Chapter 2, the Fourier transform is also a positive multiple of an exponential function in $-r^2$ itself. Although the above proposition does not apply for instance to $\phi(r) = r$, it contains two fundamental facts which will prove to be of great importance, namely

(1) The 'symbol' σ can be reformulated as

$$\sigma(\vartheta) = \sum_{j \in \mathbb{Z}^n} \hat{\phi}(\|\vartheta - 2\pi j\|), \quad \vartheta \in \mathbb{T}^n,$$

through the use of Poisson's summation formula.

(2) The $\{c_k\}_{k \in \mathbb{Z}^n}$ can be expressed by Wiener's lemma as Fourier coefficients of the reciprocal of the 2π-periodic symbol, that is

(4.1) $$c_k = \frac{1}{(2\pi)^n} \int_{\mathbb{T}^n} \frac{e^{i\vartheta \cdot k}}{\sigma(\vartheta)} \, d\vartheta, \quad k \in \mathbb{Z}^n,$$

provided the symbol has no zero. Further, these coefficients are absolutely summable.

The last formula (4.1) follows from a simple reformulation of the Lagrange conditions and from $\sigma(\vartheta) \neq 0$ for all arguments ϑ:

$$\sum_{j \in \mathbb{Z}^n} \sum_{k \in \mathbb{Z}^n} c_k \, \phi(\|j - k\|) \, e^{-ij \cdot \vartheta} = 1$$

which can be rewritten equivalently as

$$\left[\sum_{k \in \mathbb{Z}^n} c_k e^{-i\vartheta \cdot k} \right] \left[\sum_{j \in \mathbb{Z}^n} \phi(\|j\|) e^{-i\vartheta \cdot j} \right] = 1$$

and because σ does not vanish anywhere

$$\sum_{k \in \mathbb{Z}^n} c_k e^{-i\vartheta \cdot k} = \frac{1}{\sigma(\vartheta)}, \quad \vartheta \in \mathbb{T}^n.$$

We now go *backwards*, beginning by expressing the symbol $\sigma(\vartheta)$ as the series of Fourier transforms

$$\sum_{j \in \mathbb{Z}^n} \hat{\phi}(\|\vartheta - 2\pi j\|),$$

which we require to be well-defined and without a zero, and define the coefficients of the Lagrange function c_k as the symbol's Fourier coefficients. For this purpose, we consider a larger class of radial basis functions and admit any ϕ that is polynomially bounded, thus has a radially symmetric distributional Fourier

transform (e.g., Jones, 1982) $\hat{\phi}(\| \cdot \|): \mathbb{R}^n \backslash \{0\} \to \mathbb{R}$, with the properties (A1)–
(A3) which we state below. Those conditions guarantee the well-definedness of
the symbol and that it has no zero. Thus the aforementioned Fourier coefficients
exist.

To aid the analysis we briefly review some of the salient points of the distributional Fourier theory. We recall that any function that grows at most like
a polynomial of arbitrary degree (also called tempered functions) has a generalised (distributional) Fourier transform which need not be everywhere continuous, unlike conventional Fourier transforms of integrable functions (see the
Appendix). In fact a distributional Fourier transform usually has singularities
at the origin. Actually, the order of the singularity reflects the speed of the
growth of the aforementioned polynomial type at infinity. In order to define the
generalised Fourier transform we have to state a few facts about generalised
functions first.

One way to represent generalised functions or distributions is by equivalence
classes of sequences of good functions (test functions) v_j, $j \in \mathbb{N}$. The set $\mathbf{S}$
of good functions is the set of infinitely differentiable functions on $\mathbb{R}^n$ with all
derivatives (including the function itself) decaying faster than the reciprocal of
any polynomial at infinity. Therefore, g is a generalised function, represented
by a sequence with entries $v_j \in \mathbf{S}$, if there exists a sequence of good functions
v_j, $j \in \mathbb{N}$, such that

$$\int_{\mathbb{R}^n} v_j(x)\gamma(x)\,dx, \qquad j \in \mathbb{N},$$

is a convergent sequence for any good γ. As a consequence, the Fourier
transform of such a generalised function may be defined by the sequence
$\hat{v}_j \in \mathbf{S}$, $j \in \mathbb{N}$. An example of such a generalised function is Dirac's δ-
function, which is represented in one dimension by a sequence of good functions
$v_j(x) = \sqrt{j/\pi} \exp(-jx^2)$, $j \in \mathbb{N}$. For any $\gamma \in \mathbf{S}$ we then have

$$\lim_{j \to \infty} \int_{\mathbb{R}} \gamma(x)v_j(x)\,dx = \gamma(0)$$

which is also expressed by

$$\int_{\mathbb{R}} \delta(x)\gamma(x)\,dx = \gamma(0).$$

The Fourier transform of δ is then 1 because

$$\lim_{j \to \infty} \int_{\mathbb{R}} \gamma(x)\hat{v}_j(x)\,dx = \int_{\mathbb{R}} \gamma(x)\,dx.$$

More generally, generalised Fourier transforms of derivatives of the δ-function are polynomials and generalised Fourier transforms of polynomials are derivatives of the δ-function.

In fact, we need very few properties of these generalised Fourier transforms. The key fact which we shall rely on is that they exist for the radial basis functions of at most polynomial growth which we study here and that they agree for those radial basis functions with continuous functions almost everywhere (in the event, everywhere except zero). Then, we have certain properties which we demand from those almost everywhere continuous functions that we take from the literature. In addition to the part of the generalised Fourier transform that is continuous except at zero, the generalised Fourier transforms of radial basis functions often contain a term that comprises a δ-function and its derivatives. This can be removed by changing the radial basis function by adding an even degree polynomial, containing only even powers. We always assume therefore that this δ-function term is not present. Indeed, we shall learn in Sections 4.2 and 5.1 that the linear combinations of translates of radial basis functions which we study always contain coefficients which cancel out any such polynomial addition to ϕ. Therefore it is not loss of generality that already at this time, we consider the radial basis functions and their Fourier transforms modulo an even degree polynomial and derivatives of the δ-function respectively. We illustrate this by an example.

The generalised Fourier transforms can be computed by standard means using the above definition as limiting functions. An example is the generalised Fourier transform of the thin-plate spline function $\phi(r) = r^2 \log r$ in two dimensions, that is (Jones, 1982)

$$\hat{\phi}(\|x\|) = 4\pi \|x\|^{-4} + 4\pi^2(\log 2 + 1 - \gamma)\Delta\delta(x), \quad x \in \mathbb{R}^2,$$

where $\gamma = 0.5772\ldots$ is Euler's constant and Δ is the two-dimensonal Laplace operator. Therefore we consider only

$$\hat{\phi}(\|x\|) = 4\pi \|x\|^{-4}$$

as the two-dimensional generalised Fourier transform of the thin-plate spline function that we thus consider modulo a quadratic polynomial.

Another useful way of stating that a function $\hat{\phi}(\|x\|)$ with a singularity at the origin is the generalised Fourier transform (again, modulo an even degree polynomial) is by the integral identity that demands

$$\int_{\mathbb{R}^n} \hat{\phi}(\|x\|)v(x)\,dx = \int_{\mathbb{R}^n} \phi(\|x\|)\hat{v}(x)\,dx.$$

Again, this is for all test functions $v \in \mathbf{S}$. Also the test functions which are used in the above have to satisfy

$$\int_{\mathbb{R}^n} p(x)\hat{v}(x)\,dx = 0$$

for all polynomials of some order which depends on ϕ. Thus the right-hand side of the last but one display is well-defined by the fast decay of the Fourier transform of the test function and the continuity and limited growth of the radial basis function. The left-hand side is well-defined by the fast decay of v and by the above display that requires the v to have a sufficiently high order of zero at zero to meet the singularity of $\hat{\phi}$. What the correct order of the polynomials p is will be seen below.

Indeed, since for nonintegrable radial basis functions the generalised or distributional Fourier transform will usually not agree with a continuous function everywhere, the last but one display admits the singularity of the generalised Fourier transform at the origin, cancelling singularities of the same order. Therefore both sides are well-defined, absolutely convergent integrals and may be used to define generalised Fourier transforms. We will give examples of those below, having encountered the multiquadrics example already in the second chapter.

We need conditions that ensure that the interpolation problem set on the scaled lattice $h \cdot \mathbb{Z}^n$ is well-defined and uniquely solvable. These conditions are always formulated in terms of the generalised Fourier transform of the radial basis function. The three required properties are

(A1) $\hat{\phi}(r) > 0, r > 0$,
(A2) $\hat{\phi}(r) = O(r^{-n-\varepsilon}), \varepsilon > 0, r \to \infty$,
(A3) $\hat{\phi}(r) \sim r^{-\mu}, \mu \geq 0, r \to 0_+$.

In these conditions, we use the notation $\sim$ to mean that, if $f(x) \sim g(x)$, then $f(x) = \text{const.} \times g(x) + o(g(x))$ with a nonvanishing constant. Condition (A1) here is responsible for the positivity of the symbol, although nonnegativity of the Fourier transform and absence of 2π-periodic zeros would be sufficient. It is standard in the analysis of radial basis function methods to demand strict positivity, however. Indeed, almost all radial basis functions that are in use and that are mentioned in this book exhibit this property. We remark that the one exception to this rule is the radial basis functions having compact support that are treated in Chapter 6.

Of course, a function whose Fourier transform is *negative* everywhere is also acceptable, sign changes can always be absorbed into the coefficients of the approximants. We have noted already in the second chapter that the generalised Fourier transform of the multiquadric function satisfies the conditions above, object to a sign-change.

Indeed, a large class of radial basis functions exhibiting properties (A1)–(A3) above can be identified. For instance, the thin-plate spline

$$\phi(r) = r^2 \log r$$

has a generalised Fourier transform which also satisfies all conditions above except for some δ-functions.

The second condition takes care of absolute convergence of the series defining the symbol, while (A3) controls the Fourier transform's singularity (if any) at the origin. Thus, at a minimum, σ and $\frac{1}{\sigma}$ are still well-defined, except perhaps at the origin, and thus in particular the coefficients

$$c_k = \frac{1}{(2\pi)^n} \int_{\mathbb{T}^n} \frac{e^{i\vartheta \cdot k}}{\sigma(\vartheta)} \, d\vartheta, \qquad k \in \mathbb{Z}^n,$$

exist, so long as we interpret the symbol σ as *only* of the form of the series

$$\sigma(\vartheta) = \sum_{j \in \mathbb{Z}^n} \hat{\phi}(\|\vartheta - 2\pi j\|), \quad \vartheta \in \mathbb{T}^n \backslash \{0\},$$

and no longer use the standard form of the symbol as a Fourier series in $\phi(\| \cdot \|)$ we started with. We may, however, no longer use Wiener's lemma to deduce the absolute summability of the c_k and have to derive this property separately.

Note that $\hat{\phi}$'s allowed singularity at the origin does not create any harm since $\frac{1}{\sigma}$, and not σ, is decisive for the definition and the properties of the Fourier coefficients $\{c_k\}_{k \in \mathbb{Z}^n}$. Indeed, the singularities of the symbol translate only to zeros of its reciprocal. Nonetheless, the lack of decay at infinity of ϕ forces us to examine the decay of the c_k carefully, i.e. their anticipated decay for $\|k\| \to \infty$, so as to ensure the convergence of the defining series for L. As is usual with Fourier coefficients, this will depend fundamentally on the smoothness and integrability of $\frac{1}{\sigma}$ and its derivatives.

It is just as important to study the localness of L itself since we need to know what f we may introduce into the interpolating (infinite) series. Of course we can always admit compactly supported approximands f since then the approximating (interpolating) series is finite for any given x. However, we wish to interpolate more general classes of f, to which the hope is, even some polynomials will belong. This is interesting because if we can admit interpolation to (low degree) polynomials we have a chance of exactly *reproducing* certain polynomials which will help us with establishing approximation order results by local Taylor series approximations and other standard arguments.

One way of introducing smoothness into the reciprocal of the symbol, $\frac{1}{\sigma}$, in order to establish approximation orders is through strengthening conditions (A2) and (A3) by additional conditions on derivatives: we do so now and specifically require that for a fixed positive ε and a positive integer M

(A2a) $\hat{\phi}^{(\ell)}(r) = O(r^{-n-\varepsilon})$, $r \to \infty$, $\ell = 0, 1, \ldots, M$, and in particular
   $\hat{\phi} \in C^M(\mathbb{R}_{>0})$,
(A3a) $\hat{\phi}^{(\ell)}(r) \doteq r^{-\mu-\ell} + O(r^{-\mu-\ell+\varepsilon})$, $r \to 0_+$, $0 \le \ell \le M$, for a *positive*
   exponent μ.

Here, the $\doteq$ means equality up to a nonzero constant multiple which we do not state because it is unimportant to the analysis. Thus, for instance, the reciprocal $\frac{1}{\sigma}$ of the symbol is in $C^M(\mathbb{T}^n \setminus \{0\})$, and we can deduce decay estimates for c_k. In our work here we want more precise estimates that supply the dominant term and indeed we get more general and more precise results than, e.g., in Theorem 2.7. In the sequel we denote the upper part by $\lceil \mu \rceil = \min\{m \in \mathbb{Z} \mid m \ge \mu\}$.

Theorem 4.2. *Let ϕ satisfy (A1), (A2a), (A3a), $M > \lceil n + \mu \rceil$. Then, for $k \ne 0$ and for some positive ε the coefficients (4.1) satisfy*

$$|c_k| = \begin{cases} C\|k\|^{-n-\mu} + O(\|k\|^{-n-\mu-\varepsilon}) & \text{if } \mu \notin 2\mathbb{Z}, \\ O(\|k\|^{-n-\mu-\varepsilon}) & \text{otherwise.} \end{cases}$$

Proof: It follows from the form (4.1) that the c_k are the Fourier coefficients of the reciprocal of the symbol. In other words,

$$|c_k| \cdot (2\pi)^n = \int_{\mathbb{T}^n} \frac{e^{ix \cdot k}\, dx}{\hat{\phi}(\|x\|) + \sum' \hat{\phi}(\|x - 2\pi j\|)}.$$

Moreover, $\sum'$ denoting as before $\sum_{j \in \mathbb{Z}^n \setminus \{0\}}$, we have for a nonzero constant C_1

$$|c_k| \cdot (2\pi)^n = \int_{\mathbb{T}^n} \frac{e^{ix \cdot k}\, dx}{C_1 \|x\|^{-\mu} + \tilde{\phi}(\|x\|) + \sum' \hat{\phi}(\|x - 2\pi j\|)},$$

where $\tilde{\phi}(r) = \hat{\phi}(r)$ minus the dominant term in r of order $-\mu$ of (A3a). That is by cancellation

$$= \int_{\mathbb{T}^n} \frac{C_1^{-1} e^{ix \cdot k} \|x\|^\mu\, dx}{1 + C_1^{-1} \|x\|^\mu \left(\tilde{\phi}(\|x\|) + \sum' \hat{\phi}(\|x - 2\pi j\|)\right)}.$$

Now, let $\rho \colon \mathbb{R}^n \to \mathbb{R}$ be a so-called cut-off function, that is one that vanishes outside the unit ball $B_1(0)$, is one inside $B_{1/2}(0)$ and is in $C^\infty(\mathbb{R}^n)$. It is easy to construct such cut-off functions from certain ratios of exponential functions.

With this preparation, we may reduce the problem in the following fashion. As M is greater than $\lceil n + \mu \rceil$ and $\hat{\phi} \in C^M(\mathbb{R}_{>0})$, we can now apply standard methods of partial integration as explained in Chapter 2 to deduce that $(1 - \rho)$ times the integrand gives rise to Fourier coefficients which decay as $O(\|k\|^{-M}) = O(\|k\|^{-n-\mu-\varepsilon})$. This is because we have excluded the singularity of $\hat{\phi}$ from the range of integration and thereby made the integrand smooth.

Next, we may even shrink the cut-off function's support further by introducing any fixed $\hat{\varepsilon} > 0$ and by multiplying the integrand by $(1 - \rho(\frac{\cdot}{\hat{\varepsilon}}))$ for the part that contributes the $O(\|k\|^{-n-\mu-\varepsilon})$ term and by $\rho(\frac{\cdot}{\hat{\varepsilon}})$ for the rest. Hence it remains to show that for large $\|k\|$

$$(4.2) \qquad \int_{\mathbb{T}^n} \frac{e^{ix\cdot k} \, \|x\|^\mu \, \rho(\frac{x}{\hat{\varepsilon}}) \, dx}{1 + C_1^{-1} \, \|x\|^\mu \left(\widetilde{\phi}\,(\|x\|) + \sum{}' \hat{\phi}\,(\|x - 2\pi j\|) \right)}$$

is the same as a constant multiple of $\|k\|^{-n-\mu} + O(\|k\|^{-n-\mu-\varepsilon})$, where the first term in the short asymptotic expansion is *absent* if μ is an even integer, i.e. we have then only to establish $O(\|k\|^{-n-\mu-\varepsilon})$.

If $\hat{\varepsilon}$ is small enough, then we can expand one over the denominator in an absolutely convergent geometric series: so, if $\hat{\varepsilon}$ is sufficiently small,

$$\frac{1}{1 + C_1^{-1} \, \|x\|^\mu \left(\widetilde{\phi}\,(\|x\|) + \sum{}' \hat{\phi}\,(\|x - 2\pi j\|) \right)}$$

is the same as the infinite series

$$\sum_{\ell=0}^{\infty} \|x\|^{\mu\ell} (-C_1)^{-\ell} \left(\widetilde{\phi}\,(\|x\|) + \sum{}' \hat{\phi}\,(\|x - 2\pi j\|) \right)^\ell.$$

We note that the smoothness of the ℓth term in that series increases linearly with ℓ, that is, the expression for a single summand is

$$\int_{\mathbb{T}^n} e^{ix\cdot k} \, \|x\|^{(\ell+1)\mu} \, \rho\left(\frac{x}{\hat{\varepsilon}}\right) (-C_1)^{-\ell} \left(\widetilde{\phi}\,(\|x\|) + \sum{}' \hat{\phi}\,(\|x - 2\pi j\|) \right)^\ell dx,$$

for $\ell = 0, 1, \ldots$. So they contribute faster and faster decaying terms to the Fourier transform. Hence, instead of treating (4.2), we may show that

$$\int_{\mathbb{T}^n} e^{ix\cdot k} \, \|x\|^\mu \, \rho\left(\frac{x}{\hat{\varepsilon}}\right) dx = \int_{\mathbb{R}^n} e^{ix\cdot k} \, \|x\|^\mu \, \rho\left(\frac{x}{\hat{\varepsilon}}\right) dx$$

should be $\doteq \|k\|^{-n-\mu} + O(\|k\|^{-n-\mu-\varepsilon})$, where, again, the first term is omitted if μ is an even integer.

This, however, follows from the fact that the distributional Fourier transform of $\| \cdot \|^{\mu}$ is a fixed constant multiple of $\| \cdot \|^{-n-\mu}$ (Jones, 1982, p. 530) if μ is not an even integer. In the opposite case, the distributional Fourier transform is anyway a multiple of a δ-function or a derivative thereof. Thus we get $O(\|k\|^{-\ell})$ for any ℓ in that case from the above display. The theorem is proved. $\qquad\Box$

In summary, we have shown that the coefficients of L exist under the above assumptions and that

$$(4.3) \qquad L(x) = \sum_{k \in \mathbb{Z}^n} c_k \, \phi(\|x - k\|), \quad x \in \mathbb{R}^n,$$

is well-defined if $|\phi(\|x\|)| = O(\|x\|^{\mu-\varepsilon})$ for any $\varepsilon > 0$. These assumptions are true for all radial basis functions of the form (with $\varepsilon = n/2$ for instance)

$$(4.4) \qquad \phi(r) = \begin{cases} r^{2k-n} \log r, & 2k - n \text{ an even integer}, \\ r^{2k-n}, & 2k - n \notin 2\mathbb{Z}, \end{cases}$$

where $2k > n$, $\mu = 2k$ and k need not be an integer. In those cases, all M are admitted due to properties of the Fourier transform $\hat{\phi}(r)$ which is a constant times r^{-2k}. The same μ and M are suitable for shifted versions of (4.4) that are smoother, i.e. where r is replaced by $\sqrt{r^2 + c^2}$, c a positive constant, since as pointed out in Chapter 2, modified Bessel functions then occur in the transform

$$\hat{\phi}(r) \doteq K_k(cr)/(r/c)^k \sim r^{-2k}, \quad r \to 0_+,$$

which are C^{∞} except at the origin and nonzero, decay exponentially and satisfy $K_k(r) \sim r^{-k}$ for $r \to 0_+$ and positive real k. Here $k = (n+1)/2$ when multi-quadric functions are studied. Hence, in particular, the multiquadric function is included for $\mu = n + 1$. For instance, in one dimension, the Lagrange function coefficients for the multiquadric function decay at least cubically.

As a by-product of (4.1) we note the following form of L, where we use the Fourier inversion theorem. To wit, at least in the distributional sense, the Fourier transform of L can be computed as follows. Later, in Theorem 4.3, we shall see that the Fourier transform of L exists even in the classical sense, because L is absolutely integrable. For now, using distributional Fourier theory, we can write

$$\hat{L}(y) = \sum_{k \in \mathbb{Z}^n} c_k e^{-iy \cdot k} \hat{\phi}(\|y\|).$$

By virtue of the form of our symbol,

$$\hat{L}(y) = \sigma(y)^{-1}\hat{\phi}(\|y\|)$$

$$= \frac{\hat{\phi}(\|y\|)}{\sum_{\ell \in \mathbb{Z}^n} \hat{\phi}(\|y - 2\pi\ell\|)}.$$

Now, due to the properties of the radial basis function's Fourier transform, namely, in particular, (A2a), this function is absolutely integrable and can be extended to a continuous, absolutely integrable function. Therefore we can represent L as

$$(4.5) \qquad L(x) = \frac{1}{(2\pi)^n} \int_{\mathbb{R}^n} e^{ix \cdot y} \frac{\hat{\phi}(\|y\|)}{\sum_{\ell \in \mathbb{Z}^n} \hat{\phi}(\|y - 2\pi\ell\|)}\, dy.$$

This expression will be of fundamental importance to us. As a first straightforward consequence we note the following result.

Theorem 4.3.   *Under the assumptions of Theorem 4.2, L satisfies*

$$|L(x)| = O\left((1 + \|x\|)^{-n-\mu}\right);$$

in particular, L is absolutely integrable and has the Fourier transform

$$\hat{L}(y) = \frac{\hat{\phi}(\|y\|)}{\sum_{j \in \mathbb{Z}^n} \hat{\phi}(\|y - 2\pi j\|)}, \qquad y \in \mathbb{R}^n.$$

Precisely, it is true that with this Fourier transform, L satisfies

$$|L(x)| = \underline{\varepsilon}\, C\, (1 + \|x\|)^{-n-\mu} + O\left((1 + \|x\|)^{-n-\mu-\varepsilon}\right)$$

with $\underline{\varepsilon} \equiv \mu \,(\mathrm{mod}\, 2)$, $\underline{\varepsilon} \in \{0, 1\}$.

Proof: Exactly the same technique as in the previous proof can be applied and (4.5) used for the proof of decay.   □

Hence we observe that any f with polynomial growth

$$|f(x)| = O\left((1 + \|x\|)^{\mu-\varepsilon}\right),$$

with a positive constant ε, can be interpolated by

$$(4.6) \qquad s(x) = \sum_{j \in \mathbb{Z}^n} f(j) \, L(x - j), \qquad x \in \mathbb{R}^n,$$

or by

$$(4.7) \qquad s_h(x) = \sum_{j \in \mathbb{Z}^n} f(jh) \, L\left(\frac{x}{h} - j\right), \qquad x \in \mathbb{R}^n.$$

In particular, all polynomials f can be interpolated if their total degree is less than μ.

It is a remark which we shall not prove here that for radial basis functions of the form (4.4) with $k \in \mathbb{N}$, n even and less than $\frac{k}{2}$, the decay in Theorems 4.2 and 4.3 is in fact exponential (Madych and Nelson, 1990a). This is related to the fact that the Fourier transform of even powers of $\|x\|$ is supported at a point (it is the multiple of a δ-function or a partial derivative thereof) and the Fourier transform is analytic in a tube about the real axis. Now, for analytic integrable functions, the exponential decay of their Fourier transforms can be proved in a standard way by using Cauchy's theorem for analytic functions. If the function is not just analytic but entire and of exponential type with a growth condition on the real axis, then the Fourier transform is even compactly supported, but this fact is not used, and is beyond the interest of this monograph anyway.

Incidentally, similar facts (but not the full analyticity) were already employed in the proof of Theorem 4.2 and account for the presence of the ε there. (Here ε is as used in Theorem 4.2.)

In fact, it requires hardly any more work from us to show that under no further conditions, polynomials such as the above are reproduced exactly. That is a highly relevant fact to the establishment of convergence orders, when functions are approximated by their values on the scaled grids $h\mathbb{Z}^n$. We have already seen the reason for this remark in Chapter 2, where the linear polynomial reproduction properties of univariate multiquadric interpolation were used to prove – essentially – quadratic convergence of quasi-interpolation and interpolation with multiquadrics. The same features of radial basis function approximation will be established here in several dimensions and for our larger class of radial basis functions with the aid of Theorem 4.4. Thus, the polynomial reproduction property is of fundamental importance to the analysis in this chapter.

We remark that the property is in fact quite straightforward to establish as the following result shows.

Theorem 4.4.   *Under the assumptions of Theorem 4.2, it is true that we have the recovering property for interpolation*

$$(4.8) \qquad \sum_{j \in \mathbb{Z}^n} p(j) \, L(x - j) = p(x), \quad x \in \mathbb{R}^n,$$

for all polynomials p in n unknowns of total degree less than μ (μ need not be integral). Further, there is a polynomial of degree at least μ that is not recovered by the above sum, even if the infinite sum in (4.8) is well-defined for that higher degree polynomial.

Proof: Let p be a polynomial in n unknowns of degree less than μ. Then, by the Poisson summation formula, the left-hand side of expression (4.8) is, using again the standard notation

$$D = D_t = \left(\frac{\partial}{\partial t_1}, \, \frac{\partial}{\partial t_2}, \, \dots, \, \frac{\partial}{\partial t_n} \right)^T,$$

the sum of partial derivatives of an exponential times the Lagrange function's Fourier transform:

$$\sum_{j \in \mathbb{Z}^n} p(iD) \, \{e^{-ix \cdot t} \, \hat{L}(t)\}_{t = 2\pi j}.$$

We note that the multivariate polynomial p applied to iD gives rise to a linear combination of partial derivatives. The index $t = 2\pi j$ indicates that those partial derivatives are computed first, and only then does the evaluation take place. We separate the term with index $j = 0$ from the rest of the sum. Therefore, it becomes

$$(4.9) \qquad \sum_{j \in \mathbb{Z}^n \setminus \{0\}} p(iD) \, \{e^{-ix \cdot t} \, \hat{L}(t)\}_{t = 2\pi j} + \hat{L}(0) p(x).$$

We claim that the above expression is in fact $p(x)$, as demanded. This is true because (4.5) implies

$$\hat{L}(0) = 1$$

and

$$(D^\alpha \hat{L})(2\pi j) = 0 \quad \forall j \in \mathbb{Z}^n \setminus \{0\}, \ |\alpha| < \mu,$$

and finally

$$(D^\alpha \hat{L})(0) = 0 \quad \forall \alpha, 0 < |\alpha| < \mu,$$

where $\alpha \in \mathbb{Z}_+^n$. Indeed, all these properties of the Fourier transform of the Lagrange function L come from the singularity of the Fourier transform of $\hat{\phi}$ at the origin, because this singularity which occurs in the denominator of $\hat{L}$ induces the manifold zeros of $\hat{L}$ at nonzero multiples of 2π. We can for example expand $\hat{L}$ in neighbourhoods about zero and nonzero multiples of 2π to verify that the above zero properties are therefore direct consequences of condition (A3).

Thus the first part of the theorem is proved. As to the second, it is clear that the above sum (4.8) is ill-defined if μ is not an even integer and thus $L(x) \sim \|x\|^{-n-\mu}$, and when p has total degree $\geq \mu$. However, (4.8) may very well be well-defined if μ is an even positive integer and L thus decays faster. Let $p(x) = \|x\|^{\mu}$, μ an even positive integer. Thus p is a polynomial of total degree μ. In this case (4.9) contains terms like

$$(4.10) \qquad e^{-i2\pi j \cdot x} \|iD\|^{\mu} \hat{L}(2\pi j), \quad j \in \mathbb{Z}^n,$$

where the right-hand factor is a nonzero constant (it is a certain nonzero multiple of the reciprocal of the nonzero coefficient of the $r^{-\mu}$ term in $\hat{\phi}(r)$). Thus (4.8) cannot be true, and there cannot be cancellation in the sum of expressions (4.10) because of the linear independence of the exponential functions with different arguments. □

Note that this theorem implies polynomial exactness of degree $<2k$ for all examples (4.4) and their shifts; e.g. for the multiquadric function where $2k = n + 1$, we can recover all polynomials exactly if their total degree does not exceed n in $\mathbb{R}^n$. This is in itself a remarkable observation, namely that infinite linear combinations of nonanalytic (often nondifferentiable), nonpolynomial radial basis functions can recover polynomials exactly. We will come back to and use this fact often in the book.

4.2 Convergence analysis

4.2.1 Approximation orders on gridded data

It is a natural question to ask now whether the polynomial recovery enables us to deduce convergence (or synonymously approximation) orders for approximations (4.7) if f is a sufficiently differentiable target function. The answer is affirmative, as we shall see in this subsection. There are two very practical purposes of this convergence analysis. The first one is a question asked often by anyone who uses radial basis functions in practice, namely to classify the radial basis functions according to the convergence orders that can be obtained in order to know more about their usefulness in applications. The second one comes up in the very frequent situation when radial basis

functions are used as approximants within other numerical algorithms which
also have certain convergence orders. It is usual in those cases to use approximants inside of 'outer' algorithms which have at least the same convergence
orders.

As we shall see, the convergence orders are closely related to the order of the
singularity of the radial function's Fourier transform at zero. In our conditions
on ϕ we have denoted this order – which need not be integral – by μ. Indeed,
we can always show that (4.7) approximates f uniformly up to order $O(h^\mu)$
and that that is the best possible order. This approximation order is of course
only obtained for sufficiently smooth f and the smoothness we need to require
depends on the speed of L's decay, the polynomial recovery being fixed as we
have seen above. In order to demonstrate the techniques, we begin with an easy
case. That is, the simplest result along those lines is the following Theorem 4.5.
Note that in these convergence results, the fact that L satisfies the Lagrange
conditions is immaterial. We only use its decay and its polynomial recovery
properties. Therefore the results we obtain on convergence are true both for
quasi-interpolation and for interpolation where the translates of ψ and L are
used, respectively. We formulate the approximants below only with ψ and state
the (e.g. decay-) conditions as conditions on L.

Theorem 4.5.   *Let the approximant (4.6) on the multivariate integers be exact
for all polynomials of total degree less than μ and suppose*

$$|L(x)| = O\left((1 + \|x\|)^{-n-\mu-\varepsilon}\right)$$

*with $\varepsilon > \lceil \mu \rceil - \mu$. Then the scaled approximant (4.7) gives the uniform con-
vergence estimate $\|s_h - f\|_\infty = O(h^\mu)$ for all $f \in C^{\lceil \mu-1 \rceil}(\mathbb{R}^n)$ whose $\lceil \mu-1 \rceil$st
derivatives are Lipschitz-continuous.*

Proof: For any given $x \in \mathbb{R}^n$, let p be the following x-dependent (!) polynomial. To wit, we take it to be the $(\lceil \mu \rceil - 1)$st degree Taylor polynomial to f at
the point x. This Taylor polynomial

$$p(y) = f(x) + \sum_{0 < |\alpha| < \lceil \mu \rceil} \frac{D^\alpha f(x)}{\alpha!}(y - x)^\alpha$$

is recovered by (4.6) because its degree is less than μ as $\lceil \mu \rceil < \mu + 1$. It is in
particular true that

$$f(x) - s_h(x) = \sum_{j \in \mathbb{Z}^n} \left(f(x) - f(jh) \right) L\left(\frac{x}{h} - j\right)$$

is the same as

$$\sum_{j \in \mathbb{Z}^n} \Big(p(jh) - f(jh) \Big) \, L\Big(\frac{x}{h} - j\Big),$$

because

$$\sum_{j \in \mathbb{Z}^n} (jh - x)^{\alpha} \, L\Big(\frac{x}{h} - j\Big) = 0$$

for all monomials $(\,\cdot\,)^{\alpha}$ of total degree less than μ. Thus using the Lipschitz continuity condition and the decay of the function L stated in Theorem 4.3, there exist some fixed positive constants C_i, $i = 1, 2$, such that

$$(4.11) \quad |f(x) - s_h(x)| = \left| \sum_{j \in \mathbb{Z}^n} \Big(f(jh) - p(jh) \Big) \, L\Big(\frac{x}{h} - j\Big) \right|$$

$$\leq \sum_{\|x - jh\|_{\infty} \leq 2h} |f(jh) - p(jh)| \, \left| L\Big(\frac{x}{h} - j\Big) \right|$$

$$+ \sum_{\|x - jh\|_{\infty} > 2h} |f(jh) - p(jh)| \, \left| L\Big(\frac{x}{h} - j\Big) \right|.$$

This is at most

$$C_1 \sum_{\|x - jh\|_{\infty} \leq 2h} \|x - jh\|_{\infty}^{\lceil \mu \rceil}$$

$$+ C_2 \sum_{\|x - jh\|_{\infty} > 2h} \|x - jh\|_{\infty}^{\lceil \mu \rceil} \frac{h^{\mu + n}}{\|x - jh\|_{\infty}^{\varepsilon + n + \mu}}$$

which equals

$$O(h^{\mu}) + C_2 \, h^{\mu} \sum_{\|x - jh\|_{\infty} > 2h} \frac{h^n}{\|x - jh\|_{\infty}^{n + \varepsilon + (\mu - \lceil \mu \rceil)}}.$$

That is $O(h^{\mu})$, as desired. We note that we reach the final conclusion because the sum that multiplies C_1 contains only a finite number of terms, independently of h, and because the final sum is uniformly bounded as h goes to zero, $\lceil \mu \rceil - \mu$ being less than ε. □

The message from the above proof is that we need precise information about the decay of the cardinal function to obtain the best convergence results. The polynomial recovery cannot be improved upon what we have already, cf. the last result of the previous section, and hence it cannot be used to any more advantage than it is. However, better decay, that is, better localised cardinal functions, can give optimal approximation orders. Now, there are various other forms of such convergence results when L has different orders of decay. Their main purpose is to admit $\varepsilon = 0$ in the above result, but they require extra conditions on the derivatives of f. A slightly more complicated version than the previous result is

Theorem 4.6. *Let the assumptions of Theorem 4.5 hold for noninteger μ and $\varepsilon = 0$. Then (4.7) gives the error estimate*

$$\|s_h - f\|_\infty = O(h^\mu), \qquad h \to 0,$$

for all $f \in C^{\lceil \mu \rceil}(\mathbb{R}^n)$ whose $\lceil \mu \rceil$th and $(\lceil \mu \rceil - 1)$st total order derivatives are uniformly bounded.

Proof: We begin as in the previous proof and define the Taylor polynomial in the same way. Therefore we also use the polynomial exactness in the same way and consider the sum of shifts and dilates of L multiplied by $p(jh) - f(jh)$. However, then we consider three different ranges of indices in the approximant now. That is, we divide (4.11) up into an $O(h^\mu)$ term plus two further expressions, namely for positive constants C_2 and C_3

$$C_2 \sum_{2h < \|x - jh\|_\infty < 1} \|x - jh\|_\infty^{\lceil \mu \rceil} \frac{h^{\mu + n}}{\|x - jh\|_\infty^{\mu + n}}$$

plus

$$C_3 \sum_{\|x - jh\|_\infty \geq 1} \|x - jh\|_\infty^{\lceil \mu \rceil - 1 - \mu - n} h^{\mu + n},$$

where the last term is a consequence of the boundedness of the $(\lceil \mu \rceil - 1)$st order derivatives and it is, indeed, $O(h^\mu)$ as before, because $\lceil \mu \rceil - 1 < \mu$. Moreover, we bound the newly introduced second term by an integral from above:

$$\sum_{2h < \|x - jh\|_\infty < 1} \|x - jh\|_\infty^{-n - (\mu - \lceil \mu \rceil)} h^{\mu + n} \leq C_4 h^\mu \int_h^1 y^{-(\mu - \lceil \mu \rceil) - 1} \, dy$$

which is $O(h^\mu)$, because in this theorem the hypotheses give $\mu - \lceil \mu \rceil \in (-1, 0)$, as required.   □

In order to drop the positive ε in the assumptions for an integer-valued μ, much more work is required (note that the above derivation in the final display of the proof of Theorem 4.6 could give a $\log h$ term in that case unless we change the proof because $\int_h^1 y^{-1} dy = (-\log h)$). This is done in the next theorem. It is taken from a paper by Powell (1992a).

Theorem 4.7. *Let the assumptions of Theorem 4.6 hold except that now μ is an odd integer. Then (4.7) gives the error estimate*

$$\|s_h - f\|_\infty = O(h^\mu), \qquad h \to 0,$$

for h tending to zero and for all $f \in C^{\lceil \mu + 1 \rceil}(\mathbb{R}^n)$ *whose partial derivatives of total orders* $\mu - 1$, μ *and* $\mu + 1$ *are uniformly bounded.*

Proof: Let $k = \mu - 1$. Let p be the same Taylor polynomial as before. Next, let q be the polynomial of degree μ that vanishes at x of order $k + 1$, that is, it vanishes at x with all its partial derivatives of *total* degree $\leq k$ vanishing at x as well, and that also matches all partial derivatives of total degree $\mu = k + 1$ of f at this point. Thus, for all $\tilde{x}$, there are $(x, \tilde{x})$-independent constants C_i such that the following inequalities hold due to Taylor's theorem and the boundedness assumptions in the statement of Theorem 4.7:

$$|f(\tilde{x}) - p(\tilde{x})| \leq C_1 \|x - \tilde{x}\|_\infty^k,$$
$$|f(\tilde{x}) - p(\tilde{x})| \leq C_2 \|x - \tilde{x}\|_\infty^{k+1},$$
$$|q(\tilde{x})| \leq C_3 \|x - \tilde{x}\|_\infty^{k+1},$$
$$|f(\tilde{x}) - p(\tilde{x}) - q(\tilde{x})| \leq C_4 \|x - \tilde{x}\|_\infty^{k+2}.$$

Moreover, according to Theorem 4.3, there is a bounded function $\eta : \mathbb{R}^n \to \mathbb{R}$ such that for an $\varepsilon > 0$ and any x

$$|L(x) - \eta(x)(1 + \|x\|)^{-n-\mu}| = O\left((1 + \|x\|)^{-n-\mu-\varepsilon}\right).$$

We consider once more the error formula

$$f(x) - s_h(x) = \sum_{j \in \mathbb{Z}^n} \left(p(jh) - f(jh) \right) L\left(\frac{x}{h} - j\right).$$

By periodicity of the grid of centres and because we are considering uniform norms, there is no loss in generality if we assume subsequently $\|x\|_\infty \leq \frac{h}{2}$ for evaluating and estimating the error. We divide the ranges of indices in the sum much as in the previous proofs. In particular, showing that the expression

$$\sum_{\|jh\|_\infty > 1} |p(jh) - f(jh)| \left| L\left(\frac{x}{h} - j\right) \right|$$

is $O(h^\mu)$ follows precisely the same lines as before using the decay of the cardinal function. This we do not repeat here. Thus we must show that for $\|x\|_\infty \leq \frac{h}{2}$ we have this estimate for the error:

$$(4.12) \qquad \left| \sum_{\|jh\|_\infty \leq 1} \left(p(jh) - f(jh) \right) L\left(\frac{x}{h} - j\right) \right| = O(h^\mu).$$

We replace the first term in parentheses in (4.12) by $q(jh)$ which incurs an error of at most a multiple of

$$\sum_{\|jh\|_\infty \leq 1} \|x - jh\|_\infty^{k+2} \left(1 + \left\|\frac{x}{h} - j\right\|_\infty\right)^{-n-k-1}$$

$$\leq h^{k+2} \sum_{\|jh\|_\infty \leq 1} \left(1 + \left\|\frac{x}{h} - j\right\|_\infty\right)^{-n+1}$$

$$\leq C_5 h^{k+1} \int_0^1 (h + y)^0 \, dy = O(h^{k+1}).$$

We note that here we have lost a power h^n in the summand. This is due to a change in coordinates which replaces the sum in n dimensions by a univariate integral, that is by changing to polar coordinates when we have replaced the sum by an integral as in the previous proof. If we additionally replace $L(\frac{x}{h} - j)$ by $\eta(\frac{x}{h} - j)(1 + \|\frac{x}{h} - j\|)^{-n-k-1}$ in the revised form (4.12) we get an error that does not exceed a fixed constant multiple of

$$\sum_{\|jh\|_\infty \leq 1} \|x - jh\|_\infty^{k+1} \left(1 + \left\|\frac{x}{h} - j\right\|_\infty\right)^{-n-\mu-\varepsilon} \leq C_6 h^{\mu+\varepsilon} \int_0^1 (h + y)^{k-\mu-\varepsilon} dy$$

which is, as desired, $O(h^{k+1}) = O(h^\mu)$ for $h \to 0$.

Thus it remains to show that, still for $\|x\|_\infty \leq \frac{h}{2}$, it is true that

$$(4.13) \qquad \left| \sum_{\|jh\|_\infty \leq 1} (q(jh)\, \eta\left(\frac{x}{h} - j\right)\left(1 + \left\|\frac{x}{h} - j\right\|_\infty\right)^{-n-k-1} \right| = O(h^\mu).$$

We wish to make a final change in the error expression above. We want to replace q by $q(\cdot + x)$ in (4.13). What additional error does this change introduce? By the mean-value theorem and because of the bounds on q and $\|x\|_\infty$ stated above,

$$|q(jh + x) - q(jh)| \leq C_7 \cdot h(h + \|x - jh\|_\infty)^k, \qquad \|jh\|_\infty \leq 1.$$

Thus we may replace q by $q(\cdot + x)$ and get an error of at most a fixed constant multiple of the following sum:

$$\|\eta\|_\infty h \sum_{\|jh\|_\infty \leq 1} (h + \|x - jh\|_\infty)^k \left(1 + \left\|\frac{x}{h} - j\right\|_\infty\right)^{-n-k-1}$$

which is at most

$$C_8 h^{k+2} \int_0^1 (h + y)^{-2} \, dy = O(h^{k+1}),$$

and which is therefore of the order that we desire to obtain in our required result. Therefore we may make the proposed change in (4.13) without changing the final result beyond a term of $O(h^\mu)$. Further, as $q(\cdot + x)$ is a homogeneous polynomial of degree μ, μ being an odd integer, we have $q(-hj + x) = -q(hj + x)$, $\|jh\|_\infty \le 1$. Thus we bound, for $j \in \mathbb{Z}_+^n$ and $\|x\|_\infty \le \frac{h}{2}$,

$$
\left| \sum_{\|jh\|_\infty \le 1} \left(q(jh + x) \left(1 + \left\| \frac{x}{h} - j \right\|_\infty \right)^{-n-k-1} \eta \left(\frac{x}{h} - j \right) \right. \right.
$$

$$
\left. \left. + q(-jh + x) \left(1 + \left\| \frac{x}{h} + j \right\|_\infty \right)^{-n-k-1} \eta \left(\frac{x}{h} + j \right) \right) \right|.
$$

Moreover, we may factor out the function η and bound, by Hölder's inequality, by $\|\eta\|_\infty$. In addition, if we employ the estimate

$$
\left| \left(1 + \left\| \frac{x}{h} - j \right\|_\infty \right)^{-n-k-1} - \left(1 + \left\| \frac{x}{h} + j \right\|_\infty \right)^{-n-k-1} \right| \le C_9 (1 + \|j\|_\infty)^{-n-k-2}
$$

we can deduce that the above sum is in fact bounded by a fixed constant multiple of

$$
\sum_{\|jh\|_\infty \le 1} \|jh\|_\infty^{k+1} (1 + \|j\|_\infty)^{-n-k-2} = O(h^{k+1}) = O(h^\mu), \quad h \to 0,
$$

as required. The theorem is proved. $\qquad\qquad\qquad\qquad\qquad\qquad\square$

We note that for even μ, the dominant term in the expression for $L(x)$ of Theorem 4.3 does not occur, so that we always have faster decay than $(1 + \|x\|)^{-n-\mu}$. Therefore we have always Theorem 4.5 at hand which applies. It is natural now to ask two further questions that we shall discuss next.

The first one is, are the orders stated in the last three theorems saturation orders, i.e. do there exist different classes of functions (smaller than those having fixed Lipschitz smoothness) for which a larger power of the spacing h is possible (see Theorem 4.5)? And the second one is, can those orders be achieved without using interpolation?

4.2.2 Quasi-interpolation versus Lagrange interpolation

We address the second one first because the answer to it is simpler and more familiar to us, especially in view of the work we have already done in Chapter 2; it leads to the so-called quasi-interpolation, where approximants

$$
s(x) = \sum_{j \in \mathbb{Z}^n} f(j)\, \psi(x - j), \quad x \in \mathbb{R}^n,
$$

are studied for – during the rest of this section – only finite linear combinations

$$(4.14) \qquad \psi(x) = \sum_{k \in \mathbb{Z}^n} c_k \, \phi(\|x - k\|), \quad x \in \mathbb{R}^n,$$

i.e. the sequence $\{c_k\}_{k \in \mathbb{Z}^n}$ in (4.14) has finite support with respect to k. The essential feature of this quasi-interpolation, which has in fact been studied for radial basis functions much earlier than interpolation on $\mathbb{Z}^n$ (e.g. in Jackson, 1988, Rabut 1990) is that it can obtain in many cases the same polynomial recovery and (almost) the same localisation properties for ψ as Lagrange functions. Therefore, from the point of view of approximation quality, it is just as attractive as interpolation. It achieves essentially the same approximation orders and we do not need to solve linear systems for the interpolation coefficients. On the other hand, certain finite systems of equations have to be solved in order to compute the coefficients, and we shall explain this and give examples now.

These polynomial reproduction and convergence features are achieved by recalling that it is the *Fourier transform* (4.5) of the function whose shifts we use in (4.6) which helps in the proof of polynomial recovery (and, of course, just as much in the localisation as we have demonstrated in the various convergence proofs above). Now, observing that we can consider the Fourier transform of ψ like considering $\hat{L}$ above, at least in a 'generalised' sense of distributional Fourier transforms, namely in the form

$$(4.15) \qquad \hat{\psi}(t) = \hat{\phi}(\|t\|) \sum_{k \in \mathbb{Z}^n} c_k \, e^{-it \cdot k}, \qquad t \in \mathbb{R}^n,$$

we can try to mimic the essential features of $\hat{L}$, that is, in particular, its properties at the points in $2\pi \mathbb{Z}^n$.

As stated above we shall assume that the above sum in (4.15) is a trigonometric polynomial, i.e. that the $\{c_k\}_{k \in \mathbb{Z}^n}$ are compactly supported. This we do in full appreciation of the fact that thereby we exclude many important cases, which are, however, taken care of by our Lagrange functions. In other words, the only real gain we get with quasi-interpolation from our point of view is if the $\{c_k\}_{k \in \mathbb{Z}^n}$ are compactly supported because they are simpler to work with; if they are not, we can just as well use the previously demonstrated theory on Lagrange functions.

Now, given any μ which we assume to be integral, it follows from the proof of Theorem 4.4 that what we need to do is find finitely many coefficients c_k, such that $\hat{\psi}$ as stated in (4.15) satisfies the following conditions that induce the same behaviour of $\hat{\psi}$ on $2\pi \mathbb{Z}^n$ as $\hat{L}$ has. The first condition is that the integral of ψ is one or, in the Fourier domain, it has the property

$$(4.16) \qquad \hat{\psi}(0) = 1.$$

The second condition concerns the partial derivatives of total order less than μ at zero,

$$(4.17) \qquad (D^\alpha \hat{\psi})(0) = 0 \quad \forall \alpha, \; 0 < |\alpha| < \mu,$$

and the final condition is responsible for the polynomial recovery as in the proof of Theorem 4.4:

$$(4.18) \qquad (D^\alpha \hat{\psi})(2\pi j) = 0 \quad \forall j \in \mathbb{Z}^n \backslash \{0\}, \; 0 \le |\alpha| < \mu.$$

Among those three conditions, (4.17) and (4.18) are, however, simple linear conditions that can always be satisfied by choosing a large enough support set for the coefficients c_k. It has to be noticed that a zero of g at the origin forces zeros of the same order at all $2\pi j$, $j \in \mathbb{Z}^n$, by the 2π-periodicity. As a consequence, the conditions are nothing else than moment conditions on the

$$g(t) = \sum_{k \in \mathbb{Z}^n} c_k e^{-it \cdot k}, \quad t \in \mathbb{T}^n,$$

our simple trigonometric polynomial, where satisfaction of the conditions (4.16)–(4.18) essentially means $g(t) = O(\|t\|^\mu)$ in a neighbourhood of the origin.

These conditions (4.16)–(4.18) are also known as the *Strang and Fix conditions* (see, e.g., Jia and Lei, 1993) and there is very much important work on their sufficiency and necessity – in tandem with extra conditions for so-called controlled approximation order – for the establishment of polynomial reproduction and approximation order through approximation with quasi-interpolation (4.14). In particular, one distinguishes between the conditions for *polynomial preservation*, for which it is sufficient that (4.18) holds which guarantees that the quasi-interpolant to a polynomial is again a polynomial, and the conditions for the actual polynomial reproduction (4.16)–(4.18). While the meaning of polynomial reproduction is self-explanatory, we mean by polynomial preservation that the result of the application of the operator to a polynomial is a polynomial.

Results are available both when ψ is of compact support, a case particularly important for box-splines (see our third chapter and, for instance, de Boor, Höllig and Riemenschneider, 1993), and when ψ is not of compact support (Light and Cheney, 1992a) which applies to our radial basis functions and has been studied much more recently. Also, results are available for general L_p-norms and not only for uniform error estimates. Finally, there are extensions to the theory of the Strang and Fix type conditions when the approximants are not just linear combinations of one basis function ψ, but when ψ can be chosen from a finite set Ψ of basis functions. This is especially suitable when we wish to generate a whole space of multivariate piecewise polynomials of a given total degree and given smoothness by box-splines, because one box-spline and its translates alone may not be sufficient for this purpose (in contrast to univariate spline interpolation where the B-splines of one fixed degree alone are good enough). This is not needed for radial basis function approximation.

A particularly important issue in the discussion of Strang and Fix conditions is their necessity for the polynomial preservation or reproduction. This becomes especially delicate when larger sets Ψ of bases are used as described above, and this is where the aforementioned conditions on controlled approximation come in to render the conditions necessary.

All this being said, in this monograph we do not, however, especially build on this theory, because for our applications, the available approximation orders come very specifically from our conditions (A1)–(A3) on the radial basis functions which lead to Lagrange functions. The latter then lead directly to our approximations and their properties which are, as we shall argue now, optimal anyhow, and fortunately no more general theory is required.

The crux lies in satisfying (4.16) as well as (4.17) which requires μ to be even in the first place for quasi-interpolation, because otherwise the short expansion at zero, namely,

$$g(t) = \sum_{k \in \mathbb{Z}^n} c_k \, e^{-it \cdot k} \sim \|t\|^{\mu}, \quad \|t\| \to 0,$$

can never be achieved for a compactly supported coefficient sequence, whenever $\|t\|^{\mu}$, $\mu \notin 2\mathbb{Z}_+$, is not a polynomial. If $\mu \in 2\mathbb{Z}_+$, however, then all is well and (4.16)–(4.18) can be attained. Many examples are given in Jackson (1988). No further discussion of the polynomial recovery is needed therefore, because the same methods with the Poisson summation formula are employed as above in the theorems about interpolation, but we need to know how rapidly ψ decays. Generally speaking, almost the same decay rates as with cardinal interpolation can be provided. At any rate, if μ is an even integer, sufficient decay for the polynomial recovery and for obtaining convergence rates as in the theorems above can be achieved – except for a $\log h$ term in the error estimate that reflects the sometimes not-quite-as-good decay. The techniques to prove decay, however, are the same as before, albeit slightly more complicated to handle due to the extra freedom by the choice of g, quite unlike the canonical form of $\hat{L}$. We conclude that we have to concentrate on the decay of the quasi-interpolating basis functions ψ and how to get the coefficients for ψ to have sufficient decay.

To exemplify the theory of quasi-interpolation, we present just one theorem, as a central result giving all essential facts, namely, the following theorem.

Theorem 4.8. *Let ϕ satisfy (A1), (A2a), (A3a) for $\mu \in 2\mathbb{Z}_+$. Then there is a multivariate trigonometric polynomial g such that the function ψ renders (4.14) exact for polynomials of degree less than μ and has the asymptotic property*

$$|\psi(x)| = O(\|x\|^{-n-\mu})$$

for large $\|x\|$.

Note that in all our examples where μ includes an n-term, μ is integral (for example $\mu = n + 2$ for thin-plate splines) and thus that μ be even is actually a condition on the parity of n. For instance, the above result applies to multiquadrics, $\phi(r) = r$ and $\phi(r) = r^3$ in odd dimensions, and for thin-plate splines in even dimensions. This is typical for quasi-interpolation with radial basis functions.

The convergence theorems above obviously do not require the function L in (4.6) to be a Lagrange function as observed before in this section – only polynomial recovery and asymptotic decay are essential. We may therefore apply those results for quasi-interpolation instead of interpolation by virtue of Theorem 4.8. On the other hand, we benefited from the simplicity of the Lagrange formulation when we began the analysis not with quasi-interpolation but with cardinal interpolation and presented our convergence theorems. Moreover, as we shall see soon, the convergence properties attained by the interpolants are best possible at least in an asymptotic way and therefore it is natural to focus on interpolation so far.

The convergence rates which are attained by quasi-interpolation follow now from Theorem 4.6 except that the presence of an integer-valued μ forces a $-\log h$ factor into the convergence estimate. This follows from the last display in the proof of Theorem 4.6. This is the only way in which the approximation power of quasi-interpolation is weaker than that obtainable by interpolation. Disregarding this log-factor which grows very slowly indeed – in fact it is in practical applications usually not noticeable – we have the same asymptotic convergence behaviour of interpolation and quasi-interpolation.

Therefore we get without any further elaboration a convergence estimate of $\|s_h - f\|_\infty = O(h^\mu |\log h|)$ for

$$s_h(x) = \sum_{j \in \mathbb{Z}^n} f(jh)\, \psi\left(\frac{x}{h} - j\right), \quad x \in \mathbb{R}^n,$$

and sufficiently smooth approximands f, for instance those that satisfy the smoothness conditions of Theorem 4.5. In fact, sometimes the logarithmic term can even be removed due to further properties of ϕ much as in Theorems 4.6–4.7.

This, however, requires many careful extra considerations we do not undertake here, because we always have the cardinal interpolation at hand which gives the full orders without a logarithmic contribution.

We do not give a proof in all details of the result Theorem 4.8, but we indicate nonetheless how to approach the proof because it is instructive as to the functioning of the approximations. Clearly, the first necessary step is to seek suitable linear conditions on the $\{c_k\}_{k \in \mathbb{Z}^n}$ such that (4.16)–(4.18) hold. Among those, condition (4.18) is the easiest one because it is always fulfilled as long as g has a sufficiently high order zero at zero which forces zeros of the same order at all $2\pi j$, $j \in \mathbb{Z}^n$, by the 2π-periodicity. Therefore, this feature can always be achieved by letting g be a product of suitable powers of $(1 - \cos x_j)$, where x_j, $j = 1, 2, \dots, n$, are the components of $x \in \mathbb{R}^n$. However, (4.16) checks the order of the zeros we can impose, because it forces $g(t) \sim \|t\|^\mu$ at zero. The most difficult conditions are conditions (4.17), and they essentially mean that there is a gap between the $\|t\|^\mu$ term in the Taylor expansion of $g(t)$ about 0 and the next higher order term that may appear – that latter one may not be of any order lower than 2μ, so that $\hat{\phi}(\|t\|)g(t) = 1 + O(\|t\|^\mu)$ in a neighbourhood of the origin.

Alternatively, we can therefore view g's task as approximating $\frac{1}{\sigma}$ at zero (the rest is again a consequence of periodicity) up to a certain order. It suffices to achieve

$$g(t) - \frac{1}{\sigma(t)} = O(\|t\|^{2\mu}), \quad \|t\| \to 0_+,$$

including those estimates according with the derivatives on the left-hand side, to satisfy (4.16)–(4.18) if (A1), (A2a), (A3a) hold for the radial function.

This leaves us with showing the decay rate of ψ at infinity. What is used here again is the smoothness of $g(t)\hat{\phi}(\|t\|)$ to prove that its inverse Fourier transform, namely the absolutely convergent and therefore well-defined integral

$$\psi(x) = \frac{1}{(2\pi)^n} \int_{\mathbb{R}^n} e^{ix \cdot t} \hat{\phi}(\|t\|)g(t)\,dt, \quad x \in \mathbb{R}^n,$$

decays at infinity at a rate that is related to the smoothness of $\hat{\psi}$. For this, one notes that the decays of the inverse Fourier transforms of $\hat{\psi}$ and of $\hat{\phi}(\|t\|)g(t)-1$ are the same and that the latter function is smooth enough everywhere to provide quick decay of the inverse Fourier transform in the above display, except at the origin. There it is asymptotically $\sim \|t\|^\mu$ for small t. The same techniques as in Theorem 4.3 are then applied to derive the assertion of Theorem 4.8 but one has to look very carefully at the properties of g induced by the fulfilment of conditions (4.16)–(4.18). This is more difficult than the proof of Theorem 4.3 where only $\hat{\phi}(\|\cdot\|)$ and σ were involved and $\frac{1}{\sigma}$ took the task our g performs here. We omit the tedious details because they give no further insight into our particular problem.

An example presented by Jackson (1988) for the multiquadric function in three dimensions is as follows (all coefficients have to be divided by 8π for normalisation): the coefficients c_j, $j \in \mathbb{Z}^3$, $\|j\|_\infty \le 2$, are

$$c_0 = -49 - 81c^2,$$
$$c_{(\pm1,0,0)} = c_{(0,\pm1,0)} = c_{(0,0,\pm1)} = (142 + 369c^2)/12,$$
$$c_{(\pm2,0,0)} = c_{(0,\pm2,0)} = c_{(0,0,\pm2)} = (-25 - 54c^2)/12,$$
$$c_{(\pm3,0,0)} = c_{(0,\pm3,0)} = c_{(0,0,\pm3)} = (2 + 3c^2)/12,$$
$$c_{(\pm1,\pm1,\pm1)} = (-4 - 27c^2)/2,$$
$$c_{(\pm2,\pm2,\pm2)} = (1 + 72c^2)/48,$$
$$c_{(\pm2,\pm1,0)} = c_{(\pm2,0,\pm1)} = c_{(\pm1,\pm2,0)}$$
$$= c_{(0,\pm2,\pm1)} = c_{(\pm1,0,\pm2)} = c_{(0,\pm1,\pm2)} = \frac{3c^2}{4}.$$

All coefficients not specified above are set to zero.

4.2.3 Upper bounds on approximation orders

Another very important aspect of convergence analysis is to bound approximation orders from *above*, since, so far, we have just offered *lower* bounds for the obtainable approximation orders. Several central results along this line are due to Johnson (1997 and several later articles). Among other things he shows that our $O(h^\mu)$ convergence orders for approximation on grids are best possible for the class of radial basis functions we studied above, even if we choose to measure the error not in the uniform norm but in other L^p-norms, $1 \le p \le \infty$, and even if we only admit arbitrarily smooth functions as approximands into the interpolation or quasi-interpolation schemes. The proof of this result by Johnson is very advanced and beyond the scope of the book, but several examples will be given now. The examples will confirm in particular that the convergence orders on grids we had already obtained are the best possible orders.

We recall the usual notation $\check{f}$ for the inverse Fourier transform of a function

$$\check{f}(x) = \frac{1}{(2\pi)^n} \int_{\mathbb{R}^n} e^{ix \cdot t} f(t)\, dt, \qquad x \in \mathbb{R}^n,$$

see also the Appendix.

Further we recall the cut-off function ρ that has been used before in this chapter in the proof of Theorem 4.2. We remark that the next result, like many others in this book, does not require the 'radial' basis function to be radial; the result holds for general $\phi\colon \mathbb{R}^n \to \mathbb{R}$ with a measurable distributional Fourier transform $\hat{\phi}$. Nonetheless, because we especially study functions that are truly radial, we restrict the statement of the theorem to radial basis functions. In fact, the theorem is entirely expressed in terms of the Fourier transform. We recall finally the definition of band-limited functions, i.e. those functions whose Fourier transform is of compact support.

Theorem 4.9.   *Let* $1 \leq p \leq \infty$. *Let* $\hat{\phi}(\|\cdot\|)\colon \mathbb{R}^n \to \mathbb{R}$ *be continuous on* $\mathbb{R}^n \setminus \{0\}$ *and such that for some* $j_0 \in \mathbb{Z}^n \setminus \{0\}$ *and some* $\varepsilon \in (0, 1)$,

  (i) $\hat{\phi}(r) \neq 0$ *for almost all* $r \in [0, \varepsilon]$,
 (ii) *the* inverse Fourier transform

$$\left(\rho\!\left(\frac{\cdot}{\varepsilon}\right)\frac{\hat{\phi}(\|\cdot + 2\pi j_0\|)}{\hat{\phi}(\|\cdot\|)}\right)\check{}$$

is absolutely integrable for the above cut-off function ρ *with support in the unit ball,*
(iii) *for a measurable function* $\tilde{\rho}$ *which is locally bounded and for all positive* m

$$\left\|h^\mu\,\tilde{\rho} - \frac{\hat{\phi}(\|h\cdot + 2\pi j_0\|)}{\hat{\phi}(\|h\cdot\|)}\right\|_{\infty, B_m(0)} = o(h^\mu), \qquad h \to 0.$$

Then the $L^p(\mathbb{R}^n)$-*approximation order from a shift-invariant space which is spanned by using integer shifts with a scaling of* h *of* ϕ *cannot be more than* μ. *That is the error of the best approximation in the* L^p-*norm from the* L^p-*closure of*

$$\mathrm{span}\{\phi(\|\cdot/h - j\|) \mid j \in \mathbb{Z}^n\}$$

to the class of band-limited f *whose Fourier transform is infinitely differentiable cannot be* $o(h^\mu)$ *as* h *tends to zero.*

We note that the class of all band-limited f whose Fourier transform is infinitely differentiable is a class of very smooth local functions and if the L^p-approximation order to such smooth functions cannot be more than h^μ, it cannot be more than h^μ to any general (super)set of functions with lesser smoothness properties.

Examples: We consider the multiquadric function where the generalised transform $\hat{\phi}(\|x\|)$ (namely, the radially symmetric Fourier transform of the radially symmetric multiquadric function as a function of n variables) is

$$-\pi^{-1} K_{(n+1)/2}(cr)/(r/[2\pi c])^{(n+1)/2}, \quad r = \|x\| \geq 0.$$

Letting without loss of generality $c = 1$ we take

$$\hat{\phi}(r) = -\pi^{-1} K_{(n+1)/2}(r)/(r/2\pi)^{(n+1)/2}$$

in Theorem 4.9. This function is negative with a singularity of order $n + 1$ at the origin. Then (i) of Theorem 4.9 is certainly fulfiled.

Now consider, for a cut-off function ρ as stated above, the fraction

$$\rho\left(\frac{x}{\varepsilon}\right)\frac{\hat{\phi}(\|x + 2\pi j_0\|)}{\hat{\phi}(\|x\|)} = \rho\left(\frac{x}{\varepsilon}\right)\frac{\hat{\phi}(\|x + 2\pi j_0\|)}{C\|x\|^{-n-1}(1 + \|x\|^{n+1}\tilde{\phi}(x))}$$

with the same notation $\tilde{\phi}$ as in the proof of Theorem 4.2. This is the same as

$$\rho\left(\frac{x}{\varepsilon}\right)\frac{C^{-1}\hat{\phi}(\|x + 2\pi j_0\|)\|x\|^{n+1}}{1 + \|x\|^{n+1}\tilde{\phi}(x)}$$

whose inverse Fourier transform can be shown, precisely with the same techniques as in the proof of Theorem 4.2, to decay at least as fast as a constant multiple of $(1 + \|x\|)^{-2n-1}$. The latter, however, is absolutely integrable over n dimensions so that (ii) holds, as required.

We wish to show (iii) now. Let $\tilde{K}_\nu(z) = K_\nu(z)\cdot z^\nu$ which is a positive function on $\mathbb{R}_+$ without a singularity at the origin (precisely: it has a removable singularity at the origin) whenever ν is positive. Furthermore, letting in condition (iii)

$$\tilde{\rho}(x) = \frac{\tilde{K}_{(n+1)/2}(2\pi j_0)}{\tilde{K}_{(n+1)/2}(0)}\|x\|^{n+1}$$

gives the desired result (iii) for $\mu = n + 1$, namely that h^{n+1} is the highest obtainable order in any L^p-norm, confirming our earlier lower bound on the approximation rate to be best possible and thus the aforementioned saturation order.

As a second example, consider the easier case of a thin-plate spline radial basis function whose Fourier transform is

$$\hat{\phi}(r) = r^{-n-2}, \quad r > 0,$$

where for the sake of simplicity we are omitting a constant nonzero multiplying factor and δ-function terms, and again we write $\hat{\phi}$ as a multivariate function in agreement with our notation in Theorem 4.9. It is obvious that condition (i) is true. Moreover,

$$\rho\left(\frac{\cdot}{\varepsilon}\right)\frac{\hat{\phi}(\|\cdot+2\pi j_0\|)}{\hat{\phi}(\|\cdot\|)} = \rho\left(\frac{\cdot}{\varepsilon}\right)\|\cdot\|^{n+2}\|\cdot+2\pi j_0\|^{-n-2}.$$

If ε is small, then this is at a minimum in $C^{n+1}(\mathbb{R}^n)$, so that its inverse Fourier transform is of order $O(\|\cdot\|^{-n-1})$ for large argument, and it is continuous anyway since ρ is of compact support and infinitely differentiable. Thus (ii) holds as well. Moreover, let $\mu = n + 2$ and

$$\tilde{\rho}(x) = \|x\|^{n+2}\|2\pi j_0\|^{-n-2}.$$

Thus, j_0 still being nonzero, and letting $x \in B_m(0)$,

$$h^\mu \tilde{\rho}(x) - \frac{\hat{\phi}(\|h\cdot+2\pi j_0\|)}{\hat{\phi}(\|h\cdot\|)}$$
$$= h^{n+2}\|x\|^{n+2}\|2\pi j_0\|^{-n-2} - h^{n+2}\|x\|^{n+2}\|h\cdot+2\pi j_0\|^{-n-2},$$

that is, by the mean-value theorem, clearly $o(h^\mu)$ for diminishing h with $\mu = n + 2$ uniformly in all $B_m(0)$.

Hence h^{n+2} is, as expected, the best obtainable approximation order with thin-plate splines on a uniform grid which is scaled by a positive h, thus once more cementing our earlier lower bound as the best possible and the saturation order.

4.2.4 Approximation orders without Fourier transforms

It is evident that one needs to compute the Fourier transform of all the radial basis functions to which the above theorems apply in order to get the required results. This is not always easy except for our standard choices where they are mostly available from suitable tables (e.g. Jones, 1982) of Fourier transforms. Thus it is interesting and useful to avail oneself of results that no longer need these Fourier transforms and that have direct conditions on ϕ instead. One such theorem is the next result. It is highly instructive and suitable for this chapter, because it reveals a link between the completely monotonic functions of Section 2.1 and our cardinal functions.

Theorem 4.10. *Let $\hat{k}$ be a nonnegative integer and $\phi \in C(\mathbb{R}_+) \cap C^\infty(\mathbb{R}_{>0})$. Suppose that $\eta(t) := \frac{d^k}{dt^k}\phi(\sqrt{t})$, $t > 0$, is completely monotonic and satisfies the short asymptotic expansions (i.e. an expansion with a single element)*

$$\eta(t) \sim A^* t^{-\alpha}, \quad t \to \infty, \quad A^* \neq 0,$$

and

$$\eta(t) \sim A^0 t^{-\alpha'}, \quad t \to 0_+,$$

for positive and nonnegative exponents α and α' respectively. If $\hat{k} = 0$ we also assume that $\eta'(0_+)$ is bounded. If $\alpha, \alpha' \leq \frac{n}{2} - \hat{k} + 1$, $\alpha < \frac{n}{2} + 1$ and, when $\hat{k} > 0$, also $\alpha' < \hat{k}$, then there are $\{c_j\}_{j \in \mathbb{Z}^n}$ such that with $\mu = 2\hat{k} + n - 2\alpha$

(i) $|c_j| = O(\|j\|^{-n-\mu})$,
(ii) *(4.3) is a Lagrange function that satisfies*

$$|L(x)| = O\left((1 + \|x\|)^{-n-\mu} \right)$$

and all the conclusions of Theorems 4.3–4.7 are true, where the additional conditions on μ in those theorems have to be fulfilled too.

Note the absence of the $(-1)^k$ factor in the function η which is required to be completely monotonic here, in contrast with the result of Section 2.1. This is because any sign or indeed any constant factor is unimportant here for the cardinal functions, because it will cancel in the form (4.5) of the cardinal function anyhow.

We give a few examples for application of this result. For instance, $\phi(r) = r$ satisfies the assumptions of the above theorem with $\eta(t) = \frac{1}{2} t^{-1/2}$ which clearly provides the required properties of a completely monotonic function. Our μ here is $\mu = n + 1$, as before for the results that did include the Fourier transform of ϕ, and further $k = 1$, $\alpha = \alpha' = \frac{1}{2}$. A similar result holds for the thin-plate spline radial function $\phi(r) = r^2 \log r$, where $\eta(t) = \frac{1}{2} t^{-1}$ and $\mu = n + 2$, $k = 2$, $\alpha = \alpha' = 1$. Finally, the multiquadric function satisfies the assumptions too, with $\mu = n + 1$, and $\eta(t) = \frac{1}{2} (t + c^2)^{-1/2}$, $\alpha' = 0$, $\alpha = \frac{1}{2}$ and $k = 1$. In order to verify this, we note that this η is clearly completely monotonic and that it satisfies the required short asymptotic expansions at zero and for large argument. It is easy to check that the exponents for those expansions are within the required bounds.

Of course, the way this theorem is established is by showing that ϕ has a distributional Fourier transform $\hat{\phi}(\| \cdot \|)$ which satisfies the assumptions (A1), (A2a), (A3a) as a consequence of the theorem's requirements. In order to do this, one begins by noticing that the Bernstein–Widder theorem implies the function η in Theorem 4.10 has the following representation (much like the case we studied in Chapter 2), where we start with the choice $\hat{k} = 0$:

$$\eta(t) = \int_0^\infty e^{-t\beta} \, d\gamma(\beta), \quad t \geq 0.$$

Thus

$$\phi(r) = \int_0^\infty e^{-r^2\beta}\, d\gamma(\beta), \quad r \geq 0,$$

where the γ is the measure of the Bernstein–Widder theorem. We may take generalised Fourier transforms on both sides to get from the exponential's transform from Section 2.1

$$\hat{\phi}(r) = \pi^{n/2} \int_0^\infty e^{-r^2/(4\beta)}\, \frac{d\gamma(\beta)}{\beta^{n/2}}, \quad r > 0.$$

This means we take a classical Fourier transform of the exponential kernel on the right-hand side of the previous display.

When is this last step allowed so that $\hat{\phi}$ is still well-defined, i.e. continuous except at the origin? It is allowed if

(4.19) $$\int_1^\infty \frac{d\gamma(\beta)}{\beta^{n/2}} < \infty,$$

because the integral over the range from zero to one is finite at any rate due to the exponential decrease of the exponential $e^{-r^2/(4\beta)}$ in the integrand towards $\beta = 0$, and the integrand is continuous elsewhere.

In order to conclude that (4.19) is true from the properties of η which we required in the statement of the theorem, we need to understand the asymptotic behaviour of $\gamma(\beta)$, $\beta \to +\infty$. Fortunately, a so-called Tauberian theorem for Laplace transforms helps us here, because it relates the asymptotic behaviour of η at 0_+ with that of γ at infinity. We shall also see from that the reasons for the various conditions on the short asymptotic expansions of η.

Theorem 4.11. *(Widder, 1946, p. 192) If the Laplace transform*

$$\eta(t) = \int_0^\infty e^{-t\beta}\, d\gamma(\beta)$$

satisfies at zero

$$\eta(t) \sim A^0 t^{-\alpha'}, \qquad t \to 0_+,$$

with a nonzero constant factor A^0, then we have the short asymptotic expansion

$$\gamma(\beta) \sim A^0 \beta^{\alpha'} / \Gamma(\alpha' + 1), \qquad \beta \to +\infty.$$

The Γ used in Theorem 4.11 is the usual Γ-function. It is a consequence of this result that, in our case, γ remains bounded at $+\infty$, recalling that η is bounded at 0, thus $\alpha' = 0$ in Theorem 4.11, so long as $\hat{k} = 0$. Hence we may apply integration by parts to (4.19) and re-express the left-hand side of (4.19) as the sum

$$\left[\frac{\gamma(\beta)}{\beta^{n/2}}\right]_1^\infty + \frac{n}{2}\int_1^\infty \frac{\gamma(\beta)\,d\beta}{\beta^{\frac{n}{2}+1}} < \infty.$$

This integral is finite for all dimensions n, as required because our measure is bounded and because $n > 0$.

We may also apply Theorem 4.11 using the properties of $\hat\phi$ at infinity to deduce γ's asymptotic properties at the origin – only a change of variables is required for that. Indeed, since $\eta(t) \sim A^* t^{-\alpha}$, for large argument t, we get $\gamma(\beta) \sim A^* \beta^\alpha / \Gamma(\alpha + 1)$, $\beta \to 0_+$, and thus a change of variables and integration by parts yield

(4.20)
$$\hat\phi(r) = -\pi^{n/2}\int_0^\infty e^{-\beta r^2/4}\beta^{n/2}\,d\gamma(\beta^{-1})$$

$$= -\pi^{n/2}\int_0^\infty e^{-\beta r^2/4}\,d(\beta^{n/2}\gamma(\beta^{-1}))$$

$$+ \frac{2\pi^{n/2}n}{r^2}\int_0^\infty e^{-\beta r^2/4}\,d(\beta^{n/2-1}\gamma(\beta^{-1}))$$

$$\sim A^{**} r^{-n+2\alpha}, \quad r \to 0_+.$$

When this computation is performed, it always has to be kept in mind that there is necessarily a square in the argument of the exponential due to the definition of η.

For the final step in (4.20) we have applied an Abelian theorem of the following type and the fact that $\beta^{n/2-1}\gamma(1/\beta) \sim A\beta^{n/2-1-\alpha}$.

Theorem 4.12. *(Widder, 1946, p. 182) If we consider the Laplace transform*

$$\eta(t) = \int_0^\infty e^{-t\beta}\,d\gamma(\beta)$$

and the measure $\gamma(\beta)$ satisfies

$$\gamma(\beta) \sim A\beta^{\alpha'}, \qquad \beta \to +\infty,$$

with a nonvanishing A, then $\eta(t)$ satisfies

$$\eta(t) \sim A\Gamma(\alpha' + 1)t^{-\alpha'}, \qquad t \to 0_+.$$

Now we can compare the last line in (4.20) with (A3a) and set μ as defined in the statement of Theorem 4.10, recalling that still $\hat k = 0$. This is the salient technique in the proof of Theorem 4.10. The differentiability conditions come from the analyticity and the fast decay of the exponential kernel: all the $\hat{\phi}$ that occur in the above Laplace transforms are infinitely differentiable on the positive real axis because the factors of the exponential function that come from the differentiation do not change the continuity of the resulting integral, due to the exponential decay of the kernel.

Further, the radial function's asymptotic conditions on the derivatives are quite easily derived from the above form of $\hat{\phi}(r)$ with the same techniques as we have just applied, namely using the Abelian and Tauberian theorems. This is true both for its behaviour near the origin and for $r \to +\infty$. What remains, then, is to show the positivity of $\hat{\phi}$ which is straightforward due to the positivity of the exponential function and $d\gamma \geq 0$, $d\gamma$ not being identically zero. Therefore we have established all our conditions for the existence of a cardinal function. Moreover, the asserted properties of L are satisfied and the theorem is therefore proved for $\hat{k} = 0$.

When $\hat{k}$ is positive, the techniques remain almost the same, but they are combined with the techniques which are known to us from the proofs of the nonsingularity properties of Chapter 2. Indeed, we begin as in Chapter 2 by stating the derivative as

$$\frac{d^{\hat{k}}}{dt^{\hat{k}}} \, \phi(\sqrt{t}) = \int_0^\infty e^{-t\beta} \, d\gamma(\beta)$$

and integrate on both sides as in the proof of Theorem 2.3, but now $\hat{k}$ times. Then we take Fourier transforms, ignoring at first the polynomials of degree $\hat{k} - 1$ that come up through the integration. We arrive at the following Fourier transform of a radial basis function:

$$\hat{\Psi}(r) = (-1)^{\hat{k}} \, \pi^{n/2} \int_0^\infty e^{-r^2/(4\beta)} \, \frac{d\gamma(\beta)}{\beta^{n/2+\hat{k}}}, \quad r > 0.$$

Before we justify below why we could leave out the polynomial terms and why therefore the $\hat{\Psi}$ is identical to our required $\hat{\phi}$, we observe that an analysis which uses the properties of η at 0_+ and $+\infty$, and thus, by Theorems 4.11 and 4.12, the properties of γ at those two ends, leads to the conclusions stated through the short asymptotic expansions

$$\hat{\Psi}^{(\nu)}(r) \sim \widetilde{A}_\nu \, r^{-n-2\hat{k}+2\alpha-\nu}, \quad\quad r \to 0_+, \;\; \nu \geq 0,$$
$$\hat{\Psi}^{(\nu)}(r) \sim \widetilde{B}_\nu \, r^{-n-2\hat{k}+2\alpha'-\nu}, \quad\quad r \to +\infty, \;\; \nu \geq 0.$$

The coefficients in this display are suitable, ν-dependent constants. This suffices for (A2a) and (A3a), (A1) being known to be true anyhow because of the condition on α'.

Leaving out the aforementioned polynomials in the radial basis function is justified because the coefficients of the cardinal functions would annihilate them anyway when the infinite sums

$$(4.21) \qquad \sum_{j \in \mathbb{Z}^n} c_j \, \phi(\|x - j\|), \quad x \in \mathbb{R}^n,$$

are formed. This is actually a very important point, important to the understanding of the present theory, and therefore we wish to explain it in somewhat more detail. We still let $\{c_j\}_{j \in \mathbb{Z}^n}$ be the coefficients of our cardinal function that results where ϕ meets the criteria (A1), (A2a), (A3a). Thus, as we recall from above,

$$(4.22) \qquad \sum_{j \in \mathbb{Z}^n} c_j \, e^{-ix \cdot j} \sim \|x\|^{\mu}, \quad \|x\| \to 0_+,$$

where we are omitting a nonzero constant factors in front of $\|x\|$, and this is true for any μ in our condition that leads to a cardinal function. (We recall that for *quasi*-interpolation by contrast, we are only able to achieve (4.22) for μ an even positive integer, because only in that case is $\|x\|^{\mu}$ a polynomial in x, as a result of the compact support of the $\{c_j\}_{j \in \mathbb{Z}^n}$ whereupon the left-hand side of (4.22) must be a trigonometric polynomial.) Now, however, (4.22) means in particular that moment conditions

$$(4.23) \qquad \sum_{j \in \mathbb{Z}^n} c_j \, j^{\tilde{\alpha}} = 0 \quad \forall \tilde{\alpha} \in \mathbb{Z}_t^n, \ 0 \leq |\tilde{\alpha}| < \mu,$$

hold, the coefficients decaying fast enough for guaranteeing absolute convergence of the infinite series in (4.23). This is a consequence of the high order zero which (4.22) has at zero. It furthermore implies that in (4.21) any addition of a polynomial p of degree less than μ in x (or, rather, in $x - j$) to the radial function $\phi(\| \cdot \|)$ is harmless, i.e.

$$\sum_{j \in \mathbb{Z}^n} c_j \left(\phi(\|x - j\|) + p(x - j) \right), \quad x \in \mathbb{R}^n,$$

is the same as the expression (4.21). Going back to our representation of η and its $\hat{k}$-fold integration, we ask whether the polynomials of degree $2(\hat{k} - 1)$ (remember that we replace t by r^2 after the integration!) can in fact be disregarded in the vein outlined above. For that we need the inequality on the polynomial degrees

$$2\hat{k} - 2 < \mu = 2\hat{k} + n - 2\alpha,$$

or $\alpha < \frac{n}{2} + 1$ which is indeed a condition we explicitly demand in the statement of the theorem.

It is interesting to observe further that we do not need the 'full' infinite condition of complete monotonicity of η. There is another concept, called multiple monotonicity, which suffices. We shall also use it extensively in our Chapter 6, and in fact it provides a beautiful link between the work in this chapter and Chapter 6.

Definition 4.1. *The function $g \in C^{\lambda-2}(\mathbb{R}_{>0}) \cap C(\mathbb{R}_{\geq 0})$ is λ times monotonic if $(-1)^j g^{(j)}$ is ≥ 0, nonincreasing and convex for $j = 0, 1, \ldots, \lambda - 2$. If $g \in C(\mathbb{R}_{\geq 0})$ is nonnegative and nondecreasing, then it is called* monotonic.

Our claim is that all conclusions of Theorem 4.10 can be reached with the above condition replacing complete monotonicity. Naturally, a bound on λ, i.e. the minimal number of monotonicity required of the function involved, will be required as well. Radial basis functions closely related to the multiply monotonic functions will re-occur in the aforementioned chapter about compactly supported radial basis functions. The strategy is to show that the multiple monotonicity conditions plus suitable asymptotic conditions as in Theorem 4.10 lead to a radial function whose Fourier transform is such that we can invoke our previous work to deduce existence of a cardinal function. This will then again allow reproduction of polynomials of the same degrees as before and give fast enough decay to admit applications of Theorems 4.5–4.6.

For all that, we need first a characterisation of multiply monotonic functions; specifically, the proof of the following Theorem 4.14 is based on the 'Williamson representation theorem' which is in our book contained in the following statement.

Theorem 4.13. *A function g is λ times monotonic if and only if there is a nondecreasing measure γ that is bounded below with*

$$g(t) = \int_0^\infty (1 - t\beta)_+^{\lambda-1} \, d\gamma(\beta), \quad t \geq 0.$$

For the proof, see Williamson (1956), but note that the sufficiency is obvious. It is a consequence of this theorem that $g(t) = (1-t)_+^{\lambda-1}$ is the standard example for a λ times monotonic function but note that all completely monotonic functions are multiply monotonic of any order due to Definition 4.1. We have the following theorem, where λ and $\hat{k}$ are nonnegative integers.

Theorem 4.14. *Let $\lambda > 3\hat{k} + \frac{1}{2}(5n + 1)$. Let $\phi \in C^{\lambda+\hat{k}}(\mathbb{R}_{>0}) \cap C(\mathbb{R}_{\geq 0})$. Suppose that the η of Theorem 4.10 is λ times monotonic and that $(-1)^\lambda \times \beta^{\lambda+\hat{k}+\frac{n}{2}-\frac{3}{2}} \eta^{(\lambda)}(\beta^2)$ is twice monotonic. If there are real constants A^0, $A^* \neq 0$, $-\frac{1}{4} < \alpha' < \hat{k}$, $0 < \alpha < \min(1, \hat{k}) + \frac{n}{2}$, such that*

$$\eta^{(\lambda)}(t) \sim A^0 t^{-\alpha'-\lambda}, \quad t \to 0_+,$$
$$\eta^{(\lambda)}(t) \sim A^* t^{-\alpha-\lambda}, \quad t \to +\infty,$$

then all conclusions of Theorem 4.10 hold.

We observe that indeed for the application of this theorem, only finitely many conditions need to be verified. We do not give a proof of this result (see Buhmann and Micchelli, 1991, for one), but the tools are very similar to our theorem with completely monotonic functions. We can be precise about the differences in proving Theorems 4.10 and 4.14: in fact, the only extra work that is required concerns the Tauberian and Abelian theorems which have to be re-invented for Hankel transforms that come through the Fourier transforms of the compactly supported kernels in Theorem 4.14. We do not perform this generalisation because it is technical and does not help much in the understanding of the radial basis functions. Instead, we state here without proof that, for instance, the Hankel integral with the standard Bessel function J_ν and $\tilde{\Omega}_\nu(r, s) := J_\nu(rs)(s/r)^\nu s$,

$$F(r) = \int_0^\infty \tilde{\Omega}_\nu(r, \beta) g(\beta^2) \, d\beta$$

for a suitable positive ν which depends on our individual application, satisfies the asymptotic property

$$F(r) = Cr^{-2u-2\nu-2} + O(r^{-2u-2\nu-2-\varepsilon}), \qquad r \to \infty,$$

whenever

$$g(t) = C't^u + O(t^{u+\varepsilon'}), \qquad t \to 0_+.$$

This u has to be larger than $-\nu - 1$, and we require positive ε and ε'. Here C, C' are fixed positive quantities. Moreover, the Hankel transform has the property

$$F(r) = Cr^{2s-2\nu-2} + O(r^{2s-2\nu-2+\varepsilon}), \qquad r \to 0_+,$$

if

$$g(t) = C't^{-s} + O(t^{-s-\varepsilon'}), \qquad t \to \infty.$$

Similar short expansions are available also for the inverse statements which have asymptotics of g as consequences of asymptotic properties of F at zero and for large argument.

In order to explain and give at least a part of the proof, we do, however, show how the extra condition on the λth derivative of η being twice monotonic leads to the positivity of $\hat{\phi}$. To that end we begin with an auxiliary result.

Lemma 4.15. *Let g be a twice monotonic function that is not identically zero. Then, for any real $\nu \geq \frac{3}{2}$, the Hankel transform*

$$\int_0^\infty J_\nu(r\beta)\sqrt{r\beta}\, g(\beta)\beta \, d\beta, \qquad r > 0,$$

is positive provided that this integral is absolutely convergent for all positive r.

Proof: It is proved by Gasper (1975a, p. 412) that for $\nu \geq \frac{3}{2}$, for positive t and positive r the inequality

$$\int_0^\infty (1 - t\beta)_+ \beta^{3/2} J_\nu(r\beta) \, d\beta > 0$$

is satisfied. We note that we can express g in the Williamson form because it is twice monotonic, so

$$g(\beta) = \int_0^\infty (1 - t\beta)_+ \, d\gamma(t), \qquad \beta > 0,$$

for a nondecreasing measure γ that is bounded below. Further, we have for all real numbers $0 < a < b < \infty$

$$\int_a^b \int_0^\infty (1 - t\beta)_+ \beta^{3/2} J_\nu(r\beta) \, d\beta \, d\gamma(t) > 0.$$

This implies

$$\int_0^\infty \int_a^b (1 - t\beta)_+ \, d\gamma(t) \beta^{3/2} J_\nu(r\beta) \, d\beta > 0,$$

since $\int_a^b (1 - t\beta)_+ \, d\gamma(t)$ is bounded above by $g(\beta)$, while $g(\beta)\beta^{3/2} J_\nu(r\beta)$ is absolutely integrable with respect to β according to the assumptions in the statement of the lemma. We can let a and b tend to zero and infinity respectively because $\int_a^b (1 - t\beta)_+ \, d\gamma(t) \nearrow g(\beta)$ – the integrand being nonnegative – and we obtain

$$\int_0^\infty g(\beta)\beta^{3/2} J_\nu(r\beta) \, d\beta > 0,$$

for positive r and $\nu \geq \frac{3}{2}$, which completes the proof. $\square$

Now we may apply Lemma 4.15 to the Fourier transform of ϕ. We recall that this Fourier transform is the Fourier transform of η, after being integrated k times, and after replacing its argument by the square of the argument.

Before we state this Fourier transform, we wish to find a form of the $d\gamma(\beta)$ of the Williamson representation which is explicitly related to η. Specifically, define $\varepsilon_\lambda := (-1)^\lambda / \Gamma(\lambda)$; then $\varepsilon_\lambda \eta^{(\lambda)}(\beta^{-1})/\beta^{\lambda+1} \, d\beta$ is the same as the $d\gamma(\beta)$ from the characterisation of multiply monotonic functions, written in a weight form, i.e. $\gamma'(\beta)d\beta = d\gamma(\beta)$. This is straightforward to verify by direct computation and especially by λ-fold differentiation:

$$\eta(t) = \int_0^\infty (1 - t\beta)_+^{\lambda-1} \, d\gamma(\beta)$$

and therefore

$$\eta^{(\lambda)}(t) = \varepsilon_\lambda^{-1} \int_0^\infty \beta^\lambda \delta(1 - t\beta) \gamma'(\beta) \, d\beta$$

with the Dirac δ-distribution in the last display. The last displayed equation implies

$$\eta^{(\lambda)}(t) = \varepsilon_\lambda^{-1} t^{-\lambda-1} \gamma'(t^{-1})$$

which provides the required form of γ'.

Therefore we have as a representation for our radial basis function

(4.24) $$\phi(r) = \frac{(-1)^{\hat{k}} \varepsilon_\lambda}{(\lambda)_{\hat{k}}} \int_0^\infty (1 - r^2 \beta)_+^{\lambda+\hat{k}-1} \beta^{-\lambda-\hat{k}-1} \eta^{(\lambda)}(\beta^{-1}) \, d\beta,$$

where the expression $(\lambda)_{\hat{k}}$ denotes the Pochhammer symbol

$$(\lambda)_{\hat{k}} = \lambda(\lambda + 1) \ldots (\lambda + \hat{k} - 1).$$

Hence we get by Fourier integration and change of variables that $\hat{\phi}(r)$ is the same as $2^{\lambda+\hat{k}+n/2}(-1)^{\lambda+\hat{k}+1}\pi^{n/2}$ times

(4.25) $$\int_0^\infty \tilde{\Omega}_{\lambda+\hat{k}+n/2-1}(r, \beta) \eta^{(\lambda)}(\beta^2) \, d\beta$$

to which we may apply our lemma directly (and our hypothesis in the statement of the theorem) to show that no zero occurs.

With these remarks we conclude our work on the existence and convergence power of radial function interpolation and quasi-interpolation on lattices. There is only one last issue that we want to discuss in this chapter. It has to do with the stability of the computation of interpolants and it is the theme of the next section.

4.3 Numerical properties of the interpolation linear system

Many important facts are known about the norms of inverses of interpolation matrices with radial basis functions and their dependence on the spacing of the data points. This is especially relevant to applications of interpolation with radial basis functions because it admits estimates of the ℓ^p-condition numbers (mostly, however, ℓ^2-condition numbers are considered) that occur. Indeed, most readers of this book will be aware of the relationship between condition numbers of matrices and the numerical stability of the interpolation problem, i.e. the dependence of the accuracy of the computed coefficients on (rounding) errors of the function values.

For instance, if Δf_ξ is the error (rounding error, recording error, instrument error from a physical device etc.) in each f_ξ, and if furthermore $\Delta \mathbf{f}$, $\Delta \boldsymbol{\lambda}$ denote the vectors with entries Δf_ξ and the errors in the computed λ_ξ, respectively, then we have the following familiar bounds for the relative errors in the Euclidean norm:

$$\frac{\|\Delta \boldsymbol{\lambda}\|}{\|\boldsymbol{\lambda}\|} \leq \text{cond}_2(\mathbf{A}) \frac{\|\Delta \mathbf{f}\|}{\|\mathbf{f}\|},$$

where $\text{cond}_2(\mathbf{A}) = \|\mathbf{A}\|_2 \|\mathbf{A}^{-1}\|_2$ is the condition number in Euclidean norm of $\mathbf{A}$ and where the linear system $\mathbf{A}\boldsymbol{\lambda} = \mathbf{f}$ defines the $\boldsymbol{\lambda} = \{\lambda_\xi\}$ from the $\mathbf{f} = \{f_\xi\}$ with $\mathbf{A} = \{\phi(\|\zeta - \xi\|)\}_{\xi,\zeta \in \Xi}$. In a more general setting, such issues are, for instance, most expertly treated in the recent book by Higham (1996).

The work we are alluding to is particularly relevant when the data points are allowed to be scattered in $\mathbb{R}^n$, because this is the case which usually appears in practice. In addition, for scattered data, we are usually faced with more computational difficulties. In some cases, they may be circumvented by using FFT or other stable – and fast – techniques for the computation of Lagrange functions and their coefficients otherwise (see, e.g., Jetter and Stöckler, 1991).

However, the analysis is very delicate for scattered data due to potentially large discrepancies between the closest distance of the points that are used and the largest distance of neighbouring points, there being no concept of uniform distance. This is why we defer the treatment of condition numbers of interpolation matrices for scattered data points to the next chapter, and just give some very specialised remarks about multiquadrics and p-norms of inverses of interpolation matrices for *equally spaced* data here. For instance, Baxter (1992a) proves the next theorem which gives best upper bounds for the Euclidean norm of the inverse of the interpolation matrix. To get to the condition numbers from that is straightforward by standard estimates of the missing Euclidean norm of the interpolation matrix itself.

Theorem 4.16. *Suppose* $\phi(\|\cdot\|)\colon \mathbb{R}^n \to \mathbb{R}$ *has a distributional Fourier transform with radial part* $\hat{\phi} \in C(\mathbb{R}_{>0})$ *which is such that (A2), (A3) hold for* $\mu \leq n + 1$. *Further, let* $\Xi \subset \mathbb{Z}^n$ *be a finite subset of distinct points and* $\mathbf{A}$ *be the associated interpolation matrix* $\{\phi(\|\xi - \zeta\|)\}_{\xi,\zeta \in \Xi}$. *Then*

$$\|\mathbf{A}^{-1}\|_2 \leq \left\{ \sum_{k \in \mathbb{Z}^n} |\hat{\phi}(\|\pi + 2\pi k\|)| \right\}^{-1},$$

where $\pi = (\pi, \pi, \ldots, \pi)^T$, *and this is the least upper bound that holds uniformly for all finite subsets of* $\mathbb{Z}^n$.

We note that we have not explicitly demanded that the symbol which appears in disguise on the right-hand side of the display in the statement of the theorem be positive, so that the bi-infinite matrix **A** should be invertible, because then the right-hand side would be infinity anyway.

In order to show the behaviour of the multiquadric interpolation matrix as an example, we wish to apply this result for instance to the multiquadric radial basis function. The required bound is obtained from the above theorem in tandem with standard expansions of the modified Bessel function and the known Fourier transform of the multiquadric function – from our second section – because we have to estimate the reciprocal of the symbol at π. It is highly relevant to this that there is a lower bound for the Bessel function $K_{(n+1)/2}$ in Abramowitz and Stegun (1972, p. 378) that reads, for z with positive real part,

$$K_{(n+1)/2}(z) \geq \sqrt{\frac{\pi}{2z}} e^{-z}.$$

Thus, with this we can bound for the multiquadric function with positive parameter c

$$\left\{ \sum_{k \in \mathbb{Z}^n} |\hat{\phi}(\|\pi + 2\pi k\|)| \right\}^{-1} \leq \pi \left(\|\pi\| / [2\pi c] \right)^{(n+1)/2} \sqrt{\frac{2c\|\pi\|}{\pi}} e^{c\|\pi\|}.$$

Note that we have bounded the reciprocal of the symbol by the reciprocal of the term for $k = 0$ only. Therefore we cannot *a priori* be certain whether the bound we shall get now is optimal. It is reasonable to expect, however, that it is asymptotically optimal for large c because the bound in the display before last is best for all positive z.

Next we insert $\|\pi\| = \pi\sqrt{n}$ and get for the right-hand side of the last display

$$\frac{n^{(n+2)/4}\pi}{\left(\sqrt{2c}\right)^n} e^{c\sqrt{n}\pi}.$$

For instance for $n = 1$ and multiquadrics, this theorem implies by inserting $n = 1$ into the last display

$$\|\mathbf{A}^{-1}\|_2 \leq \frac{\pi}{\sqrt{2c}} e^{c\pi}.$$

This is in fact the best possible upper bound asymptotically for large c. It is important to verify that in general the exponential growth for larger c is genuine because this is the dominating term in all of the above bounds.

Indeed, Ball, Sivakumar and Ward (1992) get, for the same interpolation matrix with centres in $\mathbb{Z}^n$,

$$\|\mathbf{A}^{-1}\|_2 \geq Ce^c$$

with a positive constant C which is independent of c, i.e. we have obtained an independent verification that the exponential growth with c is genuine.

Furthermore, Baxter, Sivakumar and Ward (1994) offer estimates on such p-norms when the data are on a grid just as we have treated them in this chapter all along. The result rests on the observation that we may restrict attention to $p = 2$, but it is applied only to the Gauss-kernel. Its proof is a nice application of the so-called Riesz convexity theorem from the theory of operator interpolation and it is as follows.

Theorem 4.17. *Let* $\mathbf{A} = \{\phi(\|j - k\|)\}_{j,k\in\mathbb{Z}^n}$, *for* $\phi(r) = e^{-c^2 r^2}$, $c \neq 0$. *Then* $\|\mathbf{A}^{-1}\|_p \equiv \|\mathbf{A}^{-1}\|_2$, *for all* $p \geq 1$, *where* $p = \infty$ *is included.*

Proof: Because $\mathbf{A}$ is a Toeplitz matrix, that is a matrix whose entries are constant along its diagonals, we know already that $\{c_k\}_{k\in\mathbb{Z}^n}$ have the form (4.1) with

$$\mathbf{A}^{-1} = \{c_{j-k}\}_{j,k\in\mathbb{Z}^n}.$$

This is true because the Toeplitz matrix $\mathbf{A}$ has as an inverse the Toeplitz matrix with the said shifts of the Fourier coefficients of the symbol's reciprocal. Furthermore, we know from Theorem 4.16 that

$$(4.26) \qquad \|\mathbf{A}^{-1}\|_2 = \frac{1}{\min \sigma(x)} = \frac{1}{\sigma(\pi)} = \sum_{k\in\mathbb{Z}^n}(-1)^{k_1+\cdots+k_n}c_k.$$

Here, we use the components $k = (k_1, k_2, \ldots, k_n)$ of an n-variate multiinteger $k \in \mathbb{Z}^n$. Note that not just a bound, but an explicit expression is provided for the norm of the inverse of $\mathbf{A}$, because the bound given in Theorem 4.16 is the best uniform bound for all finite subsets of $\mathbb{Z}^n$ and therefore it is readily established that equality is attained for $\Xi = \mathbb{Z}^n$. We recall that π denotes the n-dimensional vector with π as all of its components.

Now, because of symmetry of the matrix $\mathbf{A}$, we have equal uniform and 1-norms, i.e.

$$\|\mathbf{A}^{-1}\|_\infty = \|\mathbf{A}^{-1}\|_1 = \sum_{k\in\mathbb{Z}^n} |c_k|.$$

We claim that this is the same as the ℓ^2-norm of the matrix. From this identity, the equality for all p-norms will follow from the Riesz convexity theorem from Bennett and Sharpley (1988), Stein and Weiss (1971) which we state here in this book in a form suitable to our application. For the statement, we let $(M, \mathcal{M}, \mu)$ and $(N, \mathcal{N}, \nu)$ be measure spaces, $\mathcal{M}, \mathcal{N}$ being the σ-algebras of measurable subsets and μ, ν be the respective measures on them. We let also p_t^{-1} and

q_t^{-1} be the convex combinations of p_0^{-1}, p_1^{-1}, and q_0^{-1}, q_1^{-1}, respectively, i.e.
$p_t^{-1} = t p_0^{-1} + (1 - t) p_1^{-1}$, and analogously for q_t, $0 \leq t \leq 1$.

Riesz convexity theorem. *Let T be an operator mapping a linear space of measurable functions on a measure space $(M, \mathcal{M}, \mu)$ into measurable functions defined on $(N, \mathcal{N}, \nu)$. Suppose that $\|Tf\|_{q_i} = O(\|f\|_{p_i})$, $i = 0, 1$. Then $\|Tf\|_{q_t} = O(\|f\|_{p_t})$ for all $t \in [0, 1]$, and the constant in the O-term above for q_t is the same for all t if the associated constants for the two q_i, $i = 0$ and $i = 1$, are the same.*

To prove that the above uniform and ℓ^1-norm is the same as the ℓ^2-norm of the matrix, it suffices to show that $(-1)^{k_1 + \cdots + k_n} c_k$ are always of one sign or zero for all multiintegers k, because then the above display equals (4.26).

We establish this by a tensor-product argument which runs as follows. Since

$$e^{-c^2 \|x\|^2} = e^{-c^2 x_1^2} e^{-c^2 x_2^2} \ldots e^{-c^2 x_n^2},$$

where we recall that $x = (x_1, x_2, \ldots, x_n)^T$, we may decompose the symbol as a product

$$\sigma(x) = \widetilde{\sigma}(x_1) \, \widetilde{\sigma}(x_2) \ldots \widetilde{\sigma}(x_n),$$

where $\widetilde{\sigma}(x)$ is the univariate symbol for $e^{-c^2 r^2}$, $r \in \mathbb{R}$. Hence the same decomposition is true for the coefficients (4.1), i.e. $c_k = \widetilde{c}_{k_1} \widetilde{c}_{k_2} \ldots \widetilde{c}_{k_n}$. Thus it suffices to show that $(-1)^\ell \widetilde{c}_\ell$ is always of one sign or zero in one dimension. This, however, is a consequence of the fact that any principal submatrix of $\{\phi(|j - k|)\}_{j,k \in \mathbb{Z}}$ is a totally positive matrix (Karlin, 1968) whence the inverse of any such principal submatrix exists and enjoys the 'chequerboard' property according to Karlin: this means that the elements of the inverse at the position (j, k) have the sign $(-1)^{j+k}$. This suffices for the required sign property of the coefficients. Now, however, the next lemma shows that those elements of the inverses of the finite-dimensional submatrices converge to the $\widetilde{c}_\ell$. It is from Theorem 9 of Buhmann and Micchelli (1991).

Lemma 4.18. *Let the radial basis function ϕ satisfy the assumptions (A1), (A2a), (A3a). Let*

$$\{0\} \subset \cdots \subset \Xi_{m-1} \subset \Xi_m \subset \cdots \subset \mathbb{Z}^n$$

be finite nested subsets that satisfy the symmetry condition $\Xi_m = -\Xi_m$ for all positive integers m. Moreover, suppose that for any positive integer K there is an m such that $[-K, K]^n \subset \Xi_m$. If c^m are the coefficients of the shifts of the *radial function which solve the interpolation problem*

$$\sum_{j \in \Xi_m} c_j^m \phi(\|k - j\|) + p_m(k) = \delta_{0k}, \qquad k \in \Xi_m,$$

$$\sum_{j \in \Xi_m} c_j^m q(j) = 0, \qquad q \in \mathbb{P}_n^{\mu-n},$$

with $p_m \in \mathbb{P}_n^{\mu-n}$, *then*

$$\lim_{m \to \infty} c_j^m = \frac{1}{(2\pi)^n} \int_{\mathbb{T}^n} \frac{e^{ij \cdot t} \, dt}{\sigma(t)}, \qquad j \in \mathbb{Z}^n.$$

Here, σ is the symbol associated with the radial basis function in use. So the coefficients for solutions of the finite interpolation problem converge to the coefficients of the solution on the infinite lattice.

It follows from this lemma that $(-1)^{j+k} \tilde{c}_{j-k} \geq 0$, which implies

$$(-1)^\ell \tilde{c}_\ell \geq 0.$$

It is now a consequence of the aforementioned Riesz convexity theorem that

$$\|\mathbf{A}^{-1}\|_\infty = \|\mathbf{A}^{-1}\|_1 = \|\mathbf{A}^{-1}\|_2$$

leads to our desired result, because the Riesz convexity theorem states for our application that equality for the operator norms for $p = p_1$ and $p = p_2$ implies that the norms are equal for all $p_1 \leq p \leq p_2$. Here p_1 and p_2 can be from the interval $[1, \infty]$. $\qquad\qquad\qquad\qquad\qquad\qquad\qquad\qquad\qquad\qquad$ □

Note that the proof of this theorem also gives an explicit expression, namely $\sigma(\pi, \pi, \ldots, \pi)^{-1}$, for the value not only for the 2-norm but also for the p-matrix-norm. It follows from this theorem, which also can be generalised in several aspects (see the work of Baxter), that we should be particularly interested in the ℓ^2-norm of the inverse of the interpolation matrix – which we are incidentally for other reasons because the ℓ^2-norms are easier to observe through eigenvalue computations, which can be computed stably for instance with the QR method (Golub and Van Loan, 1989). Indeed, for a finite interpolation matrix the condition number is the ratio of the largest eigenvalue in modulus of the matrix to its smallest one. We will come back to estimates of those eigenvalues for more general classes of radial basis functions and for finitely many, scattered data in the following chapter, which is devoted to radial function interpolation on arbitrarily distributed data.

4.4 Convergence with respect to parameters in the radial function

Another theorem, which is due to Baxter (1992a), Madych (1990), is the result below, which concerns the accuracy of the multiquadric radial function $\phi(r) = \sqrt{r^2 + c^2}$ interpolants. However, rather than focussing upon how the interpolants converge as the grid spacing decreases, these results examine the effect of varying the user-defined parameter c. As we have seen already in our discussion of condition numbers in Section 4.2, however, the interpolation matrix becomes severely ill-conditioned with growing c (the condition numbers grow exponentially; their exponential growth is genuine and not only contained in any unrealistic upper bound) and thus virtually impossible to use in the practical solution of the linear interpolation system. Therefore this result is largely of theoretical interest. For the statement of the result we recall once more the notion of a band-limited function, that is a function whose Fourier transform is compactly supported. We also recall in this context the Paley–Wiener theorem (Rudin, 1991) which states that the entire functions of exponential type are those whose Fourier transforms are compactly supported and which we quote in part in a form suitable to our need.

Paley–Wiener theorem. *Let f be an entire function which satisfies the estimate*

$$|f(z)| = O\left((1 + |z|)^{-N} \exp(r|\Im z|)\right), \qquad z \in \mathbb{C}^n, \; N \in \mathbb{Z}_+.$$

Then there exists a compactly supported infinitely differentiable function whose support is in $B_r(0)$ and whose Fourier transform is f. Here, $\Im z$ denotes the imaginary part of the complex quantity vector z.

With its aid we establish the following theorem.

Theorem 4.19. *Let $f \in C(\mathbb{R}^n) \cap L^2(\mathbb{R}^n)$ be an entire function as in the Paley–Wiener theorem of exponential type π, i.e. its Fourier transform is compactly supported in $\mathbb{T}^n$, so that the approximand is band-limited. Then the cardinal interpolants*

$$s_c(x) = \sum_{j \in \mathbb{Z}^n} f(j) \, L^c(x - j), \; x \in \mathbb{R}^n,$$

with cardinal Lagrange functions for multiquadrics and every positive parameter c

$$L^c(x) = \sum_{k \in \mathbb{Z}^n} d_k^c \sqrt{\|x - k\|^2 + c^2}, \; x \in \mathbb{R}^n,$$

that satisfy $L^c(j) = \delta_{0j}$, $j \in \mathbb{Z}^n$, *enjoy the uniform convergence property*

$$(4.27) \qquad \|s_c - f\|_\infty = o(1), \quad c \to \infty.$$

Proof: Let χ be the characteristic function of $[-\pi, \pi]^n$, which is one for an argument inside that cube and zero for all other arguments. Thus, since f is square-integrable and since the Fourier transform is an isometry on square-integrable functions (cf. the Appendix), the approximation error is the difference

$$(4.28) \quad s_c(x) - f(x) = \frac{1}{(2\pi)^n} \int_{\mathbb{R}^n} \sum_{k \in \mathbb{Z}^n} \hat{f}(t + 2\pi k) \left(\hat{L}^c(t) - \chi(t)\right) e^{ixt} \, dt.$$

It follows from (4.28) that $|s_c(x) - f(x)|$ is at most a constant multiple of

$$\int_{\mathbb{T}^n} |\hat{f}(t)| \sum_{k \in \mathbb{Z}^n} |\hat{L}^c(t + 2\pi k) - \chi(t + 2\pi k)| \, dt$$

which is the same as

$$\int_{\mathbb{T}^n} |\hat{f}(t)| \left(1 - \hat{L}^c(t) + \sum_{k \in \mathbb{Z}^n \setminus \{0\}} \hat{L}^c(t + 2\pi k)\right) dt$$

$$= 2 \int_{\mathbb{T}^n} |\hat{f}(t)| \left(1 - \hat{L}^c(t)\right) dt,$$

because of the form of the Lagrange function stipulated in Theorem 4.3. The same theorem and the lack of sign change of $\hat{L}^c$ (coming from the negativity of the multiquadric Fourier transform) gives the two uniform bounds

$$0 \le 1 - \hat{L}^c(t) \le 1.$$

The following proposition will evidently finish our proof of Theorem 4.19 because the problem now boils down to proving that the Lagrange function's Fourier transform is, in the limit for $c \to \infty$, pointwise almost everywhere a certain characteristic function.

Proposition 4.20. *Let* $x \in \mathbb{R}^n$. *Then* $\lim_{c \to \infty} \hat{L}^c(x) = \chi(x)$, *unless* $\|x\|_\infty = \pi$.

Proof: Let $x \notin \mathbb{T}^n$. There exists a $k_0 \in \mathbb{Z}^n \setminus \{0\}$ such that $\|x + 2\pi k_0\| < \|x\|$, and the exponential decay of $\hat{\phi}_c$ (the radial part of the Fourier transform of $\sqrt{\|\cdot\|^2 + c^2}$) provides the bounds

$$(4.29) \qquad \hat{\phi}_c(\|x\|) \le e^{-c\|x\| + c\|x + 2\pi k_0\|} \, \hat{\phi}_c(\|x + 2\pi k_0\|)$$

$$\le e^{-c\|x\| + c\|x + 2\pi k_0\|} \sum_{k \in \mathbb{Z}^n} \hat{\phi}_c(\|x + 2\pi k\|),$$

where we have now assumed without loss of generality that $\hat{\phi}_c > 0$ (we may change the sign of $\hat{\phi}_c$ without altering the cardinal function L^c in any way). Thus, using again Theorem 4.3, we get

$$0 \leq \hat{L}^c(x) \leq e^{-c\|x\| + c\|x + 2\pi k_0\|} \to 0 \quad (c \to \infty),$$

as required.

Now let $x \in (-\pi, \pi)^n \backslash \{0\}$ (the case $x = 0$ is trivial and our result is true without any further computation). Thus, for all $k_0 \in \mathbb{Z}^n \backslash \{0\}$, $\|x\| < \|x + 2\pi k_0\|$, and we have

$$\hat{L}^c(x) = \left(1 + \sum_{k \in \mathbb{Z}^n \backslash \{0\}} \frac{\hat{\phi}_c(\|x + 2\pi k\|)}{\hat{\phi}_c(\|x\|)}\right)^{-1}$$

which means that it is sufficient for us to show

$$\lim_{c \to \infty} \sum_{k \in \mathbb{Z}^n \backslash \{0\}} \frac{\hat{\phi}_c(\|x + 2\pi k\|)}{\hat{\phi}_c(\|x\|)} = 0$$

for $\|x\| < \|x + 2\pi k_0\|$. Indeed, according to (4.29), every single entry in the series in the last display satisfies the required limit behaviour. It thus suffices using (4.29) term by term and summing to show that, denoting $(1, 1, \ldots, 1) \in \mathbb{R}^n$ by $\mathbf{1}$,

(4.30) $$\sum_{\|k\| \geq 2\|\mathbf{1}\|} e^{-c\|x + 2\pi k\| + c\|x\|} \to 0 \quad (c \to \infty).$$

However, as $\|k\| \geq 2\|\mathbf{1}\|$ implies

$$\|x + 2\pi k\| - \|x\| \geq 2\pi \left(\|k\| - \|\mathbf{1}\|\right) \geq \pi \|k\|$$

for $\|x\| \leq \pi \|\mathbf{1}\|$, we get our upper bound of

$$\sum_{\|k\| \geq 2\|\mathbf{1}\|} e^{-\pi c\|k\|}$$

for the left-hand side of (4.30) which goes to zero for $c \to \infty$ and it gives the required result by direct computation. $\quad\square$

It follows from the work in Chapter 4 that the Gaussian radial basis function $\phi(r) = e^{-c^2 r^2}$ used for interpolation on a cardinal grid cannot provide any nontrivial approximation order: it satisfies all the conditions (A1), (A2a), (A3a) for $\mu = 0$ and therefore there exist decaying cardinal functions as with all the other radial basis functions we have studied in this book. However, according to Theorem 4.4, there is no polynomial reproduction with either interpolation or quasi-interpolation using Gaussians. Therefore, as we have seen above, no approximation orders can be obtained. This depends on the parameter c being fixed in the definition of the ϕ as the spacing of the grid varies. There is, however, a possibility of obtaining convergence orders of Gaussian interpolation on grids if we let $c = c(h)$ vary with h – so we get another result (of Beatson and Light, 1992) on convergence with respect to a parameter: It is the following result on quasi-interpolation as in Section 4.1 that we state without proof.

Theorem 4.21. *Let k be a natural number and ψ be a finite linear combination of multiinteger translates of the Gaussian radial basis function, whose coefficients depend on c but the number of nonzero coefficients is fixed. If $c = \sqrt{2\pi^2/(k|\log h|)}$, then there is a ψ with the above properties such that quasi-interpolation using (4.14) fulfils the error estimate for $h \to 0$ and $f \in W_\infty^k(\mathbb{R}^n)$*

$$\|f - s_h\|_\infty \leq Ch^k |\log h|^{k/2 + [(k-1)/2]} \|f\|_{k,\infty}.$$

Here, $[\,\cdot\,]$ denotes the Gauss-bracket and $W_\infty^k(\mathbb{R}^n)$ is the Sobolev space of all functions with bounded partial derivatives of total order at most k. Similarly, we recall the definition of the *nonhomogeneous* Sobolev space denoted alternatively by $H^k(\mathbb{R}^n)$ or by $W_2^k(\mathbb{R}^n)$ as

$$(4.31) \quad \left\{ f \in L^2(\mathbb{R}^n) \,\middle|\, \|f\|_k^2 := \frac{1}{(2\pi)^n} \int_{\mathbb{R}^n} (1 + \|x\|)^{2k} |\hat{f}(x)|^2 \, dx < \infty \right\}.$$

Such spaces will be particularly relevant in the next chapter.

5

Radial Basis Functions on Scattered Data

While the setting of gridded data in the previous chapter was very handy to characterise the best possible approximation orders that are obtained with interpolation or quasi-interpolation using radial basis functions, the most natural context for radial basis function approximation has always been and remains scattered data interpolation, and this is what we shall be concerned with now.

Similarly to the beginning of the last chapter, we say that the space $\mathcal{S} = \mathcal{S}_h$ of approximants which depends on a positive h, here no longer an equal spacing but a notion of distance between the data points, provides approximation order h^μ to the space of approximands in the L^p-norm over the domain Ω (often $\Omega = \mathbb{R}^n$), if

$$\mathrm{dist}_{L^p(\Omega)}(f, \mathcal{S}) := \inf_{g \in \mathcal{S}} \|f - g\|_{p,\Omega} = O(h^\mu), \qquad h = \max_{x \in \Omega} \min_{\xi \in \Xi} \|x - \xi\|,$$

for all f from the given space of approximands and no higher order may be achieved (in fact more precisely we mean, as in the previous chapter: O cannot be replaced by o in the above).

One further remark we make at the onset of this chapter, namely that in the alternative case of *quasi*-interpolation, it is hard to compute the approximants on scattered data and therefore it is desirable to choose interpolants instead of quasi-interpolation, although we do give a little result on quasi-interpolation on scattered data here as well. In fact, for applications in practice one sometimes maps scattered data to grids by a local interpolation procedure (e.g. one of those from Chapter 3), and then uses quasi-interpolation with radial basis functions on the gridded data.

We begin by extending the fundamental results from Chapter 2, some of which were quite simple and of an introductory character, to the more general setting introduced by Micchelli (1986). They are related to the unique solvability of our by-now well-known radial basis function interpolation problem. The concepts

99

we shall use here are, however, essentially the same as before, namely complete monotonicity, the Bernstein representation theorem, etc. They are all familiar from the second chapter.

We shall then continue with a *convergence analysis* of the scattered data interpolation problem. The concepts and tools are now altogether different from those of the gridded data convergence analysis. Specifically, we have now to consider boundaries and no periodicity prevails that admits the substantial use of Fourier transform techniques. Instead, certain variational properties of the radial basis functions and reproducing kernel Hilbert space theory will be applied. These concepts are the basis of the seminal contribution of Duchon (1976, 1978) and later Madych and Nelson to the convergence analysis of thin-plate splines and other radial basis functions. Though unfortunately, as we shall see, not quite the same, remarkable convergence orders as in the gridded case will be obtained in the theorems about scattered centres – with the exception of some special cases when nonstationary approximation is applied. The stationary convergence orders established in the standard theorems are typically much smaller than the orders obtainable in gridded data approximation. This is closely related to the fact that we shall prove convergence results with methods that depend on the finite domain where the interpolation takes place and which involve the boundaries of those domains, deterioration of the convergence speed on domain boundaries being a frequent phenomenon, even in simple cases like univariate splines, unless additional information (such as derivative information from the approximand f) at the boundary is introduced.

5.1 Nonsingularity of interpolation matrices

As far as the general nonsingularity results are concerned, there is the work by Micchelli which had a strong impact on the research into radial basis functions; it is the basis and *raison d'être* of much of the later work we present here.

Precisely, we recall that the unique solvability could be deduced immediately whenever the radial basis function ϕ was such that $\phi(\sqrt{t})$ is completely monotonic but not constant, since then the interpolation matrix $\mathbf{A} = \{\phi(\|\xi - \zeta\|)\}_{\xi, \zeta \in \Xi}$ is positive definite if the Ξ is an arbitrary finite subset of distinct points of $\mathbb{R}^n$. Furthermore, under certain small extra requirements, we observed it to be sufficient if the first derivative $-\frac{d}{dt} \phi(\sqrt{t})$ is completely monotonic (Theorem 2.2). That this is in fact a weaker requirement follows from the observation that if $\phi(\sqrt{t})$ is completely monotonic, then all its derivatives are, subject to a suitable sign change in the original function. Thus we may consider

ϕ such that $-\frac{d}{dt} \phi(\sqrt{t})$ is completely monotonic, or, so as to weaken the requirement further, by demanding that, for some k, $(-1)^k \frac{d^k}{dt^k} \phi(\sqrt{t})$ be completely
monotonic. If that is the case, positive definiteness of the interpolation matrix
can no longer be expected, as we have already seen even in the case when
$k = 1$, and it is, indeed, no longer the correct concept. Instead, we introduce
the following new notion of conditional positive definiteness. It has to do with
positive definiteness on a subspace of vectors. We recall that $\mathbb{P}_n^k$ denotes the
polynomial space of all polynomials of total degree at most k in n unknowns.

Definition 5.1. *A function $F: \mathbb{R}^n \to \mathbb{R}$ is conditionally positive definite (cpd)
of order k (on $\mathbb{R}^n$ but we do not always mention the dimension n) if for all finite
subsets Ξ from $\mathbb{R}^n$, the quadratic form*

$$(5.1) \qquad \sum_{\xi \in \Xi} \sum_{\zeta \in \Xi} \lambda_\xi \lambda_\zeta F(\xi - \zeta)$$

*is nonnegative for all $\lambda = \{\lambda_\xi\}_{\xi \in \Xi}$ which satisfy $\sum_{\xi \in \Xi} \lambda_\xi q(\xi) = 0$ for all
$q \in \mathbb{P}_n^{k-1}$. F is strictly conditionally positive definite of order k if the quadratic
form (5.1) is positive for all nonzero vectors λ.*

The most important observation we make now is that we can always interpolate
uniquely with strictly conditionally positive definite functions if we add polynomials to the interpolant and if the only polynomial that vanishes on our set
of data sites is zero: for later use, we call a subset Ξ of $\mathbb{R}^n$ unisolvent for $\mathbb{P}_n^{k-1}$
if $p(\xi)$ for all $\xi \in \Xi$ implies $p = 0$. The interpolant has the form

$$(5.2) \qquad \begin{cases} s(x) = \displaystyle\sum_{\xi \in \Xi} \lambda_\xi F(x - \xi) + p(x), \\ 0 = \displaystyle\sum_{\xi \in \Xi} \lambda_\xi \xi^\alpha, \quad |\alpha| < k. \end{cases}$$

We continue to use our standard multiindex notation here.

In (5.2), $p \in \mathbb{P}_n^{k-1}$, and the interpolation requirements for (5.2) are $s(\xi) = f_\xi$,
$\xi \in \Xi$, as usual. The side-conditions in (5.2) are used to take up the extra degrees
of freedom that are introduced through the use of the polynomial p.

We show now that we can interpolate uniquely with a strictly conditionally
positive definite F. Indeed, if a nonzero vector λ satisfies the above side-
conditions, then we can multiply the vector which has components $s(\zeta), \zeta \in \Xi$,
by λ on the right, and get a quadratic form with kernel $F(\zeta - \xi)$, recalling that
$\sum_{\zeta \in \Xi} p(\zeta) \lambda_\zeta = 0$. This form has to be positive because of the definition of strict
conditional positive definiteness. Therefore $s(\zeta)$ cannot be identically zero, that
is, for all $\zeta \in \Xi$, unless all its coefficients vanish. Hence the interpolant exists
uniquely. The uniqueness within the space of polynomials is assured by the

linear independence of polynomials of different degrees and by the unisolvency
of Ξ with respect to $\mathbb{P}_n^{k-1}$.

Of course, for our applications, $F(x) = \phi(\|x\|)$ is used, and now the next
result is relevant. It is due to Micchelli (1986) and generalises Theorem 2.2. Its
proof is highly instructive, much as the proof of Theorem 2.2 was, because it
shows how the side-conditions and the complete monotonicity of the derivative
of the radial basis function act in tandem to provide the nonsingularity of the
interpolation problem. Again, the proof hinges on the Bernstein representation
theorem, now applied to the derivative instead of the radial basis function itself.
Incidentally, the converse of the following result is also true (Guo, Hu and Sun,
1993), so it is in fact a characterisation, but this is not required for our work
here except in one later instance in the sixth chapter, and therefore we shall not
give a proof of that full characterisation.

Theorem 5.1. *Let ϕ be continuous and such that*

(5.3) $$(-1)^k \frac{d^k}{dt^k} \phi(\sqrt{t}), \quad t > 0,$$

*is completely monotonic but not constant. Then $\phi(\| \cdot \|)$ is strictly conditionally
positive definite of order k on all $\mathbb{R}^n$.*

Proof: The proof proceeds in much the same vein as the proofs of Theorem 2.2
and the beginning of the proof of Theorem 4.10. We call $\eta: \mathbb{R}_{>0} \to \mathbb{R}$ the
function defined through (5.3). Therefore, by the Bernstein representation
theorem,

(5.4) $$\eta(t) = \int_0^\infty e^{-\beta t} \, d\gamma(\beta), t > 0.$$

We replace the infinite integral on the right-hand side of (5.4) by an integral
from $\delta > 0$ to $+\infty$

$$\eta_\delta(t) = \int_\delta^\infty e^{-\beta t} \, d\gamma(\beta), \quad t > 0.$$

Then, for any $\varepsilon > 0$, we may integrate k times on both sides between ε and
$t = r^2$; whence we get for suitable polynomials $p_{\varepsilon,\delta} \in \mathbb{P}_1^{k-1}$ and $q_{\varepsilon,\delta} \in \mathbb{P}_1^{k-1}$,
and by multiplication by $(-1)^k$,

$$\phi_{\varepsilon,\delta}(r) := p_{\varepsilon,\delta}(r^2) + \int_\delta^\infty \{e^{-\beta r^2} - q_{\varepsilon,\delta}(\beta r^2)\} \frac{d\gamma(\beta)}{\beta^k},$$

where $\delta > 0$ is still arbitrary and where we have restricted integration over $d\gamma$
from $\delta > 0$ to $+\infty$ in order to avoid the present difficulties of integrating β^{-k}
near 0. In fact, the coefficients of the polynomials are immaterial (only their

degree is relevant to the proof in the sequel) but it turns out that for example $q_{\varepsilon,\delta}$ is essentially a Taylor polynomial of degree $k-1$ to the exponential function $e^{-\beta t^2}$. Now let $\boldsymbol{\lambda} \in \mathbb{R}^{\Xi}$ satisfy the moment conditions of (5.2). Thus, we have the following expression for a quadratic form:

$$\sum_{\xi \in \Xi} \lambda_\xi \sum_{\zeta \in \Xi} \lambda_\zeta\, \phi_{\varepsilon,\delta}(\|\xi - \zeta\|) = \sum_{\xi \in \Xi} \lambda_\xi \sum_{\zeta \in \Xi} \lambda_\zeta\, p_{\varepsilon,\delta}\,(\|\xi - \zeta\|^2)$$

$$+ \int_\delta^\infty \sum_{\xi \in \Xi} \lambda_\xi \sum_{\zeta \in \Xi} \lambda_\zeta\, \{e^{-\beta\|\xi - \zeta\|^2} - q_{\varepsilon,\delta}(\beta\|\xi - \zeta\|^2)\}\, \frac{d\gamma(\beta)}{\beta^k}.$$

We may exchange summation and integration because all sums are finite. Because of the quadratic form above with *each* sum over $\{\lambda_\xi\}_{\xi \in \Xi}$ annihilating polynomials of degree less than k, we get annihilation of both polynomials over the *double sum,* that is

$$\sum_{\xi \in \Xi} \sum_{\zeta \in \Xi} \lambda_\xi\, \lambda_\zeta\, p_{\varepsilon,\delta}(\|\xi - \zeta\|^2) = \sum_{\xi \in \Xi} \sum_{\zeta \in \Xi} \lambda_\xi\, \lambda_\zeta\, p_{\varepsilon,\delta}(\|\xi\|^2 - 2\xi \cdot \zeta + \|\zeta\|^2)$$

which is zero, and for $q_{\varepsilon,\delta}$ the same, since

$$\sum_{\xi \in \Xi} \sum_{\zeta \in \Xi} \lambda_\xi\, \lambda_\zeta \|\xi - \zeta\|^{2j} = \sum_{\xi \in \Xi} \sum_{\zeta \in \Xi} \lambda_\xi\, \lambda_\zeta (\|\xi\|^2 - 2\xi \cdot \zeta + \|\zeta\|^2)^j$$

which vanishes for all $0 \le j < k$. Therefore we get

$$\sum_{\xi \in \Xi} \sum_{\zeta \in \Xi} \lambda_\xi\, \lambda_\zeta\, \phi_{\varepsilon,\delta}(\|\xi - \zeta\|) = \int_\delta^\infty \left(\sum_{\xi \in \Xi} \sum_{\zeta \in \Xi} \lambda_\xi \lambda_\zeta\, e^{-\beta\|\xi - \zeta\|^2} \right) \frac{d\gamma(\beta)}{\beta^k}.$$

We note that the double sum (the quadratic form) in the integrand in parentheses is $O(\beta^k)$ at zero because of our side-conditions in (5.2) and by expanding the exponential in a Taylor series about zero. Hence the integral is well-defined even when $\delta \to 0$, as the first $k+1$ terms of the Maclaurin expansion of the exponential in the integrand are annihilated. Thus we may let $\delta \to 0$ and, the right-hand side being already independent of ε, we get for $\boldsymbol{\lambda} \ne 0$ and by taking limits for $\varepsilon \to 0$,

$$\sum_{\xi \in \Xi} \sum_{\zeta \in \Xi} \lambda_\xi\, \lambda_\zeta\, \phi(\|\xi - \zeta\|) = \int_0^\infty \sum_{\xi \in \Xi} \sum_{\zeta \in \Xi} \lambda_\xi\, \lambda_\zeta\, e^{-\beta\|\xi - \zeta\|^2}\, \frac{d\gamma(\beta)}{\beta^k} > 0,$$

as required, where we again use the positive definiteness of the Gaussian kernel and the fact that $d\gamma \not\equiv 0$ due to the assumptions of the theorem. $\square$

As examples we note in particular that any radial basis function of the form (4.4) satisfies the assumptions of the theorem. The k that appears in the statement of our theorem above is precisely the same k as in (4.4). We also note that we may *compose* any function of the form (4.4) with $\sqrt{r^2 + c^2}$ and still get a function that meets the assumptions of the theorem. A special case of that is the multiquadric function, and, if we start with the thin-plate spline and make this composition, the so-called shifted logarithm function $\log(r^2 + c^2)$ or the shifted thin-plate spline $(r^2 + c^2) \log(r^2 + c^2)$ results.

In retrospect we note that we have used in Theorem 2.2 the simple fact that $-\phi$ there was strictly conditionally positive of order one in all $\mathbb{R}^n$ and thus all but one of the eigenvalues of the interpolation matrix are positive. (The matrix is positive definite on a subspace of $\mathbb{R}^\Xi$ of codimension one.) Moreover, here we have that all but k eigenvalues of the matrix with the entries $\phi(\|\xi - \zeta\|)$ are positive and the associated quadratic form is positive definite over the subspace of vectors that satisfy the side-conditions of (5.2). The subspace is usually a space of small codimension. Naturally, we may no longer apply arguments such as in Theorem 2.2 when its codimension is larger than one, because we cannot control the sign of more than one eigenvalue (through the properties of the trace) that is off the set of guaranteed positive ones.

Quite in contrast with the previous chapter, there is not much more to be done for the well-posedness of the interpolation problem, essentially because all sums are finite here, but the work for establishing convergence orders is more involved. Except for the lack of periodicity of the data, the main problem is the presence of boundaries in the setting of finitely many, scattered data. It is much harder, for example, to argue through properties of Lagrange functions here, because a new, different Lagrange function must be found for every $\zeta \in \Xi$, and shifts of just one Lagrange function cannot be used. This would undoubtedly be prohibitively expensive in practice and even difficult to handle in the theory, so this is not the way to work in this case. Instead, we always consider interpolants of the form (5.2) and not Lagrange functions. Moreover, no polynomials except those added directly in (5.2) can be reproduced by expressions that are finite linear combinations of shifts of multiquadrics say. Indeed, the polynomial recovery of Chapter 4 could never have come through finite sums as the following simple demonstration shows. A finite sum

$$g(x) = \sum_{\xi \in \Xi} \lambda_\xi \|x - \xi\|, \qquad x \in \mathbb{R}^n,$$

always has derivative discontinuities at all $\{\xi\}_{\xi \in \Xi}$, even in one dimension, unless the coefficients vanish identically. However, we know from Chapter 4 that

$$1 \equiv \sum_{i \in \mathbb{Z}^n} \sum_{k \in \mathbb{Z}^n} c_k \|x - k - i\|, \qquad x \in \mathbb{R}^n,$$

for the Lagrange coefficients $\{c_k\}_{k \in \mathbb{Z}^n}$. The above expression is however analytic, and thus the derivative discontinuities must be cancelled through the vanishing sum over the c_k and other moment conditions on the c_k. That nonetheless the reproduction of constants and other polynomials is possible is of course a consequence of the infinite sums involved. They allow some of the aforementioned moment conditions on the coefficients to be true (their sum is zero etc.) without forcing them to vanish identically.

5.2 Convergence analysis

The exposition we make here is based on the early work by Duchon (1978) that was extended later on by Madych and Nelson (1990b and later references) and is fundamentally based on the work by Golomb and Weinberger (1959). Also Light and Wayne (1998), Powell (1994b), and Wu and Schaback (1993) made important contributions to the advance of this work. As to other articles on this question and on improving the rates of convergence, see also the Bibliography and the various references and remarks in the commentary on that.

Although it is not absolutely necessary, the exposition is much shorter and contains all of the salient ideas if we restrict attention to the homogeneous thin-plate-spline-type radial basis functions of the form (4.4). Therefore we shall yield to the temptation to simplify in this fashion. The usefulness of the specific class defined through (4.4) will also, incidentally, show in our penultimate chapter, about wavelets, where wavelets from such radial basis functions are relatively easy to find, especially when compared with the more difficult ones from spaces spanned by multiquadrics. We will, however, give and comment on an additional theorem at the end of the current description of the case (4.4) to explain the convergence behaviour e.g. of the multiquadric functions, so as not to omit the nonhomogeneous radial basis functions from this chapter completely. That part of the work is largely due to Madych and Nelson who were the first to realise that Duchon's approach to the question of convergence orders with homogeneous kernels can be extended to nonhomogeneous ones like multiquadrics and the so-called shifted thin-plate splines etc.

5.2.1 Variational theory of radial basis functions

The whole analysis is based on a so-called variational theory which involves interpolants minimising certain semi-norms, re-interpreting the interpolants as solutions of a minimisation problem, and which we will describe now.

To this end, we let $X \subset C(\mathbb{R}^n)$ be a linear space of continuous functions and take $(\,\cdot\,,\,\cdot\,)_*: X \times X \to \mathbb{R}$ to be a *semi*-inner product which may therefore have a nontrivial, finite-dimensional kernel but otherwise satisfies the usual conditions on inner products, i.e. it is a bilinear and Hermitian form. We call the kernel of this map $K \subset X$ and its finite dimension ℓ. In all our applications, this K will be a polynomial space, namely the null-space of a differential operator of small degree, such as the Laplacian in n dimensions (thus $K = \mathbb{P}_n^1$) or the bi-harmonic operator. As we shall see, for our functions of the form (4.4), it will always be $K = \mathbb{P}_n^{k-1}$. Sometimes, one also has to deal with pseudo-differential operators, namely when nonhomogeneous radial basis functions are used like multiquadrics, but we are not considering such radial basis functions at present for the reasons stated above and, at any rate, the kernels in those cases still have a simple form and consist of polynomials. Together with the semi-inner product there is naturally a semi-norm whose properties agree with the properties of a norm except for the finite-dimensional space on which the norm may vanish although its argument does not.

For illustration and later use at several points, we now give an example of such a semi-normed space: to wit, for any integer $k > \frac{1}{2}n$, we let our space of continuous functions be $X = D^{-k} L^2(\mathbb{R}^n)$. To explain this further, we recall the definition of

$$D^{-k} L^2(\mathbb{R}^n)$$

as the linear function space of all $f \colon \mathbb{R}^n \to \mathbb{R}$, all of whose kth total degree distributional partial derivatives are in $L^2(\mathbb{R}^n)$. That $X \subset C(\mathbb{R}^n)$ follows from the Sobolev embedding theorem which is, as quoted in the book by Rudin (1991), as follows.

Sobolev embedding theorem. *Let $n > 0$, $p \geq 0$, k be integers that satisfy the inequality*

$$k > p + \frac{n}{2}.$$

If $f : \tilde{\Omega} \to \mathbb{R}$ is a function in an open set $\tilde{\Omega} \subset \mathbb{R}^n$ whose generalised derivatives of total order up to and including k are locally square-integrable, then there is an $f_0 \in C^p(\tilde{\Omega})$ such that f and f_0 agree almost everywhere on the open set and may therefore be identified as continuous functions.

The Sobolev embedding theorem applied for $p = 0$ ensures $X \subset C(\mathbb{R}^n)$ so long as indeed $k > \frac{n}{2}$ and this will be an important condition also for the analysis of the reproducing kernel. (Strictly speaking, of course, we are identifying X with another, isomorphic space of continuous functions due to the Sobolev embedding theorem.)

In analogy to (4.31) we can also define $D^{-k}L^2(\mathbb{R}^n)$ as the space of all $f: \mathbb{R}^n \to \mathbb{R}$ such that

$$\int_{\mathbb{R}^n} \|x\|^{2k} |\hat{f}(x)|^2 dx < \infty.$$

For this, even nonintegral k are admissible.

The semi-inner product that we use in our example as stated above is

$$(f, g)_* := \int_{\mathbb{R}^n} \sum_{\substack{|\alpha|=k \\ \alpha \in \mathbb{Z}_+^n}} \frac{k!}{\alpha!} D^\alpha f(x) \overline{D^\alpha g(x)}\, dx,$$

so that $K = \mathbb{P}_n^{k-1}$ by design. Here and subsequently, factorials of vectors α are defined for any multiindex $\alpha = (\alpha_1, \alpha_2, \ldots, \alpha_n)^T$ with nonnegative integral entries, and $|\alpha| = \alpha_1 + \alpha_2 + \cdots + \alpha_n$, by

$$\alpha! := \alpha_1! \alpha_2! \cdots \alpha_n!.$$

The notation D for partial derivatives has already been introduced. The coefficients in the above come from the identity

$$\|x\|^{2k} = \sum_{|\alpha|=k} \frac{k!}{\alpha!} x^{2\alpha}.$$

We also associate the semi-norm $\|\cdot\|_*$ with the semi-inner product in a canonical way, namely $\|\cdot\|_* := \sqrt{(\cdot, \cdot)_*}$. We shall use this semi-norm and semi-inner product more often in this chapter and especially in Chapter 7.

Now, leaving our example for the moment and returning to the general setting, we let $\Omega \subset \mathbb{R}^n$ be a compact domain of $\mathbb{R}^n$ with a Lipschitz-continuous boundary and nonempty interior. We also take a finite $\Xi \subset \Omega$ which contains a K-unisolvent point-set that therefore has the property

$$p \in K \text{ and } p|_\Xi = 0 \Longrightarrow p = 0.$$

Finally we assume that $|g(x)| = O(\|g\|_*)$ for all $g \in X$ with $g(\xi) = 0, \xi \in \hat{\Xi}$, where $\hat{\Xi} \subset \Xi$ is a fixed set that also contains a K-unisolvent subset, arbitrary for the time being. This is a boundedness condition that will lead us later to the existence of a so-called reproducing kernel, a most useful concept for our work. This reproducing kernel we shall study in detail and give examples which explain the connection with our radial basis functions in the next subsection.

We observe that, by these hypotheses, the space X possesses an inner product which is an extension of our semi-inner product and which is now equipped with only a trivial kernel, that is the inner product is the sum

$$(f, g)_X = (f, g)_* + \sum_{\xi \in \hat{\Xi}} f(\xi)\, g(\xi).$$

Indeed, if f or g is in the kernel of $(\cdot, \cdot)_*$ and the first summand in the above therefore vanishes, the remaining sum vanishes if and only if f or g is identically zero. This construction works for any subset $\hat{\Xi} \subset \Xi$ that contains a K-unisolvent subset. From now on, however, we take $\hat{\Xi}$ to be a K-unisolvent subset of smallest size, that is, whose cardinality agrees with the dimension of K and that is unisolvent, i.e. the zero polynomial is the only polynomial which is identically zero on $\hat{\Xi}$.

In this fashion we can begin from a semi-inner product and semi-norm, and modify the semi-inner product so that it becomes a genuine inner product. Indeed, with this modification, the space X is in fact a Hilbert space with the induced norm. Conversely, we may consider the Hilbert space generated by a factorisation X/K, where the norm of any equivalence class $f + K$, say, $f \in X$, is defined by f's semi-norm.

We return, however, to our previous construction; the next step is that we assume from now on that the linear function space X is complete with respect to the new norm $\| \cdot \|_X$ induced by the inner product $(\cdot, \cdot)_X$ in the canonical way: we let $\| \cdot \|_X = \sqrt{(\cdot, \cdot)_X}$. With respect to this inner product, there exists a so-called *reproducing kernel* for the Hilbert space X which we have just defined with respect to $(\cdot, \cdot)_X$, as well as for a certain subspace $\hat{X}$ which we will consider below, with respect to the inner product. The linear subspace $\hat{X}$ is chosen such that the inner product and the inner product agree on that subspace. In fact $(\hat{X}, (\cdot, \cdot)_X)$ is a Hilbert space in its own right.

5.2.2 The reproducing kernel (semi-)Hilbert space

There is a close connection between the radial basis functions we study in this book and the notion of reproducing kernels. All of the radial basis functions we have mentioned give rise to such reproducing kernels with respect to some Hilbert space and/or semi-Hilbert spaces, and, more generally, the conditionally positive definite functions which occurred in the previous section give rise to reproducing kernels and Hilbert spaces and/or semi-Hilbert spaces. A semi-Hilbert space can also be made by considering a Hilbert space which is additionally equipped with a semi-norm and semi-inner product as defined above, where it is usual to require that the semi-norm of any element of the space is bounded above by a fixed constant multiple of its norm.

We define reproducing kernels as follows.

Definition 5.2. *Let X be a Hilbert space or a semi-Hilbert space of real-valued functions on $\mathbb{R}^n$, equipped with an inner product or a semi-inner product $(\cdot, \cdot)$, respectively. A reproducing kernel for $(X, (\cdot, \cdot))$ is a function $\mathbf{k}(\cdot, \cdot)$: $\mathbb{R}^n \times \mathbb{R}^n \to \mathbb{R}$ such that for any element $f \in X$ we have the pointwise equality with respect to x*

$$(f, \mathbf{k}(\cdot, x)) = f(x), \qquad x \in \mathbb{R}^n.$$

In particular, $\mathbf{k}(\cdot, x) \in X$ for all $x \in \mathbb{R}^n$.

Standard properties of the reproducing kernel are that it is Hermitian and non-negative for all arguments from X (see, for instance, Cheney and Light, 1999).

There are well-known necessary and sufficient conditions known for the linear space X to be a reproducing kernel (semi-)Hilbert space, i.e. for the existence of such a reproducing kernel $\mathbf{k}$ within the Hilbert space setting. A classical result (Saitoh, 1988, or Yosida, 1968, for instance) states that a reproducing kernel exists if and only if the point evaluation operator is a bounded operator on the (semi-)Hilbert space. If we are in a Hilbert space setting, the reproducing kernel is, incidentally, unique. In our case, a reproducing kernel exists for the *subspace* $\hat{X}$ of all functions g that vanish on the K-unisolvent subset $\hat{\Xi}$ because the condition $|g(x)| = O(\|g\|_X) = O(\|g\|_*)$ from the previous subsection gives boundedness, i.e. continuity, for the linear operator of point evaluation in that subspace $\hat{X}$.

In our example above, the boundedness of the point evaluation functional amounts to the uniform boundedness of any function g that is in $X = D^{-k}L^2(\mathbb{R}^n)$ and vanishes on $\hat{\Xi}$, by a fixed g-independent multiple of its semi-norm $\|g\|_*$. That this is so is a consequence of the two following observations, but there are some extra conditions on the domain Ω.

The first one is related to the Sobolev inequality. We have taken this particular formulation from the standard reference (Brenner and Scott, 1994, p. 32) and restricted it again to the case that interests us. We also recall, and shall use often from now on, the special notation for the semi-norm

$$|f|_{k,\Omega}^2 := \int_\Omega \sum_{\substack{|\alpha|=k \\ \alpha \in \mathbb{Z}_+^n}} \frac{k!}{\alpha!} |D^\alpha f(x)|^2 \, dx$$

as is usual elsewhere in the – especially finite element – literature, and, as a further notational simplification, $|f|_k := |f|_{k,\mathbb{R}^n}$. Accordingly, we shall sometimes use the superspace $D^{-k}L^2(\Omega)$ of $X = D^{-k}L^2(\mathbb{R}^n)$ that contains all f with finite semi-norm when restricted to the domain Ω. If f is only defined on Ω, we may always extend it by zero to all of $\mathbb{R}^n$.

Sobolev inequality.   *Let Ω be a domain in n-dimensional Euclidean space with Lipschitz-continuous boundary and let $k > n/2$ be an integer. Then there is a constant C such that for any $f \in D^{-k}L^2(\mathbb{R}^n)$ the pointwise inequality*

$$(5.5) \qquad\qquad |f(x)| \le C \sum_{j=0}^{k} |f|_{j,\Omega}$$

holds on the domain Ω, where C only depends on the domain, the dimension of the underlying real space and on k.

The second one is the Bramble–Hilbert lemma (Bramble and Hilbert, 1971). We still assume that Ω is a domain in $\mathbb{R}^n$ and that it has a Lipschitz-continuous boundary $\partial\Omega$. Moreover, from now on it should additionally satisfy an interior *cone condition* which means that there exists a vector $\xi(x) \in \mathbb{R}^n$ of unit length for each $x \in \Omega$ such that for a fixed positive $\hat{r}$ and $\vartheta > 0$,

$$\Omega \supset \{x + \lambda\eta \mid \eta \in \mathbb{R}^n, \ \|\eta\| = \hat{r}, \eta \cdot \xi(x) \ge \cos\vartheta, \ 0 \le \lambda \le 1\}.$$

Note that the radius and angle are fixed but arbitrary otherwise, i.e. they may depend on the domain. In fact, Bezhaev and Vasilenko (2001) state that the cone condition is superfluous if we require a Lipschitz-continuous boundary of the domain. We recall the definition of H^k from the end of the fourth chapter.

Bramble–Hilbert lemma.   *Let $\Omega \subset \mathbb{R}^n$ be a domain that satisfies an interior cone condition and is contained in a ball which has diameter ρ. Let F be a linear functional on the Sobolev space $H^k(\Omega)$ such that $F(q) = 0$ for all $q \in \mathbb{P}_n^{k-1}$. Suppose that the inequality*

$$|F(u)| \le C_1 \left(\sum_{j=0}^{k} \rho^{j-n/2} |u|_{j,\Omega} \right)$$

holds for u with finite semi-norm $|u|_{j,\Omega}$, $j \le k$, with a positive constant C_1 that is independent of u and ρ. Then it is true that

$$|F(u)| \le C_2 \left(\rho^{k-n/2} |u|_{k,\Omega} \right),$$

where C_2 does not depend on u or ρ.

For our application of the Bramble–Hilbert lemma we let $k > \frac{n}{2}$ be a positive integer and the operator F map a continuous function f to the value of the interpolation *error* by polynomial interpolation on $\hat{\Xi}$ with polynomials of total degree less than k. Thus, it is zero if f is already such a polynomial, as a result of $\hat{\Xi}$ being unisolvent. Therefore the first assumption of the Bramble–Hilbert lemma is satisfied. The second assumption is a consequence of the Sobolev inequality, where it is also important that f vanishes on $\hat{\Xi}$, since the interpolating polynomial for f is zero if $f \in \hat{X}$, as is the case for our setting. Therefore, the Bramble–Hilbert lemma implies

$$|f(x)| \leq C|f|_{k,\Omega} \leq C\|f\|_*,$$

as required, the constant C being independent of f. To this end, we also recall that the polynomial interpolant of degree $k - 1$ itself vanishes when differentiated k times, which is why the polynomial parts of the interpolation error expression disappear in both the middle and the right-hand side of the above inequality.

Now that we know about the existence of a reproducing kernel in our particular example (as a special case of the general radial basis function setting), we wish to identify it explicitly in the sequel. Once more, we begin here with the general case and, becoming progressively more specific, restrict to our examples for the purpose of illustrating the general case later on.

The reproducing kernel has an especially simple form in our setting. In order to derive this form, we may let, because of the condition of unisolvency of $\hat{\Xi}$, the functions p_ξ, $\xi \in \hat{\Xi}$, be polynomial Lagrange functions, i.e. such that p_ξ span K and satisfy

$$p_\xi(\zeta) = \delta_{\xi\zeta}, \qquad \xi, \zeta \in \hat{\Xi}.$$

Since we are looking for a reproducing kernel on a subspace, our reproducing kernel $\mathbf{k}$ has a particular, K-dependent form. Specifically, for our application, we can express the property of the reproducing kernel by the identity

$$(5.6) \quad f(x) = \left(f, \phi(\cdot - x) - \sum_{\xi \in \hat{\Xi}} p_\xi(x)\phi(\cdot - \xi) \right)_*, \qquad x \in \mathbb{R}^n, \quad f \in \hat{X}.$$

We have therefore the reproducing kernel of the form

$$\mathbf{k}(y, x) = \phi(y - x) - \sum_{\xi \in \hat{\Xi}} p_\xi(x)\phi(y - \xi)$$

and ϕ is a function $\mathbb{R}^n \to \mathbb{R}$ which is, notably, not necessarily itself in X. This is why ϕ does not appear itself on the right-hand side of the expression above but in the shape of a certain linear combination involving the p_ξ. This particular shape also includes the property that it vanishes on $\hat{\Xi}$. We will come to this in a moment. In the case that the function f does not vanish on $\hat{\Xi}$, then we subtract from it the finite sum of multivariate polynomials

$$\sum_{\xi \in \hat{\Xi}} f(\xi)p_\xi,$$

so that the difference does, and henceforth use instead the difference as follows:

$$(5.7) \qquad\qquad f - \sum_{\xi \in \hat{\Xi}} f(\xi)p_\xi.$$

The same principle was applied to the shift of ϕ above to guarantee that the kernel **k** vanishes on $\hat{\Xi}$. If the null-space K of the semi-inner product and of the semi-norm is trivial, no such operation is required – the sums in the preceding two displays remain empty for $K = \{0\}$. The fact that the reproducing kernel $\mathbf{k}(y, x)$ is a function of the difference of two arguments (that is, $\mathbf{k}(y, x) = \tilde{\mathbf{k}}(y - x)$) is a simple consequence of the shift-invariance of the space X (see the interesting paper by Schaback, 1997, on these issues), because any invariance properties such as shift- or rotational invariance that hold for the space X are immediately carried over to properties of the reproducing kernel **k** through Definition 5.2. The proofs of these facts are easy and we omit them because we do not need them any further in this text.

Most importantly, however, we require that the right-hand side expression of the inner product in (5.6) above is in X. We make the *hypothesis* that ϕ is such that, whenever we form a linear combination of shifts of ϕ, whose coefficients are such that they give zero when summed against any element of K, then that linear combination must be in X:

$$\sum_{\xi \in \Xi} \lambda_\xi \phi(\cdot - \xi) \in X \quad \text{if} \quad \sum_{\xi \in \Xi} \lambda_\xi q(\xi) = 0 \,\forall_{q \in K}.$$

We will show later that for the X which we have given as an example above, this hypothesis is fulfilled.

For the semi-inner product in the above display (5.6), the sum over the coefficients of the linear combination above against any $p \in K$ is indeed zero:

$$p(x) - \sum_{\xi \in \hat{\Xi}} p_\xi(x) p(\xi) = p(x) - p(x) = 0,$$

because of the Lagrange conditions on the basis elements $p_\xi, \xi \in \Xi$. Therefore, assuming our above hypothesis on ϕ being in place for the rest of this section now, the right-hand side in the inner product in the display (5.6) is in the required space X. Moreover, since $f(x)$ vanishes for $x = \zeta \in \hat{\Xi}$ on the left-hand side in the above identity (5.6), so must the expression

$$\phi(\cdot - \zeta) - \sum_{\xi \in \hat{\Xi}} p_\xi(\zeta) \phi(\cdot - \xi)$$

on the right-hand side, which it does because $p_\xi(\zeta) = \delta_{\xi\zeta}$, and therefore we have verified that it is, in particular, in $\hat{X}$.

We may circumvent the above hypothesis on ϕ which may appear strange at this point, because, as we shall see, it is fulfilled for all our radial basis functions (4.4). More specifically, given a fixed basis function ϕ, we shall construct a (semi-)Hilbert space X whose reproducing kernel has the required form (5.6).

This can be done as follows. In the case of all the radial basis functions $\phi(\|\cdot\|)$ we use in this chapter, the constructed X is the same as the above X we started with as we shall see shortly in this section. We follow especially the analysis of Schaback (1999).

To begin with, we define now a seemingly new semi-inner product for any two functions $f\colon \Omega \to \mathbb{R}$ and $g\colon \Omega \to \mathbb{R}$ that are finite linear combinations of shifts of a given general, continuous function $\phi\colon \mathbb{R}^n \to \mathbb{R}$,

$$f = \sum_{\xi \in \Xi_1} \lambda_\xi \phi(\cdot - \xi)$$

and

$$g = \sum_{\xi \in \Xi_2} \mu_\xi \phi(\cdot - \xi),$$

where Ξ_1 and Ξ_2 are arbitrary finite subsets of Ω; those sets need not (but usually do) agree. The function ϕ is required to be continuous and conditionally positive definite of order k, so that we let $K = \mathbb{P}_n^{k-1}$. Therefore we have an additional condition on the coefficients in order to define the semi-inner product which follows, that is we require that for the given finite-dimensional linear space K, the coefficient sequences must satisfy the conditions $\sum_{\xi \in \Xi_1} \lambda_\xi p(\xi) = 0$ and $\sum_{\xi \in \Xi_2} \mu_\xi p(\xi) = 0$ for all $p \in K$. Moreover, we let $L_K(\Omega)$ denote the linear space of all functionals of the form

$$\lambda\colon h \mapsto \sum_{\xi \in \Xi_1} \lambda_\xi h(\xi),$$

with the aforementioned property of the coefficients and with *any* finite set Ξ_1. Its Hilbert space completion will be $\mathcal{L}_K(\Omega)$. So λ is a linear functional on $C(\Omega)$. Now define the semi-inner product for the above two functions,

$$(5.8) \qquad (f, g)_* := (\lambda, \mu)_* := \sum_{\xi \in \Xi_1} \sum_{\zeta \in \Xi_2} \lambda_\xi \mu_\zeta \phi(\xi - \zeta),$$

where we identify functions f and g with linear functionals λ and μ in an obvious way defined through their coefficients. So functionals λ and functions f are related *via* $\lambda \mapsto f = \lambda^x \phi(x - \cdot)$, where the superscript x refers to the variable with respect to which we evaluate the functional. Note that we can always come back from one such functional to a function in x by applying it in this way to $\phi(x - \cdot)$.

As a consequence of the conditional positive definiteness of ϕ, this semi-inner product is positive semi-definite as required. Further, let $\hat{\Xi}$ still contain a K-unisolvent set and define the *modified point evaluation functional* $\delta_{(x)}$ by

$$\delta_{(x)} f = f(x) - \sum_{\xi \in \hat{\Xi}} p_\xi(x) f(\xi).$$

Here, p_ξ are Lagrange polynomials. Then the modified point evaluation functional is in $L_K(\Omega)$ for every $x \in \Omega$, which the ordinary Dirac function $\delta(\cdot - x)$, i.e. the usual function evaluation at a point x, is *not*, as it does not annihilate the polynomials in the space K unless $K = (0)$. Now let $\mathcal{X}$ be the completion of the range of

$$(L_K(\Omega), \delta_{(x)})_*, \qquad x \in \Omega,$$

that is in short, but very useful, notation the space generated by all functions from $L_K(\Omega)$ with the semi-inner product taken with $\delta_{(x)}$, always modulo K. This is a Hilbert space of functions defined for every $x \in \Omega$.

In particular, for all μ and λ from $\mathcal{L}_K(\Omega)$, we get $\mu((\lambda, \delta_{(x)})_*) = (\lambda, \mu)_*$, due to our definition of the inner product in (5.8). Here it is also relevant that the polynomials which appear in the definition of the modified point evaluation functional are annihilated by the semi-inner product

$$(\delta_{(x)}, \lambda)_* = (\delta_{(x)}, f)_* = \delta_{(x)}f = f(x) - \sum_{\xi \in \hat{\Xi}} p_\xi(x)f(\xi),$$

to which then μ has to be applied in the usual way. Thereby, a semi-inner product and a semi-norm are also defined on the aforementioned range $\mathcal{X}$. All functions in $\mathcal{X}$ vanish on our K-unisolvent point set $\hat{\Xi}$. To form X, take the direct sum of K and $\mathcal{X}$. That is the so-called 'native space'. Further, let $f \in \mathcal{X}$. Thus there exists an f-dependent functional λ such that

$$f(x) = (\lambda, \delta_{(x)})_* = (f, (\delta_{(\cdot)}, \delta_{(x)})_*)_*,$$

the second equality in this display being a consequence of the definition of $(f, g)_*$ above. That display provides the reproducing kernel property. Indeed, it is exactly the same identity as before, because, according to our initial definition of the bilinear $(\cdot, \cdot)_*$ as a semi-inner product on $L_K(\Omega)$,

$$(\delta_{(\cdot)}, \delta_{(x)})_* = \delta_{(\cdot)}^t \delta_{(x)}^z \phi(t - z),$$

where the superscript fixes the argument with respect to which the linear functional is applied. The above is the same as

$$\delta_{(\cdot)}^t \left[\phi(t - x) - \sum_{\xi \in \Xi} p_\xi(x)\phi(t - \xi) \right]$$

$$= \phi(\cdot - x) - \sum_{\xi \in \Xi} p_\xi(x)\phi(\cdot - \xi)$$

$$- \sum_{\zeta \in \Xi} p_\zeta(\cdot)\phi(\zeta - x) + \sum_{\zeta \in \Xi} p_\zeta(\cdot) \sum_{\xi \in \Xi} p_\xi(x)\phi(\xi - \zeta).$$

When inserted into the semi-inner product, the last two terms from the display disappear, since $p_\xi(\cdot) \in K$ and any such arguments in the kernel are annihilated. Therefore the two expressions for the reproducing kernel are the same.

5.2.3 Minimum norm interpolants

We claim that this reproducing kernel, which is expressed as the function ϕ minus a certain linear combination of its translates, gives rise to the 'minimum norm interpolant' for any given data $f_\xi, \xi \in \Xi$. It is no accident that the function which appears in this kernel is also denoted by ϕ, similarly to our radial basis functions, because they will form a special case of such kernels. In fact, we shall also refer – simplifying, although somewhat confusingly, but the meaning will be clear from the context – to ϕ itself as the reproducing kernel.

Proposition 5.2. *Suppose a finite set Ξ of distinct centres and the f_ξ are given. Let the assumptions about ϕ and Ω of the previous subsection hold and K be the ℓ-dimensional kernel of the above semi-inner product. Then, if Ξ contains a unisolvent subset with respect to the kernel K of the semi-norm, there is a unique s of the form*

$$s(x) = \sum_{\xi \in \Xi} \lambda_\xi \, \phi(x - \xi) + p(x), \quad x \in \mathbb{R}^n,$$

where $p \in K$ and $\sum_{\xi \in \Xi} \lambda_\xi \, q(\xi) = 0$ for all $q \in K$, such that it is the minimum norm interpolant giving $s(\xi) = f_\xi, \xi \in \Xi$, and $\|s\|_* \le \|g\|_*$ for any other such $g \in X$ which interpolates the prescribed data.

Proof: We show first that whenever s with the form in the statement of the theorem and another $g \in X$ both interpolate the prescribed data, then $\|s\|_* \le \|g\|_*$.

To this end, we first note that s belongs to X. This follows from our hypotheses on ϕ in Subsection 5.2.2. Next, we consider the following semi-inner products, recalling that $g - s$ vanishes identically on Ξ because both s and g interpolate:

$$\|g\|_*^2 - \|s\|_*^2 = (g, g)_* - (s, s)_*$$
$$= (g - s, g - s)_* + 2(g - s, s)_*$$
$$= \|g - s\|_*^2 + 2\left(g - s, \sum_{\xi \in \Xi} \lambda_\xi \, \phi(\cdot - \xi) + p\right)_*.$$

Now using the side-conditions of the coefficients λ_ξ of the interpolant, i.e.

(5.9) $$\sum_{\xi \in \Xi} \lambda_\xi q(\xi) = 0 \quad \forall q \in K,$$

together with the linearity of the semi-inner product, we get that the above is the same as the following:

$$
\|g - s\|_*^2 + 2\bigg(g - s, \sum_{\xi \in \Xi} \lambda_\xi \, \phi(\cdot - \xi) + p - \sum_{\xi \in \Xi} \lambda_\xi \sum_{\zeta \in \hat{\Xi}} p_\zeta(\xi)\phi(\cdot - \zeta)\bigg)_*
$$

$$
= \|g - s\|_*^2 + 2 \sum_{\xi \in \Xi} \lambda_\xi \bigg(g - s, \, \phi(\cdot - \xi) - \sum_{\zeta \in \hat{\Xi}} p_\zeta(\xi)\phi(\cdot - \zeta)\bigg)_*
$$

$$
+ 2(g - s, p)_*
$$

$$
= \|g - s\|_*^2 + 2 \sum_{\xi \in \Xi} \lambda_\xi \, (g(\xi) - s(\xi)) = \|g - s\|_*^2.
$$

This follows from the interpolation conditions and is indeed nonnegative, as required. Moreover, this is *strictly positive* unless $g - s$ is in the kernel K. In the latter case, however, $g = s$, because $(g - s)|_\Xi = 0$ and because Ξ contains a subset that is unisolvent with respect to K.

It remains to prove that the minimum norm interpolant exists and does indeed have the required form. This follows from the above proof of uniqueness, because the space spanned by the translates of ϕ, and also K, are finite-dimensional spaces. In fact we can also show that the finite square matrix $\{\phi(\zeta - \xi)\}_{\zeta, \xi \in \Xi}$ is positive definite on the subspace of vectors $\boldsymbol{\lambda} \in \mathbb{R}^\Xi \setminus \{0\}$ with property (5.9):

$$
\sum_{\zeta \in \Xi} \lambda_\zeta \sum_{\xi \in \Xi} \lambda_\xi \, \phi(\zeta - \xi) = (\boldsymbol{\lambda}, \boldsymbol{\lambda})_* = \bigg\|\sum_{\zeta \in \Xi} \lambda_\zeta \, \phi(\cdot - \zeta)\bigg\|_*^2
$$

$$
\geq 0 \quad (> 0 \ \text{if} \ \boldsymbol{\lambda} = (\lambda_\zeta)_{\zeta \in \Xi} \neq 0).
$$

Here we have used the above strict minimisation property, i.e. $\|g\|_* > \|s\|_*$ unless $g = s$ for all interpolating functions from X, which means that the interpolant is unique.   □

5.2.4 The power functional and convergence estimates

The previous subsection provided an alternative view of radial basis function interpolants, i.e. the so-called variational approach. Further, this development allows us to supply error estimates by introducing the notion of a power function which is well-known. This so-called power function will turn out to be closely related to a functional to evaluate the pointwise error of the interpolation. The error results which we shall obtain in Theorems 5.5 and 5.8 use the power function and the proposition below for their estimates. Additionally, the last subsection of this chapter about the so-called uncertainty principle uses the power function.

The power function is defined by the following semi-inner product of reproducing kernels, where we maintain the notation $(\,\cdot\,,\,\cdot\,)_*$ of the previous subsections:

$$P(x) = \left(\phi(\cdot - x) - \sum_{\xi \in \hat{\Xi}} p_\xi(x)\phi(\cdot - \xi), \ \ \phi(\cdot - x) - \sum_{\zeta \in \hat{\Xi}} p_\zeta(x)\phi(\cdot - \zeta) \right)_*,$$

$$x \in \mathbb{R}^n,$$

where the set $\hat{\Xi}$ is still supposed to have $\ell = \dim K$ elements and moreover to be unisolvent as demanded earlier, and where the Lagrange conditions $p_\xi(\zeta) = \delta_{\zeta\xi}$ for all $\zeta, \xi \in \hat{\Xi}$ are in place for a set of $p_\xi \in K, \xi \in \hat{\Xi}$.

Proposition 5.3. *Let X, ϕ, the set of centres and the power function be as above and in Subsection 5.2.1, s be the minimum norm interpolant of Proposition 5.2, $f_\xi = f(\xi), \xi \in \Xi$, with an f from X. Then we have the pointwise error estimate*

(5.10) $$|f(x) - s(x)|^2 \le P(x)(f, f)_*, \qquad x \in \mathbb{R}^n.$$

The inequality remains true if f is replaced by $f - s$ in both arguments of the semi-inner product on the right-hand side, where we require $f - s \in X$ instead of $f \in X$.

Proof: We will see in this proof that P serves as nothing else, as alluded to above, than a functional used for evaluating the pointwise error for our interpolant.

For proving that, we wish to re-express $f(x) - s(x)$ in terms of P and f. Note first that $f - s \in \hat{X}$ because of the interpolation conditions. Note also the inequality

$$\|f - s\|_* \le \|f\|_*$$

which is a simple application of a result from the proof of Proposition 5.2. That is, we get this from setting $g := f$ in that proof and deriving the inequality

$$\|f\|_*^2 = \|f - s\|_*^2 + \|s\|_*^2 \ge \|f - s\|_*^2$$

from $(f - s, s)_* = 0$. Using the Cauchy–Schwarz inequality 'backwards' and our previous result, we show now that the square root of the right-hand side of (5.10) is bounded below as follows:

$$
\begin{aligned}
P(x)(f, f)_* &= P(x)\|f\|_*^2 \\
&\geq P(x)\|f - s\|_*^2 = P(x)(f - s, f - s)_* \\
&= \left(\phi(\cdot - x) - \sum_{\xi \in \hat{\Xi}} p_\xi(x)\phi(\cdot - \xi), \right. \\
&\qquad \left. \phi(\cdot - x) - \sum_{\zeta \in \hat{\Xi}} p_\zeta(x)\phi(\cdot - \zeta) \right) (f - s, f - s)_* \\
&\geq \left(\phi(\cdot - x) - \sum_{\xi \in \hat{\Xi}} p_\xi(x)\phi(\cdot - \xi), f - s \right)_*^2 \\
&= (f(x) - s(x))^2,
\end{aligned}
$$

by the reproducing kernel property, due to the definition of $\{p_\xi\}_{\xi \in \Xi}$. This is the evaluation of the pointwise error by the power function.   □

If we let $\hat{\Xi}$, as in the proposition above, be a unisolvent set with ℓ elements, and if subsequently $\hat{\Xi}$ is enlarged over such an initial set, the larger set of centres being Ξ, in order to form the interpolant s, we still get the asserted inequality (5.10) of the proposition.

In order to see that, we still take those p_ξ, $\xi \in \hat{\Xi}$, to be a set of Lagrange functions for K and get that the right-hand side of (5.10) with $\hat{\Xi}$ used is a strict upper bound to the left-hand side of (5.10), Ξ being used for s. After all, the smaller a subset $\hat{\Xi}$ is, the bigger the error will be, unless it remains the same which is also possible. Thus we are simply being more modest in choosing a P with a smaller $\hat{\Xi} \subset \Xi$ to bound the error on the left-hand side by the right-hand side of (5.10) from above.

In short, (5.10) remains true even if the $\hat{\Xi}$ and p_ξ, $\xi \in \hat{\Xi}$, used for P on the right-hand side are such that $\hat{\Xi}$ is a *proper subset* of the actual data points Ξ used for the s on the left-hand side.

Now, in order to apply Proposition 5.3 and get our convergence estimates in terms of powers of h, the density measure of the centres Ξ, we choose X to be the familiar linear space $D^{-k}L^2(\mathbb{R}^n)$ when $k > n/2$. As above, there goes naturally with this space the semi-inner product $(f, g)_*$ which we shall shortly use also in a Fourier transform form by employing the standard Parseval–Plancherel formula. In view of our work on minimum norm interpolants, we shall now proceed to compute the reproducing kernel for X and its semi-inner product. There, as alluded to already, we call ϕ the reproducing kernel for simplicity – and not quite correctly – although the reproducing kernel really is the difference stated at the beginning of Subsection 5.2.2.

Proposition 5.4. *For the above $X = D^{-k}L^2(\mathbb{R}^n)$ and its semi-inner product, the reproducing kernel ϕ is a nonzero multiple of the radial basis function defined in (4.4), according to the choice of the constants n and k.*

Proof: Let f be an element of X; in particular the function f is therefore continuous by the Sobolev embedding theorem, since $k > \frac{1}{2}n$. The salient idea for the proof is to apply the Parseval–Plancherel formula from the Appendix to rewrite the expression for the semi-inner product with the reproducing kernel, namely

$$(5.11) \quad \left(f, \phi(\| \cdot - x\|) - \sum_{\xi \in \hat{\Xi}} p_\xi(x)\phi(\| \cdot - \xi\|) \right)_*$$

$$= \int_{\mathbb{R}^n} \sum_{|\alpha|=k} \frac{k!}{\alpha!} D^\alpha f(y) D^\alpha \left(\phi(\|x - y\|) - \sum_{\xi \in \hat{\Xi}} p_\xi(x)\phi(\|y - \xi\|) \right) dy,$$

in terms of the Fourier transform because then it will contain the Fourier transform of (4.4).

To begin, we assume $f \in \hat{X}$. Thus we can derive the required reproducing kernel property as follows from the Parseval–Plancherel formula applied to square-integrable functions. Expression (5.11) is a $(2\pi)^{-n}$-multiple of the displayed equation

$$\int_{\mathbb{R}^n} e^{ix \cdot t} \hat{f}(t) \sum_{|\alpha|=k} \frac{k!}{\alpha!} t^{2\alpha} \left(\|t\|^{-2k} - \sum_{\xi \in \hat{\Xi}} p_\xi(x)\|t\|^{-2k} e^{i(\xi-x)\cdot t} \right) dt$$

$$= \int_{\mathbb{R}^n} e^{ix \cdot t} \hat{f}(t) \|t\|^{2k} \|t\|^{-2k} \left(1 - \sum_{\xi \in \hat{\Xi}} p_\xi(x)e^{i(\xi-x)\cdot t} \right) dt = (2\pi)^n f(x),$$

where we use the fact that the integrals are defined in L^2 due to the Lagrange properties of p_ξ, and the continuity of f. Further, we have applied some standard Fourier theory, including the Fourier inversion formula, and the fact that $f|_{\hat{\Xi}} = 0$. Finally, we have assumed for the sake of simplicity that ϕ is scaled in such a way that

$$\hat{\phi}(\|t\|) = \|t\|^{-2k}.$$

Up to a constant multiple which is immaterial here, this is precisely the distributional Fourier transform of (4.4), as we have asserted. If f does not vanish on $\hat{\Xi}$ we proceed as in Subsection 5.2.2, that is we use (5.7) instead of f. The result now follows.   □

Since our main concern is interpolation, we can in fact always view the function ϕ and its transform $\hat{\phi}$ up to a constant multiple. This is the case because any such multiplicative constant will be absorbed into the coefficients of the unique interpolant. We have used this fact before in the analysis of Chapter 4 where we have stated that a sign-change in the radial basis functions is for this reason immaterial.

We observe that for the radial basis functions (4.4), the power function can be rephrased as

$$P(x) = -2\sum_{\xi\in\hat{\Xi}} p_\xi(x)\phi(\|x-\xi\|) + \sum_{\xi\in\hat{\Xi}} p_\xi(x)\sum_{\zeta\in\hat{\Xi}} p_\zeta(x)\phi(\|\zeta-\xi\|),$$

because, according to our original definition of the power function and because of the Parseval–Plancherel identity again,

$$P(x) = \left(\phi(\|\cdot-x\|)-\sum_{\xi\in\hat{\Xi}} p_\xi(x)\phi(\|\cdot-\xi\|),\ \phi(\|\cdot-x\|)-\sum_{\zeta\in\hat{\Xi}} p_\zeta(x)\phi(\|\cdot-\zeta\|)\right)_*$$

is the same as

$$\frac{1}{(2\pi)^n}\int_{\mathbb{R}^n}\|y\|^{-2k}\left|e^{ix\cdot y}-\sum_{\xi\in\hat{\Xi}} p_\xi(x)e^{i\xi\cdot y}\right|^2 dy,$$

where we are still scaling the radial basis function in such a way that its generalised Fourier transform is $\|\cdot\|^{-2k}$ in order to keep our computations as perspicuous as possible. Note that the integral in the last display is well-defined because the factor in modulus signs is of order $\|y\|^k$ by the side-conditions on its polynomial coefficients $p_\xi(x)$. Using the Fourier inversion formula gives finally

$$P(x) = \phi(0)-2\sum_{\xi\in\hat{\Xi}} p_\xi(x)\,\phi(\|x-\xi\|)+\sum_{\xi\in\hat{\Xi}}\sum_{\zeta\in\hat{\Xi}} p_\xi(x)\,p_\zeta(x)\,\phi(\|\zeta-\xi\|).$$

This is the same as the above, as required, since $\phi(0)=0$.

We are also now in a position to prove that if we are given a linear combination for the radial basis function from Proposition 5.4 above,

$$g(x) = \sum_{\xi\in\Xi}\lambda_\xi\phi(\|x-\xi\|)$$

whose coefficients λ_ξ satisfy the side-conditions (5.9), then $g\in X$. For settling this statement we have to prove $\|g\|_* < \infty$. Indeed, we note that after the application of the Parseval–Plancherel formula and inserting the definition of g

$$(2\pi)^n \|g\|_*^2 = \int_{\mathbb{R}^n} \|t\|^{2k} |\hat{g}(t)|^2 \, dt$$

becomes a nonzero multiple of the integral

$$\int_{\mathbb{R}^n} \|t\|^{2k} \|t\|^{-4k} \left| \sum_{\xi \in \Xi} \lambda_\xi e^{it \cdot \xi} \right|^2 dt = \int_{\mathbb{R}^n} \|t\|^{-2k} \left| \sum_{\xi \in \Xi} \lambda_\xi e^{it \cdot \xi} \right|^2 dt.$$

This integral is finite, because, again by the side-conditions required in (5.9) and by expanding the exponential function in a Taylor series about the origin, the factor in modulus signs is of order $\|t\|^k$ which gives integrability of the integrand in any finite radius ball about zero. The integrability over the rest of the range is guaranteed by the fact that $2k > n$ and by the boundedness of the term in modulus signs.

It follows by the same means that the two semi-inner products which we have considered in this section are indeed the same. Specifically, the first one we have defined in our example of thin-plate splines as used above. The second one, corresponding to our choice of thin-plate slines, was for two functions

$$f = \sum_{\xi \in \Xi_1} \lambda_\xi \phi(\| \cdot - \xi \|)$$

and

$$g = \sum_{\xi \in \Xi_2} \mu_\xi \phi(\| \cdot - \xi \|),$$

with the usual side-conditions on the coefficients λ_ξ and μ_ξ, of the symmetric form

$$(f, g)_* = \sum_{\xi \in \Xi_1} \sum_{\zeta \in \Xi_2} \mu_\zeta \lambda_\xi \phi(\|\xi - \zeta\|)$$

with the same notation for the semi-inner product as before. And indeed we can recover – by using the Fourier transforms – the same result, namely

$$(f, g)_* = \frac{1}{(2\pi)^n} \int_{\mathbb{R}^n} \frac{1}{\hat{\phi}(\|t\|)} \left[\sum_{\xi \in \Xi_1} \lambda_\xi e^{it \cdot \xi} \right] \left[\sum_{\zeta \in \Xi_2} \mu_\zeta e^{-it \cdot \zeta} \right] \hat{\phi}(\|t\|)^2 \, dt$$

$$= \lim_{\varepsilon \to 0_+} \frac{1}{(2\pi)^n} \int_{\mathbb{R}^n \setminus B_\varepsilon(0)} \left[\sum_{\xi \in \Xi_1} \lambda_\xi e^{it \cdot \xi} \right] \left[\sum_{\zeta \in \Xi_2} \mu_\zeta e^{-it \cdot \zeta} \right] \hat{\phi}(\|t\|) \, dt$$

$$=: \lim_{\varepsilon \to 0_+} I_\varepsilon(\mu, \lambda).$$

The limit exists since the side-conditions (5.9) on the coefficients λ_ξ (and the same for the coefficients μ_ξ) in the square brackets in the above display imply that the two sums therein are $O(\|t\|^k)$ each in a neighbourhood of the origin, so that the integral converges absolutely.

Moreover, it is a consequence of the definition of the generalised Fourier transform of $\phi(\|\cdot\|)$ and of the definition of the generalised inverse Fourier transform that the above is the same as

$$\sum_{\xi \in \Xi_1} \sum_{\zeta \in \Xi_2} \mu_\zeta \lambda_\xi \phi(\|\xi - \zeta\|),$$

as we shall see now. Indeed, we note that ϕ is continuous and provides for any good function $\gamma \in C^\infty(\mathbb{R}^n)$, cf. Chapter 4, with the property

$$\int_{\mathbb{R}^n} \gamma(x) p(x)\, dx = 0, \qquad p \in K,$$

the identity

$$\int_{\mathbb{R}^n} \gamma(x - \xi + \zeta)\phi(\|x\|)\, dx = \frac{1}{(2\pi)^n} \int_{\mathbb{R}^n} \hat{\gamma}(x) e^{ix \cdot (\xi - \zeta)} \hat{\phi}(\|x\|)\, dx.$$

The last integral is well-defined because $\hat{\gamma}$ has a sufficiently high order zero at the origin by virtue of the above side-condition. This equation, in turn, implies for any good γ

$$\int_{\mathbb{R}^n} \sum_{\xi \in \Xi_1} \sum_{\zeta \in \Xi_2} \mu_\zeta \lambda_\xi \gamma(x) \phi(\|x + \xi - \zeta\|)\, dx$$

$$= \frac{1}{(2\pi)^n} \int_{\mathbb{R}^n} \hat{\gamma}(x) \sum_{\xi \in \Xi_1} \sum_{\zeta \in \Xi_2} \mu_\zeta \lambda_\xi e^{ix \cdot (\xi - \zeta)} \hat{\phi}(\|x\|)\, dx,$$

where the condition in the previous display on γ can actually be waived from now on due to the property of the coefficients μ_ζ and λ_ξ. This is the same as

$$\int_{\mathbb{R}^n} \hat{\gamma}(x) \left[\sum_{\xi \in \Xi_1} \lambda_\xi e^{ix \cdot \xi} \right] \left[\sum_{\zeta \in \Xi_2} \mu_\xi e^{-ix \cdot \zeta} \right] \hat{\phi}(\|x\|)\, dx.$$

Since γ is now an arbitrary good function, the required identity

$$(f, g)_* = \lim_{\varepsilon \to 0_+} I_\varepsilon(\mu, \lambda) = \sum_{\xi \in \Xi_1} \sum_{\zeta \in \Xi_2} \mu_\zeta \lambda_\xi \phi(\|\xi - \zeta\|)$$

is a consequence of the definition of the generalised Fourier transform. That the semi-inner products are identical on the whole space now follows from the fact that the space X is defined as a completion over the space of the f and g of the above form with finite sums.

We wish to derive error estimates for those radial basis functions by considering the error on a set Ω that satisfies the above boundedness and Lipschitz requirements, i.e. boundedness, openness, Lipschitz-continuous boundary including the interior cone condition. The data sites Ξ are still assumed to stem from such Ω. The boundary of the domain in question is included in the error estimates which is one reason why many estimates of the following type are not quite as powerful as the error estimates for approximations on uniform grids. We follow the ideas of Duchon, and in particular those of the more recent article of Light and Wayne (1998) in the proof of the theorem.

Theorem 5.5. *Let Ω be as above, and let for $h \in (0, h_1)$ and some fixed $0 < h_1 < 1$, each $\Xi_h \subset \Omega$ be a finite set of distinct points with*

$$\sup_{t \in \Omega} \inf_{\xi \in \Xi_h} \|t - \xi\| \leq h.$$

Then there is a minimum norm interpolant s from $X = D^{-k}L^2(\mathbb{R}^n)$ to data on Ξ_h as in Proposition 5.3. It satisfies for all $f \in D^{-k} L^2(\Omega) \cap L^p(\Omega)$ and any $p \geq 2$, including infinity,

$$\|f - s\|_{p,\Omega} \leq Ch^{k - \frac{n}{2} + \frac{n}{p}} |f|_{k,\Omega}.$$

The constant C depends on n, k, p and on the domain, but not on the approximand f or on the spacing h or Ξ_h.

Proof: The existence of the interpolants is guaranteed for small enough h_1 because Ω has a nonempty interior, so that there are K-unisolvent subsets in $\Xi_h \subset \Omega$ if the centres are close enough together, that is when h is small enough but we will remark further on this later.

Next, we need a minimum norm interpolant f^Ω from the 'native space' X to f on the *whole* of Ω, i.e. we seek $f^\Omega \in X$ with $f^\Omega|_\Omega = f|_\Omega$ and $|f^\Omega|_k$ minimal. This will aid us in the application of our earlier results on minimum norm interpolants. The existence of this is assured in the following simple auxiliary result.

Lemma 5.6. *Let Ω be as required for Theorem 5.5. For all $f \in D^{-k} L^2(\Omega)$ there is an $f^\Omega \in D^{-k} L^2(\mathbb{R}^n)$ that satisfies $f^\Omega|_\Omega = f$ and which minimises $|f^\Omega|_k$ among all such f^Ω that meet f on Ω.*

Proof: Since Ω has a Lipschitz-continuous boundary, there is an

$$f^\Omega \in D^{-k} L^2(\mathbb{R}^n)$$

with $f^\Omega|_\Omega = f$ and $|f^\Omega|_k < \infty$, by the Whitney extension theorem (Stein, 1970, p. 181):

Whitney extension theorem. *Let $\tilde{\Omega}$ be a domain with a Lipschitz-continuous boundary in $\mathbb{R}^n$ that satisfies an interior cone condition. Let g be in the space $D^{-k}L^2(\tilde{\Omega})$. Then there exists a $\tilde{g} \in D^{-k}L^2(\mathbb{R}^n)$ that agrees with g on the domain and whose norm is at most a fixed, g-independent multiple of that of g.*

More precisely, we get that for our application with $\tilde{\Omega} = \Omega$

$$(5.12) \qquad\qquad |f^{\tilde{\Omega}}|_k \leq C\,|f|_{k,\tilde{\Omega}},$$

for a suitable, f-independent, but $\tilde{\Omega}$-dependent, constant C. We now have to take from the set of all such $f^{\tilde{\Omega}}$ one that minimises the left-hand side of (5.12). This is possible by closedness of the set over which we minimise. □

We observe that, as a consequence of the homogeneity of the semi-norm, it is possible to choose C in (5.12) $\tilde{\Omega}$-*independent* if $\tilde{\Omega}$ is a ball about t of radius $\tilde{s}$, $B_{\tilde{s}}(t)$. In other words, for this choice of $\tilde{\Omega}$, the constant C is $\tilde{s}$-independent. In order to see this, we note that we may scale f on both sides of (5.12) by composing it with the transformation ρ^{-1}, where $\rho(x) = \tilde{s}^{-1}(x - t)$ and $\tilde{\Omega} = B_{\tilde{s}}(t)$. Thus, $\tilde{\Omega}$ on the right-hand side is replaced by the $\tilde{s}$-independent $B_1(0)$, and therefore C becomes $\tilde{s}$-independent. The multiplicative powers of $\tilde{s}$ which appear on both sides of the inequality, due to the homogeneity of the semi-norm we use and through the transformation by ρ^{-1}, cancel.

The purpose of the above Lemma 5.6 is to be able to use f^{Ω} instead of f in our error estimates, because this way they become more amenable to our previous work which used the space X for error estimates, not the function space $D^{-k}L^2(\Omega)$. For the estimates, it is useful to divide the set Ω up into balls of a size that is a fixed multiple of h (stated more precisely: to cover the domain by such balls which will of course overlap), and the following result helps us with that. It is true as a consequence of the cone condition (Duchon, 1978). We do not prove this lemma here, because it is a basic result in finite element theory and not special in any way to radial basis functions. The proof can be found in Duchon's papers, in Bezhaev and Vasilenko (1993, Lemma 5.3) and in many texts on finite elements where the interior cone property is a standard requirement.

Lemma 5.7. *Let Ω satisfy the conditions of Theorem 5.5. Then there exist M, M_1 and $h_0 > 0$ such that for all $h \in (0, h_0)$ there is a finite subset $T_h \subset \Omega$ with $O(h^{-n})$ elements and*

(i) $B_h(t) \subset \Omega \quad \forall t \in T_h$,

(ii) $\bigcup_{t \in T_h} B_{Mh}(t) \supset \Omega$,

(iii) $\sum\limits_{t \in T_h} \chi_{B_{Mh}(t)}(x) \leq M_1 \ \forall x \in \Omega,$

the χ being always the characteristic function of the set that is noted in its index. The $B_{\tilde{s}}(\cdot)$ denote balls about the points given by the argument and of the radius $\tilde{s}$ given by the index.

We take now M, h and T_h as in Lemma 5.7. Let $t \in T_h$, where $0 < h < \min(h_0, h_1)$. We recall that the centres which are used for the interpolation become dense in the domain Ω with shrinking h. Therefore, through possible further reduction of h_1, we may assume that in each $B := B_{Mh}(t)$ there are at least $\ell = \dim \mathbb{P}_n^{k-1}$ unisolvent points that we cast into the set (not denoted by an index t for simplicity) $\hat{\Xi}$ and which belong to the interpolating set Ξ_h. We shall use this set $\hat{\Xi}$ now for the purpose of defining the power function and using the error estimate of Proposition 5.3. To this end, let $x \in B$, let f^Ω be an extension according to Lemma 5.6, and let further $(f^\Omega - s)^B$ be defined according to the same lemma with B replacing Ω and $(f^\Omega - s)$ replacing f in the statement of the lemma. In particular, $(f^\Omega - s)^B$ is zero at all the ξ from our set $\hat{\Xi}$ above.

Thus, according to Lemma 5.6 and (5.12), and due to Proposition 5.3,

$$(f^\Omega(x) - s(x))^2 = (f^\Omega - s)^B (x)^2$$
$$\leq P(x) \, |(f^\Omega - s)^B|_k^2$$
$$\leq C P(x) \, |f^\Omega - s|_{k,B}^2, \qquad x \in B.$$

Recall that the positive constant C in the above line may be chosen independent of the size of B, according to our earlier remarks concerning inequality (5.12).

Recall further that the definition of the power function specifically depends on the *centres* which we use in the sums that occur in the definition of P. In the current proof, we are using the notion of the power function P with respect to the $\xi \in \hat{\Xi}$ just introduced, of course. Thus, taking p-norms on the left-hand side and in the last line of the displayed equation above on the ball B, where x is the integration variable, we get the estimate below. In order to simplify notation further, we denote f^Ω simply by f from now on, because it agrees with f on the domain Ω anyway:

$$\|f - s\|_{p,B} \leq C \left(\int_B |P(x)|^{p/2} dx \right)^{1/p} |f - s|_{k,B}.$$

We immediately get a power of h from the first nontrivial factor on the right-hand side of this display by Hölder's inequality, in the new estimate

$$\|f - s\|_{p,B} \leq C h^{n/p} \|P\|_{\infty,B}^{1/2} |f - s|_{k,B}.$$

Recalling from the definition (4.4) that $\phi(0)$ vanishes and inserting the definition of the power function from the beginning of this subsection give the inequality

$$\|P\|_{\infty,B} \leq \sup_{x \in B} \left\{ 2 \left| \sum_{\xi \in \hat{\Xi}} p_\xi(x)\, \phi(\|x - \xi\|) \right| \right.$$

$$\left. + \left| \sum_{\zeta,\xi \in \hat{\Xi}} p_\xi(x)\, p_\zeta(x)\, \phi(\|\zeta - \xi\|) \right| \right\},$$

where, as before in the power function *mutatis mutandis,* the p_ξ are Lagrange polynomials at the ξ which satisfy $p_\xi(\zeta) = \delta_{\xi\zeta}$ for all ξ and ζ from $\hat{\Xi}$.

We can first bound – by Hölder's inequality again – the whole right-hand side of the last display from above by a fixed constant multiple of

$$(5.13) \qquad \max_{0 < r \leq Mh} |\phi(r)| \max_{x \in B} \left(\sum_{\zeta \in \hat{\Xi}} |p_\zeta(x)| \right)^2.$$

We begin by bounding the second factor of (5.13) because this is straightforward.

Indeed, the second factor in (5.13), including taking the maximum, is exactly the square of the Lebesgue constant for polynomial interpolation at the ξ (with polynomial degree $k - 1$ in n dimensions), i.e. the uniform norm of the polynomial interpolation operator for scattered points. As the norm of the Lagrange interpolation operator is scale-invariant, this may be bounded *independently* of h and B if the centres ξ are chosen judiciously.

This is always possible subject to a possible further reduction of h_1 – recalling that $0 < h < h_1$ – and choosing the ξ from $\hat{\Xi}$ afresh if needed, because the centres become dense in the domain with shrinking h, and using the following specific choice of ξ. We may for example first place a very fine n-dimensional square grid over the domain with a very small grid-spacing in relation to h and consider ξ from $\hat{\Xi}$ as taken directly from that grid (forming a subset thereof) or being a small perturbation of such a gridpoint, the size of the perturbation being a tiny multiple of h. This is possible for all h if $h < h_1$ is small enough. (Incidentally, it even helps to satisfy the aforementioned unisolvency condition with respect to K too if we are considering points from a grid or a small perturbation thereof, because points from a grid can be very easily chosen to be unisolvent.)

For polynomial interpolation from such square grids, however, it is evident that the norm of the Lagrange interpolation operator is bounded independently of the grid-spacing, that is, it is scale-independent, because of the uniformity requirement. If the aforementioned perturbation is sufficiently restricted, this remains true if the points are taken from a perturbed grid as outlined in this paragraph. This settles the boundedness.

The first term in (5.13) can, according to (4.4), be bounded by h^{2k-n} if n is odd, $h^{2k-n} |\log h|$ if n is even. We wish to remove the $\log h$ term now for even n by replacing $\phi(r)$ by $\phi(r) + r^{2k-n} \log \sigma$ for a suitable σ which may depend on h but not on r. This will eventually cancel the $\log h$ term. The approach is admissible for the following reasons. In short, adding a low even power of r to the radial basis function never changes the approximation because of the moment conditions (5.9) on the coefficients that annihilate such augmentations. We will demonstrate this fact by showing that the power function which occurs in the error estimate remains the same after the augmentation of the radial basis function by a low power of r; a similar analysis was already done in Chapter 4 for our cardinal Lagrange functions.

To this end, we recall that the p_ξ are Lagrange polynomials and thus we have the reproduction of monomials

$$(5.14) \qquad \sum_{\zeta \in \hat{\Xi}} p_\zeta(x) \, \zeta^{\hat{t}} = x^{\hat{t}}, \qquad \hat{t} \in \mathbb{Z}_+^n, \ |\hat{t}| < k,$$

using the standard multiinteger notation for the exponent $\hat{t}$. Keeping in mind that $2k - n$ is even, so that $\|x\|^{2k-n}$ is an even order polynomial, we get now for the power function with the augmented radial basis function the following additional contribution. We consider it first in the special case when $2k - n = 2$. Then

$$-2 \sum_{\zeta \in \hat{\Xi}} p_\zeta(x) \|x - \zeta\|^2 + \sum_{\zeta \in \hat{\Xi}} \sum_{\xi \in \hat{\Xi}} p_\zeta(x) \, p_\xi(x) \|\xi - \zeta\|^2$$

$$= -2 \sum_{\zeta \in \hat{\Xi}} p_\zeta(x) \|x - \zeta\|^2$$

$$+ \sum_{\zeta \in \hat{\Xi}} \sum_{\xi \in \hat{\Xi}} p_\zeta(x) \, p_\xi(x) \Big(\|\xi - x\|^2 - 2(\zeta - x) \cdot (\xi - x) + \|\zeta - x\|^2 \Big).$$

By the polynomial reproduction identity (5.14), the sums over the norms $\|\xi - x\|^2$ and $\|\zeta - x\|^2$ cancel, as well as the sums over $(\zeta - x) \cdot (\xi - x)$ which vanish too. This being a simple demonstration, we get similarly, with the aid of simple binomial expansions, that the above holds if $2k - n > 2$ without further consideration.

Now we may replace $\phi(r)$ in (5.13) by $\phi(r) + r^{2k-n} \log \sigma$ with $\sigma = 1/h$ in the same way without changing the approximant. This removes the undesirable

$\log h$ term when we estimate the left-hand factor of (5.13) and bound $|\phi(r)|$ from above.

In summary, we have established so far that (5.13) is bounded by a constant multiple of h^{2k-n}, and that $\|f - s\|_{p,B}$ is bounded by a fixed constant multiple of the following:

$$h^{n/p} \, h^{k-\frac{n}{2}} \, |f - s|_{k,B}.$$

To continue, we need an auxiliary result from Stein and Weiss (1971), namely the famous inequality by Young:

Young's inequality. *Let $f \in L^r(\mathbb{R}^n)$ and $g \in L^q(\mathbb{R}^n)$, with r, q and the sum $(1/q) + (1/r)$ all at least one. Then*

$$\|f * g\|_p \leq \|f\|_r \|g\|_q,$$

where $(1/p) = (1/r) + (1/q) - 1$ and $*$ denotes convolution. The same estimate is true if f and g come from the sequence spaces $\ell^r(\mathbb{Z}^n)$ and $\ell^q(\mathbb{Z}^n)$, respectively, and the norms are accordingly the sequence space norms.

Now we form the $\|f - s\|_{p,\Omega}$ by collection of the pieces $\|f - s\|_{p,B}$. Due to the choice of the Bs and Lemma 5.7, we have

$$\|f - s\|_{p,\Omega} \leq \left(\sum_{t \in T_h} \|f - s\|^p_{p,B_{Mh}(t)} \right)^{1/p}$$

$$\leq Ch^{k-\frac{n}{2}+\frac{n}{p}} \left(\sum_{t \in T_h} |f - s|^p_{k,B_{Mh}(t)} \right)^{1/p}.$$

This is at most a constant multiple of

$$\max_{x \in \Omega} h^{k-\frac{n}{2}+\frac{n}{p}} \left(\sum_{t \in T_h} \left(\sum_{\tau \in T_h} |f - s|_{k,B_{Mh}(\tau)} \chi_{B_{Mh}(t-\tau)}(x) \right)^p \right)^{1/p}.$$

Now we appeal to Young's inequality for $p \geq 2$, $r = 2$ and $1/q = (1/p) + (1/2)$. Employing Lemma 5.7 and recalling that the value of the characteristic function is always one or zero, we may bound this above by a constant multiple of

$$(5.15) \quad \max_{x \in \Omega} h^{k-\frac{n}{2}+\frac{n}{p}} \left(\sum_{t \in T_h} |f - s|^2_{k,B_{Mh}(t)} \right)^{1/2} \left(\sum_{t \in T_h} \chi_{B_{Mh}(t)}(x)^q \right)^{1/q}$$

$$\leq Ch^{k-\frac{n}{2}+\frac{n}{p}} |f - s|_{k,\Omega} \leq Ch^{k-\frac{n}{2}+\frac{n}{p}} |f|_{k,\Omega},$$

as required, the final bound being standard for an s minimising $|f - s|_{k,\Omega}$, see the proof of our Proposition 5.3. It is also evident from this proof why we have the condition $p \geq 2$, namely in order to apply Young's inequality because $p \geq 2$ follows from its requirement that $q \geq 1$ and the use of $r = 2$. □

We point out that as an alternative to Young's inequality, the end of the proof can be carried through by an application of Jensen's inequality (Hardy, Littlewood and Pólya, 1952, p. 28) which bounds a series $\sum a_i^p$ by $(\sum a_i^2)^{p/2}$ for all $p \geq 2$.

There is a neat way to improve the convergence order established above by imposing more stringent requirements on f. This is known as the 'Nitsche trick' from familiar univariate spline theory (Werner and Schaback, 1979, for instance). We demonstrate it in the book both because it provides a better convergence result closer to the results on grids, and because generally it belongs in the toolkit of anyone dealing with uni- or multivariate splines or radial basis functions anyway. It works as follows. Recall that $|\cdot|_k$ is the semi-norm otherwise called $\|\cdot\|_*$ associated with the space X of the last theorem. Now let f be any function that satisfies the *additional* smoothness condition $|f|_{2k} < \infty$, which is stronger than demanded in the statement of the last theorem, and the further condition that supp $D^{2k} f$ is a subset of Ω. This is a 'boundary condition' imposed on f. It means that all partial derivatives of f to total degree at most $2k$ are supported in Ω.

Next we recall that the semi-inner product $(s, f - s)_*$ vanishes (which is a consequence of the orthogonality of the best least squares approximation). We may deduce this from our earlier results about the minimum norm interpolant s and Proposition 5.2, namely

$$|f|_k^2 = |f - s|_k^2 + |s|_k^2,$$

and it is always true that

$$|f - s|_k^2 = |f|_k^2 - |s|_k^2 - 2(s, f - s)_*.$$

Thus for the s from the last theorem, by applying the orthogonality property of the best least squares approximation and the Cauchy–Schwarz inequality, we may now estimate

$$(5.16) \qquad |f - s|_k^2 = (f - s, f - s)_*$$
$$= (f, f - s)_*$$
$$\leq |f|_{2k} \, \|f - s\|_{2,\Omega}$$

since the support of all derivatives of total degree $2k$ of f is in Ω. We may carry on by estimating this from above by a fixed constant multiple of

$$(5.17) \qquad |f|_{2k} \, h^k \, |f - s|_k,$$

due to (5.15), i.e. the last theorem used for $p = 2$. Now, cancelling the factor $|f - s|_k$ from (5.17) and the left-hand side of (5.16), we get the estimate

$$|f - s|_k \leq C \, |f|_{2k} \, h^k.$$

This we may use once more in (5.15), now for general p, to get another factor of h^k. As a consequence we obtain the final result which we cast into

Theorem 5.8. *Let all assumptions of Theorem 5.5 hold and assume addition-ally $f \in D^{-2k} L^2(\mathbb{R}^n)$, supp $D^\alpha f \subset \Omega$ for all α with $|\alpha| = 2k$. Then, for sufficiently small h,*

$$\|f - s\|_{p,\Omega} \leq C h^{2k - \frac{n}{2} + \frac{n}{p}} \, |f|_{2k}$$

holds.

Note that for $p = \infty$ we get $O(h^{2k-n/2})$ which is better than before due to the new 'bounding conditions', but still off by a factor of $h^{-n/2}$ from the optimal result in the case of a cardinal grid. Note also that for $p = 2$ we recover $O(h^{2k})$.

We close this subsection on convergence estimates with a few remarks about the best presently available results on approximation orders for scattered data with radial basis functions of the form (4.4). To begin with, Johnson (1998b) has shown that for all $1 \leq p < 2$ the estimate of Theorem 5.8 may be retained with the exponent on the right-hand side being $2k$ only, where the only extra condition on the approximand is that the support of the approximand be in a compact set inside the interior of our domain in $\mathbb{R}^n$. Without this extra condition on the function, the best possible (and attained) L^p-approximation order to infinitely smooth functions on a domain with C^{2k} boundary for sufficiently good k is $k + 1/p$ for $1 \leq p \leq 2$ (Johnson, 2002).

The currently best contribution in the battle for optimal error estimates for all p and for scattered data comes from the paper Johnson (2000). It is to bound the interpolation error on scattered data to sufficiently smooth approximands without the support property needed in Theorem 5.8, measured in $L^p(\mathbb{R}^n)$, by $O(h^{\gamma_p})$, where in the uniform case $\gamma_\infty = k - n/2 + 1/2$ and in the least squares case $\gamma_2 = k + 1/2$ which is best possible for $p = 2$ (see also our eighth chapter).

Conversely, he has also proved in Johnson (1998a) that, for any $p \in [1, \infty]$ and the domain being the unit ball, there is an infinitely continuously differen-tiable f such that the L^p-error of the *best approximation* to f from the space of translates of the radial basis functions (4.4) including the appropriate poly-nomials, is *not* $o(h^{k+1/p})$ as $h \to 0$. This proof works by giving an example for an approximand and a simple spherical domain, i.e. with a highly smooth boundary, where the $o(h^{k+1/p})$ is not attained. As we have seen, in particu-lar, we cannot obtain any better approximation orders than $O(h^{2k})$ with our interpolation method and the error measured in the uniform norm. For other upper bounds on the convergence orders for gridded data see Chapter 4. Related results are also in Matveev (1997).

Furthermore, Schaback and Wendland (1998) have shown that if the uniform error of the radial basis function interpolant with (4.4) is $o(h^{2k})$ on any compact subset of Ω for an $f \in C^{2k}(\Omega)$ then f must be k-harmonic. In other words $\Delta^k f = 0$ on the domain Ω and with less smooth approximands f, no better error estimates than $O(h^{2k})$ are obtainable.

5.2.5 Further results

A result that applies specifically to the multiquadric function is the powerful convergence theorem below. It provides an exponential convergence estimate for a restricted class of functions f that are being interpolated. They belong to the space corresponding to our 'native space' X as before. That space X has a semi-norm which may in this context be introduced conveniently in the following fashion. It is a least squares norm weighted with the reciprocal of the radial basis function's Fourier transform, thereby becoming a semi-norm:

$$\|f\|_*^2 := \frac{1}{(2\pi)^n} \int_{\mathbb{R}^n} \hat{\phi}(\|t\|)^{-1} |\hat{f}(t)|^2 \, dt.$$

Here, $\hat{\phi}$ is as usual the generalised Fourier transform of a radial basis function which we assume to be positive. (So multiquadrics must be taken with a negative sign.)

In fact, this is precisely the same structure as before; for functions (4.4) such as thin-plate splines we weighted with $\| \cdot \|^{2k}$, namely the reciprocal of $\hat{\phi}(\| \cdot \|)$ except sometimes for a fixed factor, to get the semi-norm in Fourier transform form. Using the Parseval–Plancherel identity, we may return to the familiar form used for the semi-norm $|f|_k$ of $D^{-k}L^2(\mathbb{R}^n)$ in the previous subsections. However, here the reciprocal of the radial basis function's Fourier transform has bad properties because of its exponential increase (cf. Section 4.3 and the lower bounds on $\hat{p}$ used there), in contrast with the slow increase of the reciprocal of the Fourier transforms of the radial basis functions (4.4). They are only of polynomial growth, namely order $O(\|t\|^{2k})$. Thus we have here with the multiquadric function a least squares norm with an exponentially increasing weight in the case of the multiquadric function, the Fourier transform of the multiquadric decaying exponentially. In summary, the space of approximands X only contains very smooth functions. The following remarkable theorem with exponential convergence orders is established in Madych and Nelson (1992).

Theorem 5.9. *Let $\phi(\| \cdot \|) \in C(\mathbb{R}^n)$ be strictly conditionally positive definite of order k with a positive generalised Fourier transform, and suppose there is a v such that*

$$\int_{\mathbb{R}^n} \|t\|^{\ell} \hat{\phi}(\|t\|) \, dt \leq v^{\ell} \ell!, \qquad \forall \ell > 2k.$$

Let for a positive b_0 the set Ω be a cube of side-length at least b_0. Then, there exist positive δ_0 and $\eta \in (0, 1)$ such that for all approximands f which satisfy

$$\int_{\mathbb{R}^n} \hat{\phi}(\|t\|)^{-1} |\hat{f}(t)|^2 \, dt < \infty,$$

the interpolant to the data $\{f(\xi)\}$ at $\xi \in \Xi$, namely

$$s(x) = \sum_{\zeta \in \Xi} \lambda_\zeta \phi(\|x - \zeta\|) + p(x), \qquad x \in \mathbb{R}^n,$$

where $p \in \mathbb{P}_n^{k-1}$ and the λ_ζ satisfy the appropriate side-conditions, $K = \mathbb{P}_n^{k-1}$ being the kernel for the above semi-norm, satisfies

$$\|s - f\|_{\infty,\Omega} = O(\eta^{\delta^{-1}}), \qquad 0 < \delta < \delta_0.$$

Here, each subcube of Ω of side-length δ must contain at least one ξ.

The reason why we get in Theorem 5.9 much better convergence orders for multiquadrics even as compared with the cardinal grid case is that, first, we have a much restricted class of very smooth approximands f, due to the exponential growth of the weight function in the above integral and semi-norm, and, second, that we are here using the so-called nonstationary approach. In other words, here the scaling of the radial basis function is entirely different from that of the cardinal grid case. In the latter case we have scaled the argument of the radial basis function by the reciprocal of h (the 'stationary' case) which means, if we multiply it by the radial basis function, we multiply the parameter c by h as well:

$$\sqrt{\left(\frac{x}{h}\right)^2 + c^2} = h^{-1}\sqrt{x^2 + c^2 h^2}.$$

(The h^{-1} factor in front of the square root has no importance as it can be absorbed into the coefficients of the interpolant or approximant.) In the above Theorem 5.9, however, we do *not* scale the whole argument by the reciprocal of h. Instead we simply decrease the spacing of the centres that we use for interpolation, which leaves the parameter c untouched. In fact this is like increasing the constant c in relation to the spacing of the data sites, which gives bad conditioning to the interpolation matrix, as we can see also from our results from the fourth chapter and in the next section of this chapter.

Our observation here is confirmed by the remark that for those radial basis functions which have *homogeneous* Fourier transforms and are homogeneous themselves except for a possible log-term, where the scaling makes no difference whatsoever because it disappears into the coefficients, no such result as the above is available. Concretely, this is related to the fact that the first displayed condition of the theorem cannot be fulfilled by homogeneous radial basis functions which have homogeneous Fourier transforms.

The reason why, by contrast, multiquadrics do satisfy that condition lies, as alluded to above, in the exponential decay of their Fourier transform: indeed, the expression

$$\int_{\mathbb{R}^n} \|t\|^\ell \hat{\phi}(\|t\|) dt$$

is, using polar coordinates, bounded above by a constant multiple of the integral

$$\int_{\mathbb{R}^n} \|t\|^\ell e^{-c\|t\|} dt \leq C \int_0^\infty r^{\ell+n-1} e^{-cr} dr = C \frac{\Gamma(n+\ell)}{c^{n+\ell}}$$

which is at most a constant multiple of $\ell! c^{-n-\ell}$ for fixed dimension n, Γ being the standard Γ-function (we may take the right-hand equation in the above display as its definition if $c = C = 1$, or see Abramowitz and Stegun, 1972) and in particular $\Gamma(n+\ell) = (n+\ell-1)!$

Besides interpolation, we have already discussed the highly useful *ansatz* of quasi-interpolation which, in the gridded case, gives almost the same convergence results as interpolation on the equally spaced grid. We wish to show at this instant that for scattered data approximation also, when the centres are 'almost' arbitrarily – this will be defined precisely later – distributed, quasi-interpolation is possible. In connection with this, we shall also give a convergence result for quasi-interpolation which holds, incidentally, in the special case of gridded data too, and therefore complements our results from Chapter 4.

In practice, however, interpolation is used much more frequently when scattered data occur, for the reasons given at the beginning of this chapter, but it is worthwhile – and not only for completeness of the exposition in this book – to explain quasi-interpolation in the presence of scattered data here anyway, as even in practice some smoothing of the data for instance through quasi-interpolation may be performed when scattered data are used. We will comment further on this aspect of quasi-interpolation in Chapter 8.

So, in particular for the functions (4.4) but also for more general classes of radial basis functions, convergence results with quasi-interpolation on scattered data may be obtained without using interpolants and their optimality properties in the manner explained below. For this result, however, we require *infinite* subsets $\Xi \subset \mathbb{R}^n$ of scattered centres which satisfy the following two straightforward properties for a fixed positive constant C_0:

(B1) every ball $B_{C_0}(x)$ about any x in $\mathbb{R}^n$ contains a centre $\xi \in \Xi$,
(B2) for any L which is at least one, and any $x \in \mathbb{R}^n$, $B_L(x) \cap \Xi$ contains at most $C_0 L^n$ elements.

Condition (B1) just means that there are no arbitrarily big, empty (that is, centre-free) 'holes' in Ξ as a subset of $\mathbb{R}^n$. The fact that we exclude by condition (B1) all approximants which use only finitely many centres is of the essence. The differences between the approximation approaches and their convergence analyses lies much more in the alternatives finitely many *versus* infinitely many centres on finite *versus* infinite domains than between the alternative scattered data or gridded data. This fact is particularly clear from the convergence result below, because we may compare it with the results of the previous subsection and the theorems of Chapter 4.

Our condition (B2) is of a purely technical nature; it can always be satisfied by thinning out Ξ if necessary. After all, we are dealing with quasi-interpolation here and not with interpolation and need not use all the given centres, while with interpolation all given centres are mandatory.

In order to study the behaviour of the approximants for different sets of centres which become dense in the whole Euclidean space, we also consider sequences of centre sets $\{\Xi_h\}_{h>0}$, namely subsets of $\mathbb{R}^n$ such that each 'scaled back' $\Xi := h^{-1}\Xi_h$ satisfies (B1) and (B2) *uniformly* in C_0, i.e. C_0 remains independent of h. Therefore, in particular, $\Xi_h \subset \mathbb{R}^n$ must become dense as $h \to 0$ and so this is a useful notion for a convergence result.

It is the case, as often before, that we do not actually have to restrict ourselves to radially symmetric basis functions for this result, but in this book, for the sake of uniformity of exposition, we nonetheless state the result for an n-variate radial basis function $\phi(\|\cdot\|): \mathbb{R}^n \to \mathbb{R}$. Therefore suppose $\phi(\|\cdot\|)$ is such that it has a generalised, n-variate Fourier transform that agrees with a function except at zero, which we call as before $\hat{\phi}: \mathbb{R}_{>0} \to \mathbb{R}$. This must satisfy for an even natural number $\mu = 2m$, a nonnegative integer m_0 and a univariate function F, the following two additional properties:

(QI1) $\hat{\phi}(r) = F(r)r^{-\mu}$, $F \in C^{m_0}(\mathbb{R}) \cap C^\infty(\mathbb{R}\setminus\{0\})$,
(QI2) $F(0)$ is not zero, and for $y \to 0$, all $\gamma \in \mathbb{Z}_+$ and a positive ε,

$$\left| \frac{d^\gamma}{dy^\gamma}\left(F(y) - \sum_{\alpha=0}^{m_0} \frac{F^{(\alpha)}(0)\,y^\alpha}{\alpha!} \right) \right| = O(y^{m_0+\varepsilon-\gamma}), \qquad y > 0.$$

The condition (QI1) is nothing else than a condition about the Fourier transform's singularity at the origin. Our condition (QI2) ensures a certain smoothness of the Fourier transform of the radial basis function at zero. As a consequence of (B1)–(B2) and (QI1)–(QI2), we now have the next result, which employs the conditions we have just listed. We let K be the kernel of the differential operator $\|D\|^\mu$, where we recall that now $\mu \in 2\mathbb{Z}_+$.

Theorem 5.10. *Under the assumptions (B1)–(B2), (QI1)–(QI2), (A2a) from Chapter 4, there exist decaying finite linear combinations for a fixed positive constant C_1 of the form*

$$(5.18) \qquad \psi_\xi(x) = \sum_{\zeta \in B_{C_1}(\xi)} \mu_{\xi\zeta}\, \phi(\|x - \zeta\|), \quad x \in \mathbb{R}^n,$$

such that quasi-interpolants

$$s(x) = \sum_{\xi \in \Xi} f(\xi)\, \psi_\xi(x), \quad x \in \mathbb{R}^n,$$

are well-defined for all $f \in \mathbb{P}_n^{m_0} \cap K$ and recover those f. Moreover, let f be $q := \min(\mu, m_0 + 1)$ times continuously differentiable and let it have q and $q + 1$ total order bounded partial derivatives. Then, for sets Ξ_h as defined in the paragraph before last, approximants

$$s_h(x) = \sum_{\xi \in \Xi_h} f(\xi)\, \psi_{h^{-1}\xi}(h^{-1}x), \quad x \in \mathbb{R}^n,$$

satisfy the error estimate

$$\|f - s_h\|_\infty = O(h^q |\log h|), \quad h \to 0.$$

When using this result, it should be noted that the conditions on the data and the scaled back sets of data are relatively weak conditions and thus many different distributions of data points are admitted in the above theorem.

It should also be observed that this theorem works for $\Xi = \mathbb{Z}^n$, $\Xi_h = (h\mathbb{Z})^n$, and admits a typical convergence result for quasi-interpolation on regular grids with spacing h as a special case. This nicely complements our work of Chapter 4.

The central idea in the proof of this theorem is, as in polynomial recovery and convergence proofs in the fourth chapter, to construct suitable, finite linear combinations (5.18) that decay sufficiently fast to admit polynomials into the approximating sum. They must, like their counterparts on equally spaced grids, give polynomial reproduction. These are the two difficult parts of the proof while the convergence order proofs are relatively simple, and, at any rate, very similar to the proofs in the cardinal grid case for interpolation of Chapter 4 in Section 4.2. We do not supply the rest of the proof here as it is highly technical, relying heavily on our earlier results on gridded data, see Buhmann, Dyn and Levin (1995) but below we give a few examples for the theorem's applications.

For instance, the radial basis functions of the class (4.4) satisfy all the assumptions (QI1)–(QI2) for a nonzero constant F and $m = k$ in (QI1)–(QI2), m_0 being an arbitrary positive integer. Therefore using the quasi-interpolants of Theorem 5.10, all polynomials f from the kernel K of the differential operator Δ^k are recovered, that is, all polynomials of total degree less than $\mu = 2k$. The resulting convergence orders are $O(h^{2k}|\log h|)$ as the 'spacing' h goes to zero. This should be compared with the $O(h^{2k})$ convergence in the cardinal grid case of the previous chapter.

Multiquadrics satisfy the assumptions for odd dimension n, and $m = (n + 1)/2$, $\mu = n + 1$, as we recall from Subsection 4.2.3

$$F(r) = -\pi^{-1}(2\pi)^m \widetilde{K}_m(cr), \quad r > 0,$$

$m_0 = 2\mu - 1$, because F has an expansion near the origin so that $F(r)$ is

$$a_0 + \sum_{j=1}^{2m-1} a_j r^{2j} + \sum_{j=0}^{m-1} b_j r^{2m+2j} \log r + O(r^{2n+1}), \quad r \to 0_+,$$

where the a_js and b_js are suitable nonzero constants. The result is that all polynomials of total order at most n are recovered and the convergence orders that can be attained are $O(h^{n+1}|\log h|)$. Again, a comparison with the cardinal grid case of Chapter 4 is instructive and should be looked at by the reader.

5.3 Norm estimates and condition numbers of interpolation matrices

5.3.1 General remarks

An important and mathematically appealing part of the quantitative analysis of radial basis function interpolation methods is the analysis of the ℓ^2-condition number of the radial basis function interpolation matrix. The crux of the work of this section is in bounding below in modulus the smallest eigenvalue of the symmetric matrix in question (often positive definite or coming from a conditionally positive definite (radial basis) function), so that we can give an upper bound on the ℓ^2-norm of the inverse of the matrix. What remains is an estimate on the ℓ^2-norm of the matrix itself, which is always simple to obtain.

We have already motivated the search for those bounds in the penultimate section of Chapter 4. As an application, the bounds are also highly relevant in the application of conjugate gradient methods for the numerical solution of linear systems, because there are estimates for the rate of convergence of such (often preconditioned) conjugate gradient methods which depend crucially on those condition numbers in the Euclidean norm. The conjugate gradient method and a closely related Krylov subspace method will be stated and used in Chapter 7 when we write about implementations.

In this section, we are interested in both lower and upper bounds to the condition numbers of interpolation matrices which will provide us with negative and positive answers for the stability of computations and, for instance, speed of convergence of conjugate gradients. The lower bounds for the condition numbers are essentially there in order to verify if the upper bounds are reasonable (ideally, they should of course be best possible) or not, i.e. whether they are correct up to a constant, say, whether there are orders of magnitude differences between the upper bounds and the lower estimates. It turns out that the ones which are known up to now usually are reasonable, but sometimes when we have exponentially growing estimates, as occur for example for multiquadrics, upper and lower bounds differ by a power and by constant factors. There, as with multiquadrics for example, the constants, especially the multiquadric's parameter c, in the exponent also play a significant rôle.

Before embarking, it has to be pointed out that the following analysis applies to the nonstationary case, when the basis functions are not scaled as they are in Chapter 4, namely by $1/h$. When a suitable scaling is applied in the stationary case, most of the remarks below about the condition numbers are not really relevant, because the matrices have condition numbers that can usually be worked out easily and can be independent of h.

5.3.2 Bounds on eigenvalues

Surprisingly, several of the results below give bounds on the aforementioned lowest eigenvalues that are *independent* of the number of centres; instead they depend solely on the smallest distance between adjacent centres. The latter quantity divided by two is termed the 'separation radius' q of the data sites ξ, and this will appear several times in the rest of this chapter. On the other hand, it is not surprising that the bounds tend to zero as that distance goes to zero as well. They must do, because eventually, i.e. for coalescing points, the interpolation matrix must become singular, as it will then have at least two identical rows and columns.

Lemma 5.11. *Let $\phi(\|\cdot\|)\colon \mathbb{R}^n \to \mathbb{R}$ be strictly conditionally positive definite of order zero or one and, in the latter case, suppose further that $\phi(0)$ is nonpositive. If for all $d = (d_\xi)_{\xi \in \Xi} \in \mathbb{R}^\Xi$,*

$$\sum_{\xi \in \Xi} \sum_{\zeta \in \Xi} d_\xi \, d_\zeta \phi(\|\xi - \zeta\|) \geq \vartheta \sum_{\xi \in \Xi} |d_\xi|^2 = \vartheta \times \|d\|^2,$$

and ϑ is positive, then the inverse of the invertible interpolation matrix $\{\phi(\|\xi - \zeta\|)\}_{\xi,\zeta \in \Xi}$ *is bounded above in the Euclidean matrix norm by* ϑ^{-1}.

The proof of this lemma is simple; it relies on exactly the same arguments as we have applied in Chapters 2 and 5 (e.g. Theorem 5.1) of this book when we have shown nonsingularity in the above cases – in this case here, we only need to take $\vartheta \times \|d\|^2$ as the uniform lower bounds on the positive or conditionally positive quadratic forms that occur there. In fact, the proof is entirely trivial for order zero, because the definition of eigenvalues immediately provides the desired result which is then standard for positive definite quadratic forms. This lemma is central to this section because the radial basis functions which are conditionally positive definite of order one or zero provide the most important examples. Therefore we now wish to apply this auxiliary result to our radial basis functions, notably to the negative of the ubiquitous multiquadric function, that is $\phi(r) = -\sqrt{r^2 + c^2}$.

Thus, for any radial basis function $\phi \in C([0, \infty))$ to be such that the expression

$$-\frac{d}{dt} \phi(\sqrt{t}), \quad t > 0,$$

is completely monotonic, we note the following straightforward condition. It is necessary and sufficient that there exists a nondecreasing measure μ such that for some positive ε

$$\phi(r) = \phi(0) - \int_0^\infty \frac{1 - e^{-r^2 t}}{t} \, d\mu(t), \quad r \geq 0,$$

(5.19)

$$\int_\varepsilon^\infty \frac{d\mu(t)}{t} < \infty, \quad \int_0^\varepsilon d\mu(t) > 0.$$

For the sufficiency of the form in the display above, we only need to differentiate once and apply the Bernstein representation theorem which shows that the above indeed is completely monotonic when differentiated once.

Further, the above representation is necessary: we outline the argument as follows. For showing that, we apply once again the Bernstein representation theorem to the derivative of ϕ in the display before (5.19) and integrate once, the first of the two extra conditions in (5.19) being responsible for the existence of the integral after the integration of the exponential function therein. We also know that it is necessary that the measure μ is nondecreasing and nonconstant, and as a consequence it is in particular necessary that for some positive ε

$$\int_0^\varepsilon d\mu(t) > 0,$$

as required in (5.19).

We recall as an important example that thus the above representation (5.19) applies to the multiquadric function; it is conditionally positive definite of order one if augmented with a negative sign, i.e. $\phi(r) = -\sqrt{r^2 + c^2}$, $c \geq 0$. We wish to use Lemma 5.11 and (5.19) which imply that it suffices, in order to establish a lower bound ϑ for the smallest eigenvalue in modulus, simply to find a lower bound on the positive definite quadratic form with kernel $e^{-r^2 t}$, i.e. to prove an inequality of the form

$$(5.20) \qquad \sum_{\xi \in \Xi} \sum_{\zeta \in \Xi} d_\xi d_\zeta e^{-t \|\xi - \zeta\|^2} \geq \vartheta(t) \sum_{\xi \in \Xi} |d_\xi|^2 = \vartheta(t) \|d\|^2.$$

Then, ϑ in Lemma 5.11 can be taken from (5.19) and (5.20) by integration against the measure

$$\vartheta = \int_0^\infty \frac{\vartheta(t)}{t} \, d\mu(t) > 0,$$

because this is how the radial basis function is defined using the representation (5.19). When the above integral exists, the (with respect to r) *constant* terms $\phi(0)$ and $1/t$ disappear because we require the sum of the components of the coefficient vector $\{d_\xi\}$ to vanish. It is therefore, in the next step of the analysis of the condition number, necessary to find a suitable $\vartheta(t)$ for (5.20). This we do not do here in detail, but we state without proof that a suitable $\vartheta(t)$ is provided by Narcowich and Ward (1991) and what its value is. It is their work that was instructive for the presentation here. They define, for a certain constant δ_n which depends on the dimension of the underlying space but whose value is immaterial otherwise, and for the aforementioned separation radius q of points ξ and ζ from Ξ,

$$(5.21) \qquad q = \frac{1}{2} \min_{\xi \neq \zeta} \|\xi - \zeta\|,$$

the quantity $\vartheta(t)$ for the desired estimate (5.20) as a constant multiple of

$$(5.22) \qquad t^{-\frac{n}{2}} q^{-n} e^{-\delta_n^2/(qt)}, \quad t > 0.$$

Note that we are leaving out a constant factor which only depends on the dimension, our bounds being dimension-dependent in all sorts of ways anyhow; what is most interesting to us is always the asymptotic behaviour with respect to parameters of the radial basis function and possibly the number and spacing of centres.

This gives a ϑ through integration as above, whose reciprocal bounds the required ℓ^2-matrix-norm by integrating (5.22) with respect to $d\mu$. Further, it is easy to bound the ℓ^2-condition number now, since one can bound the ℓ^2-norm of the matrix $\mathbf{A} = \{\phi(\|\xi - \zeta\|)\}_{\xi,\zeta\in\Xi}$ itself for instance by its Frobenius norm

$$\|\mathbf{A}\|_F := \sqrt{\sum_{\xi,\zeta\in\Xi} \phi(\|\xi - \zeta\|)^2}.$$

This, in turn, can for example be bounded above by the following product which uses the cardinality of the centre-set:

$$|\Xi| \times \max_{\xi,\zeta\in\Xi} |\phi(\|\xi - \zeta\|)|.$$

Here, $|\Xi|$ is still the notation for the finite cardinality of the set of centres Ξ which we use.

We give a few examples of the quantitative outcome of the analysis when the radial basis function is the multiquadric function and its parameter c is set to be one. Then the measure $d\mu$ is defined by the weight function that uses the standard Γ-function (Abramowitz and Stegun, 1972),

(5.23) $$d\mu(t) = \frac{e^{-c^2 t}\, t^{\alpha-1}}{\Gamma(\alpha)} dt.$$

We let, for the multiquadric function, $c = 1$, $n = 2$ and in order to simplify notation further

$$p := q^2/(1 + \sqrt{1 + q^2/4}) \sim \frac{q^2}{2}, \qquad q \to 0,$$

for small q. Then the Euclidean matrix norm of the inverse of the interpolation matrix is at most $24e^{(12/p)}/p$, as follows from (5.22) and the above. This is asymptotically

$$\frac{48}{q^2} e^{(24/q^2)}, \qquad q \to 0.$$

For $n = 3$ one gets $36\, e^{(16/p)}/p$ and asymptotically

$$\frac{72}{q^2} e^{(32/q^2)}, \qquad q \to 0.$$

For $q \to \infty$ one gets in two and three dimensions the asymptotic bounds $12/q$ and $18/q$, respectively. Indeed, one expects quickly – here exponentially – growing bounds for $q \to 0$ because naturally the matrix becomes singular for smaller q and the speed of the norm's growth is a result of the smoothness of ϕ, the multiquadric being infinitely smooth. Conversely, for large q, one expects to get bounds that are similar to the bound for $c = 0$, because in that case, c is very small compared with q. Indeed, an important result of Ball (1992) gives an upper bound that is a constant multiple of $\sqrt{n}/q$ on the size of the inverse of the interpolation matrix for $c = 0$.

These bounds have been refined in Ball, Sivakumar and Ward (1992) to give for multiquadrics, general c and all n the upper bound for the Euclidean norm of the inverse of the matrix

$$(5.24) \qquad q^{-1}\exp(4nc/q),$$

multiplied by a universal constant that only depends on the dimension. This again matches nicely with the aforementioned result due to Ball. We note in particular the exponential growth of this bound when q diminishes. We note also its important dependence on the parameter c in the exponent.

5.3.3 Lower bounds on matrix norms

Other *lower* bounds on the ℓ^2-norms of the inverse of the interpolation matrix have been found by Schaback (1994). Because they confirm our earlier upper bounds, they give an idea of the typical *minimum loss* of significance we must expect in the accuracy of the coefficients for general right-hand sides which are due to large condition numbers and the effects of rounding errors. We give three examples of the results of this work below.

Theorem 5.12. *Let $\hat{q}$ be the maximal distance $\max_{\xi,\zeta\in\Xi}\|\xi-\zeta\|$ between the finite number of data points $\Xi \subset \Omega$. Let $m = |\Xi|$ be the finite cardinality of the set of centres.*

(i) *For $\phi(r) = \sqrt{r^2+c^2}$, the interpolation matrix satisfies the asymptotic bound*

$$\|A^{-1}\|_2 \geq C\frac{\exp\big(c[(\tfrac{1}{2}n!m)^{1/n}-\tfrac{1}{2}]/\hat{q}\big)}{m}, \qquad m \to \infty.$$

(ii) *Let $\phi(r) = r^{2k}\log r$ and n be even. Then the interpolation matrix satisfies the asymptotic bound*

$$\|A^{-1}\|_2 \geq Cm^{\frac{2k-1}{n}-1}, \qquad m \to \infty.$$

(iii) *Let $\phi(r) = r^{2k+1}$, and let n be odd. Then the interpolation matrix satisfies the asymptotic bound*

$$\|A^{-1}\|_2 \geq Cm^{\frac{2k+1}{n}-1}, \qquad m \to \infty.$$

We observe, e.g. by an application of Stirling's formula, that the bound in (i) compares favourably with the upper bound on the matrix norm given above, except for a power of q which has little effect in comparison with the exponential growth of the upper and lower bounds as $\hat{q}$ diminishes. Also (iii) compares quite well with the stated upper bounds for $k = 0$, although we are usually interested in $2k \pm 1 > n$.

We remark that the above statements show that the minimum norm becomes larger with a larger number of centres ($m \to \infty$). This is, of course, no contradiction to the results in the previous subsection, because the minimal separation distance becomes smaller with a larger number of data in the same domain and the $\hat{q}$ above is the maximal, not the minimal distance between centres.

It has been noted already that, while the upper bounds on ℓ^2-matrix-norms of the inverse of interpolation matrices above do not depend on the cardinality of Ξ, the bounds on the condition numbers do, as they must according to Buhmann, Derrien and LeMéhauté (1995). This is also because the bounds on the norm of the interpolation matrix itself depend on the number of centres. An example is given by the following theorem which states a lower bound on the condition number; it applies for instance to the multiquadric radial basis function.

Theorem 5.13. *If* $-\phi$ *is conditionally positive definite of order one and* $\phi(0) \geq 0$, *then the* ℓ^2-*condition number of the interpolation matrix with centres* Ξ *is bounded* below *by* $|\Xi| - 1$.

Proof: The ℓ^2-condition number is the same as the – in modulus – largest eigenvalue divided by the smallest eigenvalue. We note that ϕ is necessarily nonnegative. Now, because the interpolation matrix has only nonnegative entries, the largest eigenvalue is bounded from *below* by

$$\min_{\zeta \in \Xi} \sum_{\xi \in \Xi} \phi(\|\zeta - \xi\|) \geq \Big(|\Xi| - 1\Big)\phi(2q),$$

the separation distance q still having the same meaning as before. Moreover, the smallest eigenvalue can be bounded from *above* by $\phi(2q) - \phi(0) \leq \phi(2q)$. This is because we may apply Cauchy's interlacing theorem to the two-by-two principal submatrix of the interpolation matrix which has entries $\phi(0)$ on the diagonal and $\phi(2q)$ elsewhere. Concretely, this means that the smallest eigenvalue is bounded by $\phi(2q)$, which gives the result, recalling our lower bound on the largest eigenvalue from the above display, namely $(|\Xi| - 1)$ times $\phi(2q)$. $\qquad\qquad\square$

However, this Theorem 5.13 only applies when $-\phi$ is a conditionally positive definite matrix of order one; for other radial basis functions (Narcowich, Sivakumar and Ward, 1994) are able to give ℓ^2 (actually, the statement is more generally for ℓ^p, $1 \leq p \leq \infty$) bounds on condition numbers that only depend on q, as defined in (5.21). A typical result is as follows.

Theorem 5.14. *Let ϕ be conditionally positive definite of order zero and the separation radius q be fixed and positive, let D_q be the collection of all finite subsets of centres Ξ of $\mathbb{R}^n$ whose separation radii are at least q. Then* sup $\|A\|_p$ *over all $\Xi \in D_q$ is finite for all $1 \leq p \leq \infty$, where A is as usual the interpolation matrix for the given radial basis function and centres.*

There is another highly relevant feature of radial basis function interpolation to scattered data which we want to draw attention to in the discussion in this book. We have noted the importance of the norm estimates to the interpolation matrices and their inverses above, and of course we know about the importance of the convergence estimates and the approximation orders therein. What we wish to explain now, at the end of this chapter, is that there is a remarkable relationship between the convergence order estimates that have been presented earlier in this chapter and the norm estimates above (precisely: the upper bounds on ℓ^2-norms of inverses of interpolation matrices). It is called the uncertainty principle.

5.3.4 The uncertainty principle

This relationship which we will demonstrate now was pointed out first and termed the 'uncertainty principle' in Schaback (1993). It is as follows. In order to explain it, we shall still call the interpolation matrix $\{\phi(\|\xi - \zeta\|)\}_{\xi,\zeta\in\Xi}$, for centres Ξ, the matrix A. For simplicity, we avoid any polynomial term in the interpolant, that is we restrict ourselves to conditionally positive definite ϕ of order zero, where there is no such term. Thus if $\lambda \in \mathbb{R}^\Xi$ is such that the interpolation conditions expressed in matrix form

$$(5.25) \qquad A\lambda = \{f_\xi\}_{\xi\in\Xi} =: \mathbf{f}$$

hold, it is always true that by multiplication on the left by the vector λ^T on both sides of the last display

$$(5.26) \qquad \lambda^T A\lambda = \lambda^T \mathbf{f}.$$

We know that the real eigenvalues of the symmetric positive definite matrix A are bounded away from zero by a suitable quantity. We have seen such bounds in Subsection 5.3.1. We take such a positive bound for the eigenvalues and call it ϑ. Thus we have in hand a positive real ϑ which provides the inequality

$$\vartheta\|\lambda\|^2 \leq \lambda^T A\lambda,$$

for all vectors $\lambda \in \mathbb{R}^\Xi$. Therefore, we get from this and from (5.26) the important inequality

$$\|\lambda\| \cdot \|\mathbf{f}\| \geq \lambda^T \mathbf{A} \lambda = \lambda^T \mathbf{f} \geq \vartheta \|\lambda\|^2.$$

Moreover, letting $\lambda \neq 0 \neq \|\mathbf{f}\|$ and dividing by the positive number $\vartheta \|\lambda\|$, we get

(5.27) $$\|\lambda\| \leq \vartheta^{-1} \|\mathbf{f}\|.$$

Now we recall the definition of the power function. Forgetting for the moment its use for error estimates, it can be taken for any of the radial basis functions to be the expression

$$P(x) = \phi(0) - 2 \sum_{\xi \in \Xi} p_\xi(x) \phi(\|x - \xi\|)$$

(5.28)
$$+ \sum_{\xi \in \Xi} \sum_{\zeta \in \Xi} p_\xi(x) p_\zeta(x) \phi(\|\xi - \zeta\|),$$

where the p_ξ may be *arbitrary functions* (i.e. not necessarily polynomials) so long as we have not yet in mind the aforementioned use for error estimates. For example, $p_\xi(x)$ may be the full Lagrange functions, i.e. linear combinations of $\phi(\| \cdot - \xi\|)$ which provide the conditions

$$p_\xi(\zeta) = \delta_{\xi\zeta}, \quad \xi, \zeta \in \Xi.$$

In fact, in the section about convergence of this chapter, we have taken the p_ξ only from the kernel of the associated semi-inner product. The properties which we use here remain true, however, if the p_ξ are Lagrange functions on the whole of Ξ. Moreover, we recall the important fact that by its definition through the inner product notation at the beginning of this chapter, the power function is always nonnegative. This is unchanged even if we no longer restrict p_ξ to K. Thus, letting the new $|\Xi| + 1$ vector $\underline{p}_x$ be

$$\underline{p}_x = \left(1, (-p_\xi(x))_{\xi \in \Xi}\right)^T \in \mathbb{R} \times \mathbb{R}^\Xi \setminus \{0\}$$

and

$$\mathbf{A}_x = \begin{pmatrix} \phi(0) & \left(\phi(\|x - \xi\|)\right)_{\xi \in \Xi} \\ \left(\phi(\|x - \xi\|)\right)_{\xi \in \Xi}^T & \mathbf{A} \end{pmatrix},$$

we get the quadratic, x-dependent form

$$P(x) = \underline{p}_x^T \mathbf{A}_x \underline{p}_x$$

as an alternative form of the power function above. Let $x \notin \Xi$. Now, for each $\mathbf{A}_x$ there exists a $\vartheta_x > 0$ that fulfils the same bound as ϑ does for $\mathbf{A}$ because $\mathbf{A}_x$ is obtained very simply from $\mathbf{A}$ by amending it by a new row and a new column which represent the augmentation of Ξ by x to $\Xi' := \Xi \cup \{x\}$. Consequently, with ϑ_x replacing ϑ while at the same time $\mathbf{A}_x$ replaces $\mathbf{A}$, we get from the above lower eigenvalue estimate (5.27) now

$$(5.29) \qquad \vartheta_x \|\tilde{\boldsymbol{\lambda}}\|^2 \leq \tilde{\boldsymbol{\lambda}}^T \mathbf{A}_x \tilde{\boldsymbol{\lambda}}, \quad \tilde{\boldsymbol{\lambda}} = (\hat{\lambda}, \lambda)^T \in \mathbb{R} \times \mathbb{R}^\Xi,$$

for all $x \notin \Xi$. Next, it follows from setting in the above display $\tilde{\boldsymbol{\lambda}} := \underline{p_x}$ and from computing its Euclidean norm that

$$\vartheta_x \left(1 + \sum_{\xi \in \Xi} p_\xi^2(x) \right) \leq P(x),$$

because $\|\underline{p_x}\|^2 = 1 + \sum_{\xi \in \Xi} p_\xi^2(x)$. In particular, we get the so-called uncertainty principle below. It was proved in the original article by Schaback for general conditionally positive definite functions, and we state it so here.

Proposition 5.15. *If ϕ is any strictly conditionally positive definite radial basis function and $P(x)$, ϑ_x are defined through (5.28) and (5.29), respectively, then*

$$P(x)\, \vartheta_x^{-1} \geq 1,$$

which is the uncertainty principle.

If ϕ is conditionally positive definite of positive order, we can use the work of Subsection 5.3.2 to establish a bound (5.29). Recalling that ϑ_x is a lower positive bound on the smallest eigenvalue of the extended interpolation matrix $\mathbf{A}_x$, thus $1/\vartheta_x \geq \|\mathbf{A}_x^{-1}\|_2$, and recalling that bounding $P(x)$ above is a most important intermediate step in our convergence estimates, cf. our proof of Theorem 5.5 for instance, we can see that upper bounds on ℓ_2-norms of inverses of $\mathbf{A}$ and associated convergence orders are deeply interrelated through the uncertainty principle.

Concretely, if ϑ_x is big – thus our norm estimate is small which is the desirable state of affairs – then $P(x)$ cannot be small as well. This is why this is called the uncertainty principle in Proposition 5.15. Indeed, in most cases we use bounds of the following form, where the first one is for the convergence proofs, and we have encountered it already in this chapter in different shape, e.g. when bounding (5.13):

$$P(x) \leq \mathbf{F}(h),$$
$$\vartheta_x \geq \mathbf{G}(q),$$

with continuous and decreasing $\mathbf{F}$ and $\mathbf{G}$ for small arguments $h \to 0$ and $q \to 0$. These two functions are often monomials (see Table 5.1).

Table 5.1

Radial basis function $\phi(r)$	$\mathbf{F}(h)$	$\mathbf{G}(h)$
r^{2k-n}, $2k - n \notin 2\mathbb{Z}_+$, $2k > n$	h^{2k-n}	h^{2k-n}
$r^{2k-n} \log r$, $2k - n \in 2\mathbb{Z}_+$	h^{2k-n}	h^{2k-n}
$(r^2 + c^2)^{\alpha}$, $\alpha \notin \mathbb{Z}_+$,	$e^{-\tilde{\delta}/h}$	$h^2 e^{-24/h^2}$ $(2\alpha = 1,\ n = 2)$,
$\quad \alpha > -\frac{1}{2}n, c > 0$		$h\, e^{-4n/h}$ $(2\alpha = c = 1)$,
		$h^{2\alpha}\, e^{-12.76cn/h}$

Typical functions $\mathbf{F}$ and $\mathbf{G}$ that correspond to some of our radial basis functions are given in Table 5.1, where e.g. the first two expressions for $\mathbf{G}$ for multiquadrins are already known from Subsection 5.3.2 and constants that are independent of r and h are omitted from $\mathbf{F}$ and $\mathbf{G}$ in order to increase the readability of the table, taken from the paper by Schaback (1995a). The $\tilde{\delta}$ is a positive constant, independent of h.

We may simplify this even further, if we make a choice of centres such that we have for h and q and some $\delta > 0$

$$q \geq \frac{1}{2}h\delta.$$

Then one can establish the bounds

$$\mathbf{G}\left(\frac{1}{2}\delta h\right) \leq \vartheta_x \leq P(x) \leq \mathbf{F}(h)$$

for all $x \in \Omega$. In closing this chapter, we remark that our analysis above also provides bounds on the Lagrange functions mentioned before, namely

$$1 + \sum_{\xi \in \Xi} p_{\xi}(x)^2 \leq \frac{P(x)}{\vartheta_x} \leq \frac{\mathbf{F}(h)}{\mathbf{G}\left(\frac{1}{2}\delta h\right)}.$$

This means in particular that they cannot grow too quickly (in other words, too badly) in regions where there are sufficiently many regularly (not gridded, but 'sufficiently close' to a grid) distributed centres ξ, because then q is not too small in relation to h, and thus δ in the penultimate display need not be very small.

6
Radial Basis Functions with Compact Support

6.1 Introduction

This chapter deals with radial basis functions that are compactly supported, quite in contrast with everything else that we have encountered before. In fact the constructions, concepts and results developed in this chapter are closely related to the piecewise polynomial B- and box-splines of Chapter 3 and the finite elements of the well-known numerical methods for the numerical solution of (elliptic) differential equations. The radial basis functions we shall study now are particularly interesting for those applications. Compactly supported radial basis functions are particularly appealing amongst practitioners.

They can be used to provide a useful, mesh-free and computationally efficient alternative to the commonly used finite element methods for the numerical solution of partial differential equations.

All of the radial basis functions that we have considered so far have global support, and in fact many of them do not even have isolated zeros, such as the multiquadric function for positive c. Moreover, they are usually increasing with growing argument, so that square-integrability and especially absolute integrability are immediately ruled out. In most cases, this poses no severe restrictions since, according to the theory of Chapter 5, we can always interpolate with these functions. We do, however, run into problems when we address the numerical treatment of the linear systems that stem from the interpolation conditions, as we have seen in the discussion of condition numbers in the previous two chapters and as we shall see further on in Chapter 7. On the other hand, we can form quasi-interpolating basis functions ψ (not to be confused with our growing radial basis functions ϕ from which they stem), and the ψ decay quickly. This is a change of basis as is the further use of the Lagrange functions L that we have studied in Chapter 4 and that decay quickly too. Sometimes, we can get very substantial, fast decay and the spaces for the radial basis functions contain bases which are 'essentially' locally supported. The importance of this idea is nicely illustrated by the case of polynomial splines where we use the B-spline basis to render the approach practicable.

A similar thought was important for our convergence proofs and it plays an important rôle in the design of the fast algorithms of the next chapter.

The above substantive arguments notwithstanding, there are applications that demand genuine local support which none of our radial basis functions studied up to now can provide, in spite of their otherwise highly satisfactory approximation properties. One reason for the requirement of compact support may be that there are such masses of data, which need to be interpolated, that even exponential or quick algebraic decay is not sufficient to localise the method well enough so as to provide stable and fast computations. Further applications, e.g. from meteorology, require the approximants to vanish beyond given cut-off distances.

Among the many practical issues associated with choosing a suitable data fitting scheme is the need to update the interpolant often with new data at new locations, and this may need to be done fast or in 'real time' for applications. Updating interpolants for radial basis functions with global support, however, requires significant changes to the nonsparse interpolation matrix – one complete (nonsparse) row and column are added each time a new centre comes up – whereas interpolation matrices with compactly supported basis functions are usually banded and admit easy 'local' changes in the sparse matrix when centres are added, for example. Also, in standard finite element applications, compact support for the test functions with which the inner products are formed is usually required in order that the resulting stiffness matrices are banded and that the inner products which form their entries are easily computable, either explicitly or by a quadrature rule. A typical finite element application results from a Galerkin *ansatz* to solve linear or even nonlinear elliptic PDEs numerically, where the given differential equation is reformulated in a weak form that contains inner products of the test functions as outlined in the last section of Chapter 2. In this case, the quadrature of the integrals which form the inner products is the computational 'bottle neck', because the integrals have to be computed accurately, and there are many of them in a high-dimensional stiffness matrix. If the test functions are compactly supported, many of the inner products vanish and therefore the stiffness matrix whose entries are the inner products becomes banded.

Finally, we may be dealing with exponentially growing data that do not fit into our algebraically localised (i.e. with algebraically decaying basis functions) quasi-interpolating methods or radial basis function interpolation schemes because their fast increase influences the approximants everywhere.

In the search for compactly supported radial basis functions, it is important to remember that we do not want to give up the remarkable nonsingularity results of Chapters 2 and 5 which are still a fundamental reason for studying and using radial basis functions. It turns out that there is no need to give up those desirable properties even while remaining in our general setting of (conditionally) positive definite functions. Therefore we address first the question what kind of results we may expect in this direction.

It follows from the Bernstein representation theorem that a compactly supported univariate function cannot be completely monotonic so long as trivial cases such as constant functions are excluded. This is so because a *globally* supported kernel, that is one on the whole n-dimensional Euclidean space, is – if exponentially decaying – integrated with respect to a nonnegative measure (not identically zero), in the Bernstein representation theorem, to form and characterise a completely monotonic function. In some sense, the 'most local' completely monotonic function is the exponential function, i.e. the kernel in the Bernstein representation, which results if the measure appearing in the representation formula is a point measure. And so a compactly supported radial basis function cannot be (conditionally) positive definite on all $\mathbb{R}^n$. Therefore, for a start, *complete* monotonicity cannot be the correct concept in this chapter about compactly supported radial basis functions.

Also, as we have seen, the property of conditional positive definiteness of positive order is closely related to the Fourier transform of the radial basis function having a singularity at the origin; specifically, the converse of Theorem 5.1 – the proof can be found in the literature, cf. Guo, Hu and Sun (1993) – which states that conditional positive definiteness of the interpolation matrix for our radial basis function requires that the derivative of some order is completely monotonic, subject to a suitable sign-change. Further, the proof of Theorem 4.10 shows that the complete monotonicity of the kth derivative of a univariate function $\phi \colon \mathbb{R}_+ \to \mathbb{R}$ implies that the generalised Fourier transform of the n-variate function $\phi(\| \cdot \|)$ possesses an algebraic singularity of order k at the origin.

Now, since every continuous radial basis function of compact support is absolutely integrable, its Fourier transform is continuous, and, in fact, analytic and entire of exponential type, a notion which we have already encountered in the last section of Chapter 4. This rules out conditional positive definiteness of nonzero order for compactly supported radial basis functions. Thus we are now looking for functions that give rise to positive definite interpolation matrices for certain fixed, but not all, dimensions n, and are not completely monotonic. (We shall see in the sequel that instead *multiple monotonicity* of a finite order is the suitable notion in the context of compactly supported radial basis functions.)

We remark that since it is not possible to construct compactly supported radial basis functions that are positive definite on $\mathbb{R}^n$ for all n, as a result of our remarks above, we get positive definite functions only for some dimensions, restricted by an upper bound.

We also include a short study of the convergence of interpolants with compactly supported radial basis functions in a section of this chapter. It is there where most of the discussion in the approximation theory and radial basis function community about the usefulness of the radial basis functions with compact support enters, because the known convergence results are not nearly as good as those observed for the globally supported ones in the previous two chapters. In particular, there is a significant trade-off between the size of the support of the radial basis functions relative to the spacing of the centres – or the number of centres within – and the estimated error of the interpolant. This may also mean that special attention may need to be given to the condition numbers of the matrix of interpolation when methods are used in practice.

We are now in a position to introduce the new concept of positive definite compactly supported radial basis functions. Specifically, we present two seemingly different classes of radial basis functions that have compact support and discuss their properties.

6.2 Wendland's functions

A particularly interesting part of the work in this direction is due to Wendland, initiated in part by Wu (1995a, b) who was the one of the first approximation theorists to study positive definiteness of compactly supported radially symmetric matrices. Indeed, Wendland demonstrated that certain piecewise polynomial functions have all the required properties. They usually consist of only two polynomial pieces. Note that the n-variate radially symmetric $\phi(\| \cdot \|)$ is not piecewise polynomial, however, so we are not dealing with spline spaces here unless the dimension is one.

Moreover, for Wendland's class of basis functions, we can identify the minimal degree of these pieces such that a prescribed smoothness and positive definiteness up to a required dimension are guaranteed. We have already mentioned in the introduction to this chapter that the positive definiteness is dimension-dependent and this clearly shows in the construction. We will describe his approach in some detail now.

For the Wendland functions, we have to consider radial basis functions of the form

$$(6.1) \qquad \phi(r) = \begin{cases} p(r) & \text{if } 0 \le r \le 1, \\ 0 & \text{if } r > 1, \end{cases}$$

with a univariate polynomial p. As such they are supported in the unit ball, but they can be scaled when used in applications. The Fourier transform in n dimensions is, due to a standard formula for radially symmetric functions (see, e.g., Stein and Weiss, 1971), the univariate transform

$$(6.2) \qquad \hat{\phi}(r) = \frac{2\pi^{n/2}}{\Gamma(n/2)} \int_0^1 \phi(s) s^{n-1} \Omega_n(rs) ds, \qquad r \geq 0.$$

Here,

$$\Omega_n(x) = \Gamma\left(\frac{n}{2}\right)\left(\frac{x}{2}\right)^{-n/2+1} J_{n/2-1}(x).$$

As always, the letter J denotes the standard Bessel function of the order indicated by its index (Abramowitz and Stegun, 1972, p. 358). This is related to $\tilde{\Omega}_\nu$ of Chapter 4 by $\Omega_n(x) = \Gamma(\frac{n}{2})2^{\frac{n}{2}-1}\tilde{\Omega}_{\frac{n}{2}-1}(x, 1)$. We only integrate over the unit interval because of the support size $[0, 1]$ of expression (6.1).

It is well-known that this Fourier transform *as a function in n dimensions*, i.e. the expression (6.2), is positive if (6.1) is coupled with the univariate polynomial $p(r) = (1 - r)^\ell$ and an exponent $\ell \geq [n/2] + 1$ (Askey, 1973). Beginning with this particular choice, Wendland's more general radial basis functions are constructed by integration of this univariate function. This is suitable because of the surprisingly simple result we state next. The result uses the notation $\mathcal{I}f(r)$ for the integral of f in the form

$$\int_r^\infty sf(s) ds, \qquad r \geq 0,$$

if the integral exists for all nonnegative arguments.

Lemma 6.1. *If the transformed* $(\mathcal{I}\phi)(\|\cdot\|)$ *is absolutely integrable as a function of n variables, then the radial part of its Fourier transform is the same as the radial part of the* $(n + 2)$-*variable Fourier transform of* $\phi(\|\cdot\|)$, *i.e.*

$$\hat{\phi}(r) = \frac{2\pi^{n/2+1}}{\Gamma(n/2+1)} \int_0^1 \phi(s) s^{n+1} \Omega_{n+2}(rs) ds$$

$$= \frac{4\pi^{n/2+1}}{n\Gamma(n/2)} \int_0^1 \phi(s) s^{n+1} \Omega_{n+2}(rs) ds, \qquad r \geq 0.$$

The last line follows from the well-known multiplication formula for the Γ-function

$$\Gamma(x + 1) = x\Gamma(x)$$

(Gradshteyn and Ryzhik, 1980, p. 937). The proof of this auxiliary result about the introduced operator $\mathcal{I}$ is an easy consequence of direct computation with a straightforward exercise in computing Fourier transforms and the expression (6.2). In fact, as Castell (2000) points out, it can be deduced from very general principles, where the operator $\mathcal{I} = \mathcal{I}^1$ is a special case of the operator

$$\mathcal{I}^\kappa f(r) = \int_r^\infty (s^2 - r^2)^{\kappa-1} s f(s) \, ds, \qquad r \geq 0,$$

where κ is a positive constant. For $\kappa = 1$, this is the above definition of $\mathcal{I}$. Now, Castell (2000) shows by using the expansion for the Bessel function from Gradshteyn and Ryzhik (1980, p. 959)

$$\Omega_n(x) = \Gamma\left(\frac{n}{2}\right) \sum_{k=0}^\infty \frac{(-1)^k \left(\frac{x}{2}\right)^{2k}}{k! \Gamma\left(\frac{n}{2} + k\right)}$$

that

$$\Omega_{n-2}(x) = \left(\frac{1}{n-2}\right) \mathcal{I}^1 \Omega_n(x).$$

From this, the lemma may be derived. Moreover, we get a one-step recurrence if we use $\mathcal{I}^{1/2}$ instead (Castell, 2000).

In view of this lemma, we define $\phi_{n,k}(r) = \phi(r) = \mathcal{I}^k(1 - r)_+^\ell$, where $\mathcal{I}^k$ marks the $\mathcal{I}$-operator applied k times. This is clearly of the general form (6.1). Furthermore, we assert that ϕ is positive definite on $\mathbb{R}^n$ and $2k$ times continuously differentiable if $\ell = [n/2] + k + 1$. At this point, this is easy to establish. Indeed, the Fourier transform is nonnegative and positive on a set of nontrivial measure by Lemma 6.1 and by the remark we made above about the Fourier transform of $(1 - r)_+^\ell$. The differentiability follows directly from its definition as a k-fold integral and from the definition of ℓ. The following two results claim uniqueness of this function of compact support and a representation with recursively computable coefficients. Their proofs can be found in the work of Wendland (1995).

Proposition 6.2. *There is no nonzero function of the form (6.1) that is $2k$ times continuously differentiable and conditionally positive definite of order zero and has smaller polynomial degree than the above p.*

Proposition 6.3. *We have the representation*

$$\mathcal{I}^k(1 - r)_+^\ell = \sum_{m=0}^k \beta_{m,k} r^m (1 - r)_+^{\ell+2k-m}$$

with recursively computable coefficients

$$\beta_{j,k+1} = \sum_{m=j-1}^k \beta_{m,k} \frac{[n+1]_{m-j+1}}{(\ell + 2k - m + 1)_{m-j+2}}$$

and $\beta_{00} = 1$. *Here* $(\cdot)_k$ *is the Pochhammer symbol* $(q)_k = q(q+1)\ldots(q+k-1)$ *and* $[q]_k = q(q-1)(q-2)\ldots(q-k+1)$.

We give several examples for radial basis functions of this type for various dimensions. Examples are for

$n = 1$ and $k = 2$ the radial basis function $\phi(r) = (1-r)_+^5 (8r^2 + 5r + 1)$,

$n = 3$ and $k = 3$ the radial basis function $\phi(r) = (1-r)_+^8 (32r^3 + 25r^2 + 8r + 1)$ and

$n = 5$ and $k = 1$ the radial basis function $\phi(r) = (1-r)_+^5 (5r + 1)$.

The advantage of these radial basis functions is that they have this particularly simple polynomial form. Therefore they are easy to use in practice and popular for PDE applications. Their Fourier transforms are nonnegative but not identically zero which implies the positive definiteness of the interpolation matrix for distinct centres. This fact follows from Bochner's theorem which is stated and used below.

6.3 Another class of radial basis functions with compact support

In another approach we require compactly supported radial basis functions with positive Fourier transform, and pursue an altogether different route from the one used by Wendland by defining the functions not recursively, but by an integral that resembles a continuous convolution. We obtain a class of functions some of which are related to the well-known thin-plate splines. Indeed, the functions we get initially are only once continuously differentiable like thin-plate splines, but we will later on present an extension which enables us to obtain other functions of compact support with arbitrary smoothness. This is especially important in connection with the convergence result which we shall prove for the radial basis functions of compact support, because, as in the convergence theorems established in the previous two chapters, the rate of convergence of the interpolant to the prescribed target function is related to the smoothness of the approximating radial basis function.

The reason for seeking compactly supported analogues of thin-plate splines is that we wish to employ their properties in applications, e.g., when differential equations are solved by radial basis function methods (for the importance of this issue see our remarks in the last section of Chapter 2 and in the introduction to this chapter).

The radial basis functions we seek are generally represented by the convolution-type integrals

$$(6.3) \qquad \phi(\|x\|) = \int_{\|x\|^2}^{\infty} \left(1 - \|x\|^2/\beta\right)^{\lambda} g(\beta)\, d\beta, \qquad x \in \mathbb{R}^n,$$

where g is from the space of compactly supported continuous functions on $\mathbb{R}_+$. The resulting expression is clearly of compact support, the support being the unit ball as in the previous section. It is scalable to other support sizes as may be required in applications. Further, we shall require that g be nonnegative. Similar radial basis functions that are related to *multiply monotonic* functions and their Williamson representation were already considered in Chapter 4, but they were not of compact support since the weight function g was not compactly supported. However, they also gave rise to nonsingular interpolation matrices.

In principle, we can admit radial basis functions whose support radii depend on the centres $\phi_\xi(r) = \phi(r/\delta_\xi)$, but then the interpolation matrix $\mathbf{A} = \{\phi_\xi(\|\zeta - \xi\|)\}$ is no longer symmetric and may be singular. Indeed if Ξ consists of zero and $m - 1$ centres on a circle of radius two, say, and the centres on the circle have equal angular distance, then singularity can occur; for this we place $m - 1$ of the basis functions about the points on the circle and we scale the one about the origin by arranging its support and its coefficient in such a way that we have a nontrivial approximant which is identically zero on Ξ.

It is easy to see by the Williamson representation theorem that (6.3) is multiply monotonic because by a change of variables, with nonnegative weight function g, ϕ can be reformulated as

$$\phi(r) = \int_0^{r^{-2}} \left(1 - r^2\beta\right)^{\lambda} g(1/\beta)\beta^{-2}\, d\beta, \qquad r \geq 0.$$

Therefore, we note already at this point that (6.3) is such that $\phi(\sqrt{r})$ is $(\lambda - 1)$ times monotonic (see Definition 4.1) by appealing to Theorem 4.13 as mentioned. This justifies our observation in the introduction to this chapter, namely that radial basis functions with compact support are related to multiply monotonic functions.

We require $\hat{\phi} > 0$ everywhere; indeed, this allows an application of Bochner's theorem to deduce immediately the positive definiteness of the interpolation matrix. Alternatively, $\hat{\phi} < 0$ for negative definiteness. The compact support, and therefore the integrability of ϕ and the existence of a smooth Fourier transform, are guaranteed automatically by the compact support of g in (6.3).

Pertinent is therefore the following result of Bochner (Widder, 1946). We state it only in the specific, slightly modified way which is suitable for our application, but in fact it is available as a more general characterisation of all nonnegative definite distributions, which have to be the Fourier transform of a nonnegative measure.

Bochner's theorem. *If the Fourier transform of a continuous $F \colon \mathbb{R}^n \to \mathbb{R}$ is positive, then the symmetric matrix with entries $F(\xi - \zeta)$, $\xi, \zeta \in \Xi$, is positive definite for all finite sets of distinct points $\Xi \subset \mathbb{R}^n$. Conversely, if a continuous $F \colon \mathbb{R}^n \to \mathbb{R}$ is such that all finite matrices with entries $F(\xi - \zeta)$, $\xi, \zeta \in \Xi$, are nonnegative definite, then F must be the Fourier transform of a nonnegative finite-valued Borel measure.*

We have no recursive formula as in the previous section but wish to compute the Fourier transform directly. As has been used in this book before, in the sequential computations, the symbol $\doteq$ means equality up to a generic positive (multiplicative) constant whose exact value is immaterial to us in the analysis. The following evaluation of $\hat{\phi}$ follows the same lines as the analysis at the end of Chapter 4 concerning multiply monotonic functions. We begin with

$$\hat{\phi}(\|x\|) \doteq \int_0^\infty \int_0^{\sqrt{\beta}} (1 - s^2/\beta)^\lambda s^{n-1} \Omega_n(s\|x\|) \, ds \, g(\beta) \, d\beta.$$

This is, by integrating the truncated power multiplied by the Bessel function, substituting the expression for Ω_n into the integral, according to equation (11.4.10) of Abramowitz and Stegun with $\mu = \frac{1}{2}n - 1$, $\nu = \lambda$,

$$\int_0^{\pi/2} J_\mu(z \sin t) \sin^{\mu+1} t \cos^{2\nu+1} t \, dt = \frac{2^\nu \Gamma(\nu + 1)}{z^{\nu+1}} J_{\mu+\nu+1}(z).$$

Making changes of variables $\|x\|/\sqrt{\beta} \mapsto z$, $s \mapsto \sqrt{\beta} \sin t$ allows us to deduce that $\hat{\phi}(\| \cdot \|)$ is a fixed constant multiple of

$$\|x\|^{-\lambda - n/2} \int_0^\infty J_{\lambda+n/2}\left(\|x\|\sqrt{\beta}\right) \beta^{n/4 - \lambda/2} g(\beta) \, d\beta.$$

This again becomes, by a change of variables, a constant multiple of the integral

$$\|x\|^{-n-2} \int_0^\infty J_{\lambda+n/2}(t) t^{n/2 - \lambda + 1} g\left(t^2 \|x\|^{-2}\right) dt.$$

Finally, by including our particular weight function g we get an expression for the required Fourier transform,

$$\hat{\phi}(\|x\|) \doteq \|x\|^{-n-2-2\mu\nu} \int_0^{\|x\|} J_{\lambda+n/2}(t) t^{n/2 - \lambda + 1} \left(\|x\|^{2\mu} - t^{2\mu}\right)^\nu dt.$$

As we will see below in the proof of Theorem 6.4, this weight function g is sufficient for the Fourier transform of the radial basis function to be always nonzero, where certain additional conditions on the various exponents are involved. Specifically, we use as additional conditions $0 < \mu \le \frac{1}{2}$, that ν

be at least one, and the following extra requirement which depends on the dimension:

$$\lambda \geq \max\left(\frac{1}{2}, \frac{1}{2}(n-1)\right).$$

For example we can choose for feasible parameters $\mu = \frac{1}{2}$ and $\nu = 1$, which leads to the applicable basis functions given below.

In view of the above analysis and recalling Bochner's theorem, we have the following Theorem 6.4, where the positivity of the Fourier transform has still to be settled in the proof following.

Theorem 6.4. *For $g(\beta) = (1 - \beta^\mu)_+^\nu$, $0 < \mu \leq \frac{1}{2}$ and $\nu \geq 1$, the Fourier transform of the radial basis function of compact support stated above is everywhere well-defined and strictly positive, whenever λ satisfies the condition in the display above. Therefore the associated symmetric interpolation matrix $\{\phi(\|\xi - \zeta\|)\}_{\xi,\zeta \in \Xi}$ is positive definite for all finite sets of distinct points Ξ in n-dimensional Euclidean space.*

Proof: We have already computed what the Fourier transform of the radial basis function is up to a positive constant multiple. The results of Misiewicz and Richards (1994, Corollary 2.4, using items (i), (ii) and (v)), and the *strict* inequalities (1.1) and (1.5) in the paper Gasper (1975a) provide the following result which we shall summarise here in Lemma 6.5 and shall subsequently use in our computations.

Lemma 6.5. *Let $0 < \mu \leq \frac{1}{2}$, $\nu \geq 1$; the so-called Hankel transform, which is defined as*

$$\int_0^x (x^{2\mu} - t^{2\mu})^\nu t^\alpha J_\beta(t)\, dt,$$

is positive for all positive reals x if α and β satisfy any of the following three conditions:

(i) $\frac{3}{2} = \alpha \leq \beta$,

(ii) $-\frac{1}{2} \leq \alpha = \beta \leq \frac{3}{2}$,

(iii) $\alpha = \alpha_0 - \delta$, $\beta = \beta_0 + \delta$ where α_0, β_0 satisfy either (i) or (ii) above and δ is nonnegative.

We continue with the proof of Theorem 6.4. Now, for the case $n > 1$ of Theorem 6.4, we use (i) and (iii) and set here $\alpha_0 = \frac{3}{2}$, $\alpha = \frac{1}{2}n - \lambda + 1$, that is $\delta = \lambda + \frac{1}{2} - \frac{1}{2}n$, and $\beta = \lambda + \frac{1}{2}n$. So we see that our result is true for $\lambda \geq \frac{1}{2}(n-1)$ – this is in order that the constant δ be nonnegative – provided that $\beta_0 = \lambda + \frac{1}{2}n - \delta \geq \frac{3}{2}$. For the latter it suffices that n is at least two, recalling our condition on λ.

In the univariate case we apply the result from Lemma 6.5 that the above display is positive whenever $-\frac{1}{2} \le \alpha_0 + \delta = \beta_0 - \delta \le \frac{3}{2}$, again for a nonnegative δ. Thus we get that $\delta = \lambda - \frac{1}{2}$ which has to be nonnegative. In other words, $\lambda \ge \frac{1}{2}$. Moreover, we have a condition that $\alpha_0 + \delta = 1 \ge -\frac{1}{2}$ which is always true, and $\beta_0 - \delta = 1 \le \frac{3}{2}$ which is easily seen to be true as well. $\qquad\square$

An example to illustrate this result is provided by the choices $\lambda = \nu = 1$, $\mu = \frac{1}{2}$, which give

$$\phi(\|x\|) = \begin{cases} \frac{1}{3} + \|x\|^2 - \frac{4}{3}\|x\|^3 + 2\|x\|^2 \log\|x\| & \text{if } 0 \le \|x\| \le 1, \\ 0, & \text{otherwise,} \end{cases}$$

by simple calculation directly from (6.3).

Another example is provided by the choices $\lambda = 1$, $\nu = 4$, $\mu = \frac{1}{2}$, which give $\phi(\|x\|)$ as the function

$$\phi(\|x\|) = \begin{cases} \frac{1}{15} + \frac{19}{6}\|x\|^2 - \frac{16}{3}\|x\|^3 + 3\|x\|^4 - \frac{16}{15}\|x\|^5 \\ \qquad + \frac{1}{6}\|x\|^6 + 2\|x\|^2 \log\|x\| & \text{if } 0 \le \|x\| \le 1, \\ 0, & \text{otherwise.} \end{cases}$$

We recall that these functions can be scaled for various support sizes. We also note that the resulting basis functions are continuously differentiable as multivariate functions. Indeed, possessing one continuous derivative is the best possible smoothness that can be obtained from the results of Theorem 6.4, due to the nature of the Euclidean norm inside the truncated power in the definition of our radial basis functions.

There is an extension to this theory which shows that positive definiteness of the interpolation matrices prevails if we implement certain smoother radial basis functions of compact support instead. The proofs are more involved but nonetheless follow the same ideas as the proof of Theorem 6.4.

Theorem 6.6. *Let n, μ and g be as above, $\nu \ge 1$, and suppose λ and $\varepsilon > -1$ are real numbers with*

$$\lambda \ge \begin{cases} 1, & \varepsilon \le \frac{1}{2}\lambda, & \text{if } n = 1, \text{ or} \\ \frac{1}{2}, & \varepsilon \le \min\left(\frac{1}{2}, \lambda - \frac{1}{2}\right), & \text{if } n = 1, \text{ and} \\ \frac{1}{2}(n-1), & \varepsilon \le \frac{1}{2}\left(\lambda - \frac{1}{2}(n-1)\right), & \text{if } n > 1. \end{cases}$$

Then the radial basis function

$$\phi(\|x\|) = \int_{\|x\|^2}^{\infty} \left(1 - \|x\|^2/\beta\right)^{\lambda} \beta^{\varepsilon} g(\beta)\, d\beta, \qquad x \in \mathbb{R}^n,$$

has an everywhere positive Fourier transform. It therefore gives rise to positive definite interpolation matrices for finitely many distinct centres in n-dimensional

Euclidean space. Moreover, those radial basis functions satisfy $\phi(\| \cdot \|) \in$ $C^{1+\lceil 2\varepsilon \rceil}(\mathbb{R}^n)$.

Proof: We apply the same arguments as before, and therefore shall be brief here; we note in particular that according to Lemma 6.5 (iii) the Fourier transform of ϕ is still positive within the stated range of ε, for the following reasons.

(1) The α in the proof of Theorem 6.4 may be replaced by $\alpha = \frac{1}{2}n - \lambda + 1 + 2\varepsilon$, and Lemma 6.5 states that the positivity of the Fourier transform prevails, so long as we replace δ by $\delta - 2\varepsilon$ and δ remains nonnegative.
(2) That we still require the latter to be nonnegative gives rise to the specific conditions on ε in the statement of the theorem, i.e. that for $n > 1$ the quantity ε must be in the range

$$\varepsilon \le \frac{1}{2}\left(\lambda - \frac{1}{2}(n - 1)\right).$$

(3) The analysis for $n = 1$ requires several special cases. Briefly, conditions (i) and (ii) of Lemma 6.5 show that the conditions $\lambda \ge 1$ in tandem with $\varepsilon \le \frac{1}{2}\lambda$ are sufficient for our demands. By contrast condition (ii) alone admits $\lambda \ge \frac{1}{2}$, together with $\varepsilon \le \min(\frac{1}{2}, \lambda - \frac{1}{2})$. On the other hand, no further comments are supplied here for the rest of the proof of positivity of the Fourier transform. It is left as an exercise to the reader.

The next fact we need to prove is the smoothness of our radial basis function *as a multivariate, radially symmetric function.*

A way to establish the asserted smoothness of the radial basis function of compact support here is as follows. It uses results from the chapter before last. Specifically, the Abelian and Tauberian results after the statement of Theorem 4.14 in Chapter 4 can be applied to the Fourier transform of our radial basis function of compact support. They give for the choice $u = \varepsilon - \lambda$ that this Fourier transform satisfies $\hat{\phi}(r) \sim r^{-n-2-2\varepsilon}$ for $r \to \infty$. Therefore, recalling that $\hat{\phi}$ is continuous everywhere else and in particular uniformly bounded in a neighbourhood of the origin, we have $\hat{\phi}(r) \sim (1 + r)^{-n-2-2\varepsilon}$. We can employ the Fourier inversion formula from the Appendix for absolutely integrable and rotationally invariant $\hat{\phi}$,

$$\phi(r) = \frac{2^{1-n}}{\Gamma(n/2)\pi^{n/2}} \int_0^\infty \hat{\phi}(s)s^{n-1}\Omega_n(rs)ds, \qquad r \ge 0,$$

to deduce that we may differentiate the radial basis function of compact support fewer than $1 + 2\varepsilon$ times and still get a continuous function, each of the differentiations giving a factor of order s as a multiplier to the Fourier transform $\hat{\phi}(s)$ in the above display.

We now use our conditions on v and ε to verify the required conditions for the application of the analysis of Chapter 4. Specifically, the condition $u = \varepsilon - \lambda > -n/2 - \lambda - 1$ from Chapter 4 is satisfied since we demand $\varepsilon > -1$ because this immediately implies $\varepsilon - \lambda > -\lambda - \frac{1}{2}n - 1$, together with $n > 0$. The theorem is proved. $\qquad\square$

6.4 Convergence

The convergence properties with these radial basis functions are satisfactory but not spectacular and are not, in particular, becoming better with the dimension n as with most of the other radial basis functions we have seen so far. One typical result that shows convergence is given below. We note immediately that no universal constant scaling of the radial basis functions is suitable when it is of compact support. Indeed, if the data are much further apart than the support size of the compactly supported radial basis functions, we will get a useless approximation to the data, although it always interpolates, the interpolation matrix being a nonsingular diagonal matrix. Conversely, if the data are far closer to each other on average than the support size of the radial basis function, we lose the benefits of compact support. Thus, the scaling should be related to the local spacing of the centres. Therefore, one has to apply a variable scaling which also shrinks with the distance of the data sites. This fact is represented in the following typical result, although it only uses a scaling which is the same for all centres and does not differ locally for different centres. We recall the definition of $D^{-k}L^2(\mathbb{R}^n)$ for nonintegral k from Subsection 5.2.1.

Theorem 6.7. *Let ϕ be as in the previous theorem. Let Ξ be a finite set in a domain Ω with the fill-distance h defined in Theorem 5.5. Finally, let s be the interpolant to $f \in L^2(\Omega) \cap D^{-n/2-1-\varepsilon}L^2(\mathbb{R}^n)$ of the form (for a positive scaling parameter δ)*

$$s(x) = \sum_{\xi \in \Xi} \lambda_\xi \phi\left(\delta^{-1}\|x - \xi\|\right), \qquad x \in \mathbb{R}^n,$$

satisfying the standard interpolation conditions $(s - f)|_\Xi = 0$. Then the convergence estimate

$$\|f - s\|_{\infty,\Omega} \leq Ch^{1+\varepsilon}(1 + \delta^{-n/2-1-\varepsilon})$$

holds for $h \to 0$.

Proof: The interpolant exists because of Theorem 6.6. The first observation we make is that $\phi_\delta(\| \cdot \|)$ is the reproducing kernel for the Hilbert space with inner product defined by the integral for square-integrable f and g as usual

$$(f, g) = \frac{1}{(2\pi)^n} \int_{\mathbb{R}^n} \frac{1}{\hat{\phi}_\delta(\|t\|)} \hat{f}(t)\overline{\hat{g}(t)}dt.$$

Here ϕ_δ denotes the scaled radial basis functions $\phi(\cdot/\delta)$. The terminology follows the description of the fifth chapter, except that the above is a proper (i.e. not semi-) inner product.

This reproducing kernel property is actually much easier to show than in the previous chapter, because our radial basis function here is absolutely integrable and positive definite due to the positivity of its Fourier transform. Therefore, the weight function in the above integral is well-defined and positive. Indeed, let f be square-integrable and let it have finite Hilbert space norm

$$\|f\|^2_{\phi_\delta} = \frac{1}{(2\pi)^n} \int_{\mathbb{R}^n} \frac{1}{\hat{\phi}_\delta(\|t\|)} |\hat{f}(t)|^2 \, dt,$$

as induced by the inner product. This is nothing other than an L^2-norm with positive continuous weight. Then, we get by a standard argument using the Parseval–Plancherel theorem (see Appendix) applied to the inner product $\left(f, \phi_\delta(\|x - \cdot\|)\right)$ the pointwise identity

$$\left(f, \phi_\delta(\|x - \cdot\|)\right) = f(x),$$

because both f and the radial basis function are square-integrable and continuous.

The analysis by Wu and Schaback (1993) performed with straightforward modifications, which is similar to our proof of Theorem 5.5, provides the error estimate for our setting on the right-hand side,

(6.4) $\|f - s\|_{\infty,\Omega} \le C(h/\delta)^{1+\varepsilon} \|f\|_{\phi_\delta}.$

We recall where the two factors in (6.4) come from. The first factor in (6.4) comes, as in the proof of Theorem 5.5, from the bound on the power function, and the second factor comes from Proposition 5.3 which is now applied to the Hilbert space X of all functions with finite norm, equipped with inner product $(\cdot, \cdot)$ as above and the reproducing kernel $\mathbf{k}(x, y) = \phi_\delta(\|x - y\|)$. There is no additional polynomial term as in Section 5.2, because the radial basis functions we use in this chapter give rise to positive definite interpolation matrices, i.e. they are conditionally positive definite of order zero.

Now, a bound on the second factor on the right-hand side of (6.4) will lead to the desired result. It can be bounded as follows. We use that $\hat{\phi}(0)$ is a positive quantity and that $\hat{\phi}(r) \sim r^{-n-2-2\varepsilon}$ for $r \to \infty$ from the proof of Theorem 6.6. We therefore we have the estimate $\hat{\phi}(r) \sim (1+r)^{-n-2-2\varepsilon}$ for all r.

Next, we have, recalling the definition of ϕ_δ and its Fourier transform as can be computed by the rules specified in the Appendix, the following integral:

$$(6.5) \qquad (2\pi)^n \delta^n \|f\|_{\phi_\delta}^2 = \int_{\mathbb{R}^n} \frac{1}{\hat{\phi}(\delta \|t\|)} |\hat{f}(t)|^2 \, dt.$$

This can be bounded above uniformly by a fixed constant positive multiple of

$$(6.6) \qquad \|f\|_2^2 + \delta^{n+2+2\varepsilon} \int_{\mathbb{R}^n} \|t\|^{n+2+2\varepsilon} |\hat{f}(t)|^2 \, dt.$$

The first norm in (6.6) denotes the standard Euclidean function norm, as we recall. We note in passing that this estimate implies that our reproducing kernel Hilbert space X is a superset of

$$L^2(\mathbb{R}^n) \cap D^{-n/2-1-\varepsilon} L^2(\mathbb{R}^n).$$

In summary, this gives the error estimate

$$\|f - s\|_{\infty,\Omega} \leq C(h/\delta)^{1+\varepsilon} (\delta^{-n/2} + \delta^{1+\varepsilon}).$$

This provides the required result. $\qquad\qquad\qquad\qquad\qquad\qquad\qquad\qquad\square$

The same convergence analysis can be carried through for the radial basis functions of compact support of Wendland. One available result from Wendland (1998) is the following.

Theorem 6.8. *Let ϕ be the unique radial basis function of the form (6.1) as stated in Proposition 6.3 for the integer k at least one. Let $s = n/2 + k + 1/2$ and Ξ be a finite set in a domain Ω with the fill-distance h as defined in Theorem 5.5. Let, finally, s be the interpolant to $f \in D^{-s} L^2(\mathbb{R}^n)$ of the standard form*

$$s(x) = \sum_{\xi \in \Xi} \lambda_\xi \phi(\|x - \xi\|), \qquad x \in \mathbb{R}^n,$$

with the interpolation conditions $(s - f)|_\Xi = 0$. Then the uniform convergence estimate

$$\|f - s\|_{\infty,\Omega} \leq C h^{k+1/2}$$

holds for $h \to 0$ and a fixed h-independent, but f-dependent, positive constant C.

We note that here the number of centres in each of the supports of the shifted radial basis functions grows.

6.5 A unified class

As always in mathematics, we aim to unify results and cast separate classes of functions into one class wherever that is possible. It turns out that it is for the class of Wendland's radial basis functions of compact support and those of the subsection before last. Indeed, it is remarkable, then, that Wendland's functions may also be interpreted as certain special cases of our functions of Section 6.3. This can be demonstrated as follows.

To wit, we note that if we apply the differentiation operator which is inverse to $\mathcal{I}$ from Section 6.2,

$$\mathcal{D}f(r) = -\frac{1}{r}f'(r), \qquad r \geq 0,$$

to our radial basis functions of Theorem 6.6 (see also Castell, 2000), it gives for the choice $\mu = \frac{1}{2}$ and any integral λ

$$\mathcal{D}^\lambda \phi(r) = \lambda! 2^\lambda \int_{r^2}^1 \beta^{\varepsilon-\lambda}(1 - \sqrt{\beta})^\nu \, d\beta, \qquad 0 \leq r \leq 1,$$

the unscaled radial basis functions being always zero outside the unit interval in this chapter. Therefore, if we apply the differentiation operator once again and evaluate the result explicitly, we get for $\mathcal{D}^{\lambda+1}\phi(r)$

$$\mathcal{D}^{\lambda+1}\phi(r) = \lambda! 2^{\lambda+1} r^{2\varepsilon-2\lambda}(1-r)_+^\nu = \lambda! 2^{\lambda+1}(1-r)_+^\nu, \qquad 0 \leq r \leq 1,$$

for $\varepsilon = \lambda$. Now, in order to get Wendland's functions, we let $\lambda = k - 1$, $\nu = [\frac{1}{2}n] + k + 1$ and recall that, on the other hand, we know that the radial basis functions $\phi_{n,k}$ of Wendland are such that

$$\mathcal{D}^k \phi_{n,k}(r) = (1-r)_+^{[\frac{1}{2}n]+k+1}, \qquad 0 \leq r \leq 1,$$

because $\mathcal{D}$ is the inverse operator to $\mathcal{I}$. By comparison of the two preceding displays, we see that indeed Wendland's functions may be written in the general form studied in the section before last, except that there are certain multiplicative constants which are irrelevant. We point out, however, that the proofs of Section 6.3 about the positivity of the Fourier transforms $\hat{\phi}$ require ranges of the λ and ν which do not contain the choices we need for Wendland's functions above. Therefore his proofs of positive definiteness of those functions are still needed.

7
Implementations

One of the most important themes of this book is the implementation of radial basis function (interpolation) methods. Therefore, after four chapters on the theory of radial basis functions which we have investigated so far, we now turn to some more practical aspects. Concretely, in this chapter, we will focus on the *numerical solution* of the interpolation problems we considered here, i.e. the computation of the interpolation coefficients. In practice, interpolation methods such as radial basis functions are often required for approximations with very large numbers of data sites ξ, and this is where the numerical solution of the resulting linear systems becomes nontrivial in the face of rounding and other errors. Moreover, storage can also become a significant problem if $|\Xi|$ is very large, even with the most modern workstations which often have gigabytes of main memory.

Several researchers have reported that the method provides high quality solutions to the scattered data interpolation problem. The adoption of the method in wider applications, e.g. in engineering and finance, where the number of data points is large, was hindered by the high computational cost, however, that is associated with the numerical solution of the interpolation equations and the evaluation of the resulting approximant.

In the case where we intend to employ our methods in a collocation scheme in two or more dimensions in order to solve a differential or integral equation numerically with a fine discretisation, for instance, it is not at all unusual to be faced with 10^5 or 10^6 data sites where collocation is required. The large numbers of centres result here from the number of elements of square or cubic discretisation meshes or other higher-dimensional structures.

7.1 Introduction

For most of the radial basis functions which we have encountered up to now, solving the interpolation equations

$$\mathbf{A}\boldsymbol{\lambda} = \mathbf{f},$$

where $\mathbf{f} = \{f_\xi\}_{\xi\in\Xi}$, $\boldsymbol{\lambda} = \{\lambda_\xi\}_{\xi\in\Xi}$, as usual $\mathbf{A} = \{\phi(\|\xi - \zeta\|)\}_{\xi,\zeta\in\Xi}$, with direct methods, such as Gauss-elimination, requires as many as $O(|\Xi|^3)$ operations and is therefore computationally prohibitively expensive for the above mentioned sizes of the Ξ. (This is so because there is no sparsity in the interpolation matrix $\mathbf{A}$ unless the radial basis functions are of compact support as in the previous chapter, when $\mathbf{A}$ is a band-matrix, with a certain amount of sparsity. This helps to ease the numerical burden, but in most cases is not sufficient due to large band-widths.) If $\mathbf{A}$ is positive definite, a Cholesky method may be applied which requires less computational effort but even this reduced work will be too much if there are 5000 points, say, or more. Incidentally, if $\mathbf{A}$ stems from a *conditionally* positive definite radial basis function, it can be preconditioned to become positive (semi) definite. This will be relevant in the last section of this chapter.

In summary, it is a standard, useful approach to apply iterative methods for solving the resulting large linear system instead of direct algorithms. This is especially suitable when the methods are used within other iterative schemes, for instance for the numerical solution of partial differential equations, so approximate solutions of the interpolation problem are acceptable if their accuracy is within a specified range, according to the accuracy of the method inside which the radial basis function approximations are used.

In this chapter, we will describe three efficient such types of approach to solving the linear systems by iteration. The approaches are responsible both for the fast solution of the aforementioned linear systems of equations and for the fast evaluation of the approximants. The two issues are closely related.

The first approach uses the fact that the radial basis function interpolants for, say, thin-plate splines (but for instance also multiquadrics), are, in spite of their global support, in fact acting locally. This is so because there exist, as we have seen especially in the case when the data are located on a square integer grid, quickly decaying Lagrange functions in the radial basis function spaces for interpolation, or other decaying functions for quasi-interpolation. Thus we are entitled to expect that we may in some sense *localise* the computations of the interpolant's coefficients $\boldsymbol{\lambda}$ as well. This motivates the basic idea of the first approach, which we shall call after its inventors the BFGP

(Beatson–Faul–Goodsell–Powell) method and of which there is an alternative implementation in the shape of a Krylov subspace method; it is as follows.

We *assume* that the coefficients of the required interpolant depend almost exclusively on the nearby scattered data sites, then decompose and solve the problem accordingly and derive an iterative method that reduces the residuals quickly in each step. The type of approach that is employed here is a *domain decomposition* approach, in the language of algorithms for the computation of numerical solutions of differential equations. Indeed, it subdivides the spatial domain of the problem containing the full data set Ξ into small subdomains with few centres (small subsets of Ξ) inside each, where the problem can be solved more easily because it is much smaller. The idea differs from the usual domain decomposition algorithms, however, in that the different subdomains in which the centres we use reside are not disjoint – or even 'essentially disjoint', i.e. overlapping by small amounts.

In this, as well as in the next section, we focus on the important special case of thin-plate spline interpolation. We note, however, that the methods can, in principle, be applied to all radial basis functions (4.4) and their shifts such as multiquadrics, shifted logarithms $\log(r^2 + c^2)$ for positive parameters c, etc.

A second route we pursue is a so-called multipole method, an approach which has been fostered in the form that is suitable for radial basis functions mainly by Powell, Beatson, Light and Newsam. Originally, however, the basic idea is due to Greengard and Rokhlin (1987) who used the multipole method to solve integral equations numerically when they have globally supported kernels that lead to large, densely populated discretisation matrices. Thereby they want to reduce the computational complexity of particle simulations. The problems which come up when numerical solutions of the linear systems are computed with those matrices are the same as the ones we face with our collocation matrices, namely the absence of sparsity, large condition numbers etc. The multipole method we describe in the third section relies on a decomposition (and thus structuring) of the data in a similar way to that in the BFGP algorithm, but it does not use the locality of the approximants or any Lagrange functions as such. Moreover, the sets into which the centres Ξ are split, are usually disjoint and in a grid-like structure.

Instead of the Lagrange function's locality, the fast multipole method uses the fact that the radial basis functions like thin-plate splines are, except at the origin, expandable in infinite Laurent series. For numerical implementation, these series are truncated after finitely many terms, the number of retained terms being determined by the accuracy which we wish to achieve. The idea is then to group data sites within certain local domains together and approximate all thin-plate spline terms related to those centres as a whole, by a single short asymptotic expansion. This reduces the computational cost substantially, because the thin-plate spline terms related to each centre need no longer be computed one by one. This is particularly pertinent in the thin-plate spline case, for example, because of the logarithmic term that makes this one-by-one evaluation prohibitively expensive. The same is true for taking square roots when multiquadrics are in use.

This *ansatz* can be used in the manner outlined above whenever we wish to evaluate the approximant s at an x which is far away from the aforementioned group of centres. Otherwise, a direct evaluation is applied at a relatively small cost, because it is only needed for the few nearby centres, few meaning typically about 30, in implementations.

Bringing these ideas together, we are availing ourselves of a scheme that admits fast evaluation of large sums of thin-plate splines in one operation per group of centres. In addition we recall that algorithms such as conjugate gradient methods or, more generally, Krylov space methods can be applied to solve $\mathbf{A}\boldsymbol{\lambda} = \mathbf{f}$ efficiently, if we can evaluate matrix by vector multiplications fast; the latter is equivalent to evaluating the approximant

$$s(\zeta) = \sum_{\xi \in \Xi} \lambda_\xi \phi(\|\zeta - \xi\|)$$

at vectors of points ζ, because $\{s(\zeta)\}_{\zeta \in \Xi} = \mathbf{A}\boldsymbol{\lambda}$. The details are described below.

In fact, as we have indicated at the beginning of this section, even if the computation of the interpolation coefficients is not the issue, efficient and numerous evaluation of the approximants may very well be. For instance, it is the case when the interpolant, using only *a few* centres ξ, has to be *evaluated* very often on a fine multidimensional lattice. This may be required for comparing two approximants on that lattice or for preprocessing data that will later be approximated by another method which is restricted to a square grid. Moreover, this may be needed for displaying a surface on a computer screen at high resolution. In fact, even the BFGP and Krylov space methods require inside the algorithm especially fast multiplications of interpolation matrix times coefficient vector for computation of the residuals, thus fast multipole methods are also used in conjunction with the first algorithm we present in this chapter. This is particularly relevant if we wish to use the BFGP method in more than two dimensions (although at present implementations are mostly used only up to $n = 4$ – Cherrie, Beatson and Newsam, 2000).

The third approach we offer is that of directly preconditioning the matrix $\mathbf{A}$ with a preferably simple preconditioning matrix $\mathbf{P}$, which is quite a standard idea. Standard iterative methods such as a Gauss–Seidel or conjugate gradient method can then be applied to $\mathbf{A}$. However, it is normal that ill-conditioning occurs, sometimes severely so, as we have learned in Chapters 4 and 5, and the speed of convergence of the conjugate gradient method depends very sensitively on the condition number. Moreover, often the symmetric matrix $\mathbf{A}$ is itself positive definite not on the whole of $\mathbb{R}^\Xi$ but only on a subspace, a hyperplane for instance. So the radial basis function is *conditionally* positive definite, and indeed a preprocessing is required to move the spectrum of $\mathbf{A}$ to the positive real half-line anyhow. As such it is unsuitable to apply standard methods to solve our linear system without extra work. In many cases, however, we have enough information about the properties of ϕ to find suitable preconditioning matrices for $\mathbf{A}$. For instance, as we have learned in this book already, in many cases $\phi(\| \cdot \|)$ is the fundamental solution of an elliptic, radially invariant differential operator, such as the Laplace operator or the bi-harmonic operator, i.e. that operator applied to $\phi(\| \cdot \|)$ gives a multiple the Dirac δ-function. Therefore we may approximate those operators by discretisation and obtain decaying linear combinations of shifts of $\phi(\| \cdot \|)$, much like the linear combinations that occurred in our discussion of quasi-interpolation in the fourth chapter. Dyn, Levin and Rippa were the first to attempt such schemes successfully, both on square grids and later on scattered data sites. They gave guidance to preconditioning thin-plate spline and multiquadric interpolation matrices. Explicit expressions for the preconditioning coefficients are presented that constitute scattered data discretisations of bi-harmonic differential operators.

In fact, in many cases finding a suitable preconditioner is the same as finding coefficients of quasi-interpolating basis functions ψ that are finite linear combinations of translates of our radial basis functions. This is not surprising, as in both cases we wish to get a localisation of our globally acting radial basis functions although localisation *per se* does not automatically mean a better conditioning of the interpolation matrix $\mathbf{A}$ unless we achieve diagonal dominance.

In the simplest cases, however, both the coefficients of the ψ and the entries of a preconditioning matrix are coefficients of a symmetric difference scheme and lead to better condition numbers of $\mathbf{A}$ and better localisation simultaneously. This is straightforward to establish in the case of a grid of data sites (as is quasi-interpolation in this setting) but it may be less suitable if the data are unstructured. Again, the details are given below.

7.2 The BFGP algorithm and the new Krylov method

This algorithm is, as already outlined above, based on the observation that there are local, sufficiently quickly decaying cardinal functions for radial basis function interpolants for ϕ from a class of radial basis functions. This class contains all radial basis functions where decaying cardinal functions for gridded data were identified in Chapter 4 and, in particular, all ϕ of the form (4.4) and their shifts, such as thin-plate splines or multiquadrics, are included. The data Ξ we use now are allowed to be arbitrarily distributed in Euclidean space $\mathbb{R}^n$ and the algorithm computes an interpolant at those data sites to given function values, as we are familiar with in this book. Of course, the arbitrariness of the distribution must be slightly restricted by the condition that Ξ contains a K-unisolvent subset.

7.2.1 Data structures and local Lagrange functions

The first requirement is to structure the data into groups of about $q = 30$ sites at a time which are relatively close together. The number 30 (sometimes up to 50) is used successfully for q in current thin-plate spline implementations for two or three dimensions. Further, we assume that there is a number q^* of data with which we can solve the interpolation system directly and efficiently; this is certainly the case when q^* is also of the order of 30 points for instance but it may be of the order of a few hundred. Often, one chooses $q = q^*$. In the simplest case, we can use a standard elimination procedure for solving the linear system such as Gauss-elimination or Cholesky factorisation if the matrix is positive definite (see our remarks at the beginning of the previous section, and the final section of this chapter).

Since the following method is iterative, it is helpful now to enumerate the finite number of data (which are still assumed to be distinct) as $\Xi = \{\xi_1, \xi_2, \ldots, \xi_m\}$.

There are various useful ways to distribute the centres in Ξ into several nondisjoint smaller sets. One is as follows. For each $k = 1, 2, \ldots, m - q^*$, we let the set $\mathcal{L}_k$ consist of ξ_k and those q different points among ξ_ℓ, where $\ell = k+1, k+2, \ldots$, which minimise the Euclidean distances $\|\xi_k - \xi_\ell\|$ among all $\xi_{k+1}, \xi_{k+2}, \ldots, \xi_m$. If there are ties in this minimisation procedure, they are broken by random choice. The set $\mathcal{L}_{m-q^*+1}$ contains the remaining q^* points, that is we let

$$\mathcal{L}_{m-q^*+1} = \{\xi_{m-q^*+1}, \xi_{m-q^*+2}, \ldots, \xi_m\}.$$

As an additional side-condition we also require each of the sets of points to contain a K-unisolvent subset in the sense of Section 5.2, according to which radial basis function is in use, K being the kernel of the semi-inner product associated with the radial basis function ϕ. For instance, for two dimensions and the thin-plate splines, K is the space of linear polynomials and we require an additional linear polynomial. This may alter our above strategy slightly. That this can be done, however, for each of our sets of nearby points is only a small restriction on the Ξ which are admitted. For higher-dimensional problems we may have to increase the proposed number $q = 30$ points for each of the local sets $\mathcal{L}_k$ when K has dimension more than 30, but for example for thin-plate splines in two dimensions, $q = 30$ will be sufficient.

One suitable selection is to choose the last dim K points to be unisolvent, or renumber the points in Ξ accordingly. Then they are included in *all* $\mathcal{L}_k$. Three points like that are sufficient when thin-plate splines are used in two dimensions.

Next, we introduce the central concept of *local Lagrange functions* L_k^{loc} associated with $\mathcal{L}_k$. They are linear combinations

$$(7.1) \qquad L_k^{\mathrm{loc}}(x) = \sum_{\xi \in \mathcal{L}_k} \lambda_{k\xi} \phi(\|x - \xi\|) + p_k(x), \qquad x \in \mathbb{R}^n,$$

with $p_k \in K$ and with the usual side-conditions on the coefficients not stated again here, which satisfy the partial cardinality conditions within the $\mathcal{L}_k$

$$(7.2) \qquad\qquad L_k^{\mathrm{loc}}(\xi_k) = 1, \qquad L_k^{\mathrm{loc}}\,|_{\mathcal{L}_k \setminus \{\xi_k\}} = 0.$$

Because the cardinality $|\mathcal{L}_k|$ is relatively small, each of the sets of coefficients $\lambda_{k\xi}$ can be computed directly and efficiently by a standard method. This is done in advance, before the iterations begin, and this is always the way we proceed in the practical implementation of the algorithm.

The basic idea is now to approximate the interpolant at each step of the main iteration by a linear combination of such local Lagrange functions, in lieu of the true Lagrange functions which would have to satisfy the full Lagrange conditions on the whole set Ξ of centres for the given data and which would, of course, lead to the correct interpolant without any iteration.

Using these approximate Lagrange functions during the iteration, we remain in the correct linear space $\mathcal{U}$ which is, by definition, spanned by the translates $\phi(\| \cdot -\xi\|), \xi \in \Xi$, and contains additionally the semi-norm kernel K. We note that the space $\mathcal{U}$ contains in particular the sought interpolant s^*. Because, in general, it is true that the local Lagrange functions satisfy

$$L_k^{\mathrm{loc}}\,|_{\Xi \setminus \mathcal{L}_k} \neq 0,$$

this produces only a fairly rough approximation to the Lagrange functions. The reduction of the error has to be achieved iteratively by successive correction of the residuals. Therefore, the corrections of the error come from recursive application of the algorithm, where at each iteration the *same* approximate Lagrange functions are employed. We note, however, the remarkable fact that it is not even required that $|L_k^{\mathrm{loc}}|$ restricted to $\Xi \setminus \mathcal{L}_k$ be small. Instead, as Faul (2001) shows in a re-interpretation of the algorithm we shall present below, the semi-inner products $(\tilde{L}_j^{\mathrm{loc}}, \tilde{L}_k^{\mathrm{loc}})_*$ should be small for $j \neq k$ for efficiency of the algorithm. Using Faul's approach we may view the algorithm as a Jacobi iteration performed on a positive definite Gram-matrix with such entries which is expected to converge faster, of course, if the off-diagonal elements are small. Here, $\tilde{L}_j^{\text{loc}}$ are the local Lagrange functions normalised to have semi-norm one, that is

$$\tilde{L}_j^{\text{loc}} = \frac{L_j^{\text{loc}}}{\|L_j^{\text{loc}}\|_*}.$$

We recall that such a Gram-matrix is always nonnegative definite, and if the generating functions are linearly independent and not in the kernel of the semi-inner product, it is nonsingular. This is the case here, as the local Lagrange functions $\tilde{L}_j^{\text{loc}}$, $j = 1, 2, \ldots$, are never in K unless there are as many points as or fewer than dim K, and they are linearly independent by the linear independence of the translates of the radial basis functions and the cardinality conditions.

We remark already at this point that the algorithm we describe in this section will be an $O(m \log m)$ process. This is because we require approximately m of the local Lagrange functions, each of which requires a small, *fixed* number of operations to compute the coefficients – this number is small and independent of m – and because the evaluation of the residuals which is needed at each iteration normally requires $O(m \log m)$ computer operations with a suitable scheme being used for large m, such as the one related to the particle methods presented in the next section. In fact, those methods require only $O(\log m)$ operations for the needed function evaluation but the set-up cost for the Laurent expansions uses up to $O(m \log m)$ operations.

7.2.2 Description of the algorithm

Throughout this section we let s denote the current numerical approximation to the exact required radial basis function interpolant s^*; this s is always in the set $\mathcal{U}$.

To begin with, we let $j = 1$, $s_1 = s = 0$. Each step of the algorithm within one sweep replaces $s_j = s$ by $s + \eta_k$, $k = 1, 2, \ldots, m - q^*$, η_k being a suitable correction that depends on the current residuals $(s - f)|_\Xi$ and that is defined below. We then finish each full iteration (also called a 'sweep') of the algorithm by replacing the sum

$$\tilde{s} = s + \eta_1 + \eta_2 + \cdots + \eta_{m-q^*}$$

by $s_{j+1} = \tilde{s} + \tilde{\eta}$. Here $\tilde{\eta}$ is the radial basis function interpolant to the residual $f - \tilde{s}$ at the final q^* points that may be computed either with a direct method such as Cholesky or preconditioned Cholesky or by applying perhaps one of the alternative preconditioning methods that will be described in detail in Section 7.4. Thus one full 'sweep' of the algorithm which replaces s_j by s_{j+1}

will be completed and constitutes a full iteration. We are now in a position to present the details of the algorithm *within* each iteration step $j = 1, 2, \ldots$.

The interpolation conditions we demand for the final, additional term $\tilde{\eta}$ are

$$(7.3) \qquad \tilde{\eta}(\xi) = f_\xi - \tilde{s}(\xi), \qquad \xi \in \mathcal{L}_{m-q^*+1}.$$

This final correction term $\tilde{\eta}$ is derived by standard radial basis function interpolation at q^* points, and this part of the scheme remains identical in all variants of the BFGP method. However, we now turn to the key question of how the other η_k are found so that they provide the desired iterative correction to the residuals.

The simplest way is to take for each index k the update

$$(7.4) \qquad \eta_k(x) = \left(\sum_{\xi \in \mathcal{L}_k} \left(f_\xi - s(\xi) \right) \lambda_{k\xi} \right) \phi(\|x - \xi_k\|).$$

This is the shift of the radial basis function $\phi(\|x - \xi_k\|)$ times that coefficient which results from the definition of the local Lagrange functions when η_k is being viewed as part of a cardinal interpolant to the residuals on $\mathcal{L}_k$. The residuals that appear in (7.3)–(7.4) can be computed highly efficiently through the method presented in Section 7.3. We remind the reader that all coefficients $\lambda_{k\xi}$ are worked out before the onset of the iteration and therefore need no further attention.

In order to understand the method better, it is useful to recall that *if* the $\mathcal{L}_k$ were all Ξ, i.e. $q = |\Xi|$, and we set, for simplicity, $q^* = 0$, and we also use, for illustration, $K = \{0\}$, then (7.4) would lead immediately to the usual interpolant on all of the data, because the coefficients λ would be the ordinary cardinal functions' coefficients. Thus from (7.2) and from the facts that

$$\sum_{\xi \in \Xi} s(\xi) L_\xi(x) = s(x),$$

$$\sum_{\xi \in \Xi} f_\xi L_\xi(x) = s^*(x),$$

recalling that the L_ξ are the full Lagrange functions, we get the following expression in a single step of the algorithm, all sums being finite:

$$s(x) + \sum_{k=1}^{m} \sum_{\xi \in \Xi} \left(f_\xi - s(\xi) \right) \lambda_{k\xi} \phi(\|x - \xi_k\|)$$

$$= s(x) + \sum_{\xi \in \Xi} \left(f_\xi - s(\xi) \right) \sum_{k=1}^{m} \lambda_{k\xi} \phi(\|x - \xi_k\|)$$

$$= s(x) + \sum_{\xi \in \Xi} \left(f_\xi - s(\xi) \right) L_\xi(x) = s^*(x),$$

where we have used that $\lambda_{k\xi_j} = \lambda_{j\xi_k}$ when $\mathcal{L}_k = \Xi$ for all k. This is because the interpolation matrix $\mathbf{A} = \{\phi(\|\zeta - \xi\|)\}_{\zeta,\xi \in \Xi}$ is symmetric and so is its inverse which contains the coefficients $\lambda_{k\xi_j}$ of the full cardinal functions as entries.

This computational algorithm, as it is described, uses only an approximation to the cardinal functions and therefore only an approximate correction to the residual can be expected at each step, the iterations proceeding until convergence. Under what circumstances convergence

$$|s_j(\xi) - f_\xi| \to 0$$

for $j \to \infty$ can be guaranteed is explained further below.

At any rate, it turns out to be computationally advantageous to update at each stage with (7.5) below instead of (7.4), that is the modified expression for the following second variant of the BFGP algorithm:

$$(7.5) \qquad \eta_k(x) = \frac{1}{\lambda_{k\xi_k}} L_k^{\text{loc}}(x) \sum_{\xi \in \mathcal{L}_k} \left(f_\xi - s(\xi)\right)\lambda_{k\xi}.$$

We note, however, that (7.5) of course contains the term (7.4). It is important to observe that the coefficient $\lambda_{k\xi_k}$ by which we divide is positive because by the analysis of Section 5.2, $\lambda_{k\xi_k} = (L_k^{\text{loc}}, L_k^{\text{loc}})_* > 0$ in the semi-inner product introduced in that section.

Indeed, since the Lagrange conditions $L_k^{\text{loc}}(\xi_j) = \delta_{jk}$ hold for $\xi_j \in \mathcal{L}_k$ and L_k^{loc} has the form (7.1), we have

$$(7.6) \qquad 0 < \|L_k^{\text{loc}}\|_*^2 = (L_k^{\text{loc}}, L_k^{\text{loc}})_*$$

$$= \left(\sum_{\xi \in \mathcal{L}_k} \lambda_{k\xi}\phi(\| \cdot -\xi\|), \sum_{\zeta \in \mathcal{L}_k} \lambda_{k\zeta}\phi(\| \cdot -\zeta\|)\right)_*,$$

where the inequality follows from $L_k^{\text{loc}} \notin K$ since K's dimension is much lower than that of $\mathcal{U}$ if trivial cases are excluded. The far right-hand side of the last display is the same as

$$\sum_{\xi \in \mathcal{L}_k} \sum_{\zeta \in \mathcal{L}_k} \lambda_{k\xi}\lambda_{k\zeta}\phi(\|\zeta - \xi\|)$$

$$= \sum_{\xi \in \mathcal{L}_k} \lambda_{k\xi}\left(\sum_{\zeta \in \mathcal{L}_k} \lambda_{k\zeta}\phi(\|\zeta - \xi\|) + p_k(\xi)\right) = \lambda_{k\xi_k},$$

according to the analysis in Section 5.2 and in particular the reproducing kernel property of ϕ. We have also employed the fact that $p \in K$ for any polynomial that is added to the sum in the last but one display and used the side-conditions on the coefficients. It is useful to recall a consequence from the work in the fifth chapter; namely that the following lemma holds.

Lemma 7.1. *Let* Ξ *contain a* K-*unisolvent set and let* ϕ, $\mathcal{U}$ *and the semi-inner product* $(\,\cdot\,,\,\cdot\,)_*$ *be as above. Then any scalar product* $(t_1, t_2)_*$ *satisfies*

$$(t_1, t_2)_* = \sum_{\xi \in \Xi} \hat{\lambda}_\xi t_2(\xi),$$

as long as

$$t_1(x) = \sum_{\xi \in \Xi} \hat{\lambda}_\xi \phi(\|x - \xi\|) + q(x),$$

with the usual side-conditions on the coefficients, $q \in K$, *and* t_2 *also in* $\mathcal{U}$.

The same arguments and Lemma 7.1 imply that we may simplify and cast the sum of $\eta_1, \eta_2, \ldots, \eta_{m-q^*}$ into one operator by using the convenient form

$$\eta: g \mapsto \sum_{k=1}^{m-q^*} \frac{L_k^{\mathrm{loc}}}{\lambda_{k\xi_k}} (L_k^{\mathrm{loc}}, g)_*,$$

so

$$f - s \mapsto \eta(f - s) = \sum_{k=1}^{m-q^*} \frac{L_k^{\mathrm{loc}}}{\lambda_{k\xi_k}} (L_k^{\mathrm{loc}}, f - s)_*$$

$$= \sum_{k=1}^{m-q^*} \frac{L_k^{\mathrm{loc}}}{\lambda_{k\xi_k}} (L_k^{\mathrm{loc}}, s^* - s)_*,$$

using $(L_k^{\mathrm{loc}}, s^*)_* = (L_k^{\mathrm{loc}}, f)_*$ by the interpolation conditions and by our above Lemma 7.1.

Hence it follows that s_j is replaced by $s_{j+1} = s_j + \eta(s^* - s_j) + \tilde{\eta}$ for each sweep of the algorithm. Such an operator we will also use later, in the subsection 7.2.4 in another algorithm.

Powell (1997), who has found and implemented most of what we explain here, presents, among others, the computational results shown in Table 7.1 using the update technique demonstrated above. The method was tested on four different test problems A–D. The centres are scattered in two different problems A and B. In the third problem C, track data situated on two straight line segments, the endpoints of which are chosen randomly, are used. The fourth problem D is the same as C, except that 10 tracks are used instead of 2. In each case, the tracks must have length at least one. The centres are always in the unit square. The thin-plate-spline approximant is used in each case and the accuracy required is 10^{-10}. The iteration counts are within the stated ranges as shown in Table 7.1. For the algorithm, the updates (7.5) are used.

It is important to note that in calculating the residuum at each sweep j of the single iteration, the same s is used, namely the *old* approximation to our

Table 7.1

Centres	q	Problem A	Problem B	Problem C	Problem D
500	10	31–41	45–51	9–44	31–44
500	30	6–7	7	4–8	7–13
500	50	5	5–6	4–6	5–8
1000	10	43–52	54–67	16–68	39–57
1000	30	7	7–11	5–9	8–13
1000	50	5–6	5–8	4–6	6–8

required interpolant. The algorithm was found to be actually more robust if for each k the already updated approximant is used, i.e. the η_{k-1} is immediately added to s before the latter is used in η_k. This slight modification will be used in our convergence proof below because it leads to guaranteed convergence. Other methods to enforce convergence are introduction of line searches (see Faul, 2001) or the Krylov subspace method described below. The version (7.5) above, however, gives a faster algorithm and converges almost as fast as the more 'expensive' one with continuous updates at each stage. The saving in the operations count which we make with this faster version is by a factor of q. Incidentally, hybrid methods have also been tested where, for example, the old s is used for the first $[k/2]$ stages and then an update is made before continuing.

As an aside we mention that an improvement of the convergence speed of the above method in a multigrid way (Beatson, Goodsell and Powell, 1995) may be obtained by finding a nested sequence of subsets

$$\Xi_{j+1} \subset \Xi_j, \quad j = 1, 2, \ldots, \ell - 1, \qquad \Xi_1 = \Xi, \quad |\Xi_\ell| > q,$$

that is such that each Ξ_{j+1} contains about half the number of elements that Ξ_j does. Moreover, for each j, a collection of subsets $\{\mathcal{L}_k^{(j)}\}_{k=1}^{|\Xi_j|-q^*}$ and the associated local Lagrange functions are found in the same way as described before, where each $\mathcal{L}_k^{(j)} \subset \Xi_j$ contains approximately the same number of elements, still q say. There is also a set $\mathcal{L}_{|\Xi_j|-q^*+1}^{(j)}$ containing the remaining points. Then the algorithm is applied exactly in the same fashion as before, except that at each stage η_k and $\tilde{\eta}$ depend on Ξ_j. Thus we have a double iteration here: in computing the η_k we have a loop over k and then we have a loop over j for the Ξ_j. Finally this algorithm can be repeated up to convergence, convergence meaning, as always in this section and in the next, that the maximum of the residuals $s(\xi) - f_\xi$ goes to zero (in practice: is smaller in modulus than a prescribed tolerance). This strategy can lead to improved convergence if the residuals contain smoother components. The smoother components may be removed using fewer centres than the full set Ξ, while the initial updating with the full set removes any very high frequency components from the error. This works like a fine to coarse sweep of the well-known multigrid algorithm. For details see Powell (1996 and 1997). We leave this aside now and turn to the question of convergence of the algorithm in the following subsection.

7.2.3 Convergence

For the sake of demonstration, we shall prove convergence of one of the versions of the BFGP method in this subsection for all radial basis functions of the form (4.4). In some cases each full iteration of the scheme e.g. for thin-plate splines in two dimensions leads in implementations to a reduction of the residual by a factor of as much as about 10. Therefore the numerical results provide

$$\max_{\xi} |f_\xi - s_{j+1}(\xi)| \approx 0.1 \max_{\xi} |f_\xi - s_j(\xi)|,$$

e.g. Powell (1994a).

The convergence proof is as follows. It uses the implementation of the method we have presented where, in the individual updates η_k, each time the most recent s, i.e. the one already updated by η_{k-1}, is used. We have pointed out already that this makes the method more robust in implementations, although more expensive, and indeed we establish its guaranteed convergence now. In this description, we recall the notation s for the current approximant to the 'true' interpolant s^*. Due to the immediate update of the residuals we need to introduce a second index k for each stage within each sweep j.

The first thing we note is the fact that the semi-norm induced by the semi-scalar product of Chapter 5 $\|s - s^*\|_*$ is always decreasing through each update of the algorithm. Indeed, at each stage of the algorithm except the last one, where an explicit solution of a small system takes place to compute $\tilde{\eta}$, $\|s - s^*\|_*^2$ is replaced by

$$\|s^* - s_{j,k-1} - \vartheta L_k^{\mathrm{loc}}\|_*^2$$

where the factor ϑ is defined through (7.5) as

(7.7) $$\vartheta = \frac{1}{\lambda_{k\xi_k}} \sum_{\xi \in \mathcal{L}_k} \left(f_\xi - s_{j,k-1}(\xi)\right) \lambda_{k\xi}.$$

The s is now doubly indexed; as above, the index j denotes the full sweep number while the index k is the iteration number within each sweep. Note that if ϑ vanishes, then the algorithm terminates. We see that here the updates are performed immediately within each sweep. It is elementary to show that the quadratic form in the last but one display is least for exactly that value (7.7) of ϑ, because this choice gives the minimum by the quotient

$$\vartheta = \frac{(s^* - s_{j,k-1}, L_k^{\mathrm{loc}})_*}{\|L_k^{\mathrm{loc}}\|_*^2}$$

due to $\|L_k^{\mathrm{loc}}\|_*^2 = \lambda_{k\xi_k} > 0$, as we have noticed already, and due to the identity

$$
\begin{aligned}
(s^* - s_{j,k-1}, L_k^{\mathrm{loc}})_* &= \sum_{\xi \in \mathcal{L}_k} \lambda_{k\xi}\left(s^*(\xi) - s_{j,k-1}(\xi)\right) \\
&= \sum_{\xi \in \mathcal{L}_k} \lambda_{k\xi}\left(f_\xi - s_{j,k-1}(\xi)\right)
\end{aligned}
$$

which follows from Lemma 7.1.

In order to complete the proof of the reduction of the semi-norm, we need only to look at the last stage of the algorithm now. We recall that the final stage of the iteration adds to $\tilde{s}$ a solution $\tilde{\eta}$ of a small linear system to obtain the new s. With this in mind, it is sufficient to prove that the following orthogonality condition holds (for this standard fact about best least squares approximation cf., e.g., Section 8.1):

$$(s^* - \tilde{s} - \tilde{\eta}, \tau^*)_* = 0$$

for all linear combinations

(7.8)
$$\tau^*(x) = \sum_{\xi \in \mathcal{L}_{m-q^*+1}} \tau_\xi \phi(\|x - \xi\|) + p(x),$$

$p \in K$ being a polynomial as required by the theory of Section 5.2, linear in the paradigmatic thin-plate spline case. We call the set of all such sums T^*. Under these circumstances it then follows that $\tilde{\eta}$ out of all functions from T^* minimises the semi-norm $\|s^* - \tilde{s} - \tilde{\eta}\|_*$:

$$\tilde{\eta} = \mathrm{argmin}_{\tau^* \in T^*} \|s^* - \tilde{s} - \tau^*\|_*,$$

So $\tilde{\eta}$ is the argument where the minimum is attained. Now, the orthogonality claim is true as a consequence of Lemma 7.1 and

$$(s^* - \tilde{s} - \tilde{\eta}, \tau^*)_* = \sum_{\xi \in \mathcal{L}_{m-q^*+1}} \tau_\xi\left(s^*(\xi) - \tilde{s}(\xi) - \tilde{\eta}(\xi)\right),$$

the τ_ξ being the above coefficients of the shifts of the radial basis function in τ^* in (7.8). The right-hand side of the above display vanishes because all the terms in brackets are zero due to the interpolation conditions stated in (7.3).

In summary, $\|s_{j-1,k-1} - s^*\|_*^2$ is always decreasing through each sweep of the algorithm. Therefore, this quadratic form converges to a limit, being bounded from below by zero anyway. Moreover, it follows that specifically

$$(s^* - s_{j,k-1}, L_k^{\text{loc}}) \to 0, \qquad j \to \infty, \ \forall k \le m - q^*,$$

from

$$\|s^* - s_{j,k-1} - \vartheta L_k^{\text{loc}}\|_*^2 = \|s^* - s_{j,k-1}\|_*^2 - \frac{(s^* - s_{j,k-1}, L_k^{\text{loc}})_*^2}{\|L_k^{\text{loc}}\|_*^2},$$

because of the positivity and boundedness of the denominator in the above display on the right-hand side. Here the parameter ϑ is the same as in (7.7).

At the end of the kth iteration, the current approximation to the interpolant is $s = s_j$, and in particular $(s^* - s_j, L_1^{\text{loc}})_* \to 0$ as j increases. From this it follows that $(s^* - s_j, L_2^{\text{loc}})_* \to 0$ for $j \to \infty$ since at the beginning of the $(j + 1)$st iteration,

$$s_{j,1} = s_j + L_1^{\text{loc}}(s^* - s_j, L_1^{\text{loc}})_*/\lambda_{1,\xi_1}$$

and thus

$$\left(s^* - s_j - L_1^{\text{loc}}(s^* - s_j, L_1^{\text{loc}})_*/\lambda_{1,\xi_1}, L_2^{\text{loc}}\right)_* \to 0.$$

Alternatively, by Lemma 7.1,

$$\sum_{j=\ell}^{m} \lambda_{j\xi_\ell}\left(s^*(\xi_j) - s_k(\xi_j)\right) \to 0, \qquad k \to \infty,$$

for all $\ell \le m - q^*$. From this, it follows that we have pointwise convergence for all centres, remembering that $s_k(\xi_j) = s^*(\xi_j)$ for all $j > m - q^*$ anyhow and remembering that $\lambda_{\ell\xi_\ell}$ in the above display is always positive.

In summary, we have established the following theorem of Faul and Powell (1999a).

Theorem 7.2. *Let the BFGP algorithm specified above generate the sequence of iterations $s_0, s_1, \ldots$. Then $s_k \to s^*$ as $k \to \infty$ in the linear space $\mathcal{U}$, where s^* denotes the sought interpolant.*

The method which we have described above extends without change (except for the different semi-inner product with different kernel K) to multiquadric interpolation and there is hardly any change in the above convergence proof as well. Moreover, work is under way at present for practical implementation of cases when $n > 2$ (Cherrie, Beatson and Newsam, 2000, for example).

7.2.4 The Krylov subspace method

A more recent alternative to the implementation of the above BFGP method makes use of a Krylov subspace method by employing once again the semi-norm

$\| \cdot \|_*$ used in the previous subsection and in Section 5.2. It uses basically the same local Lagrange functions as the original BFGP method but the salient idea of the implementation has a close relationship to conjugate gradient (it is the same as conjugate gradients except for the stopping criterion and the time of the updates of the elements of K in the interpolant) and optimisation methods. There are practical gains in efficiency in this method. Also, it enjoys guaranteed convergence in contrast with the algorithm according to (7.5) which is faster than the one shown above to converge, but may itself sometimes fail to converge.

We continue to let $\mathcal{U}$ be the space of approximants spanned by the translates of the radial basis functions $\phi(\| \cdot -\xi \|)$, $\xi \in \Xi$, plus the aforementioned polynomial kernel K of the semi-inner product. Additionally $\mathcal{U}_j$ denotes the space spanned by

$$(7.9) \qquad\qquad \eta(s^*), \eta(\eta(s^*)), \ldots, \eta^j(s^*),$$

where $s^* \in \mathcal{U}$ is still the required interpolant and $\eta: \mathcal{U} \to \mathcal{U}$ is a prescribed operator whose choice is fundamental to the definition and the functioning of the method. We shall define it below. (It is closely related to the η of the subsection before last.) The main objective of any Krylov subspace method is to compute in the jth iteration $s_{j+1} \in \mathcal{U}_j$ such that s_{j+1} minimises $\| s - s^* \|_*$ among all $s \in \mathcal{U}_j$, where we always begin with $s_1 = 0$. Here, $\| \cdot \|_*$ is a norm or semi-norm which corresponds to the posed problem, and it is our semi-norm from above in the radial basis function context.

We have the following three assumptions on the operator η which must only depend on function evaluations on Ξ:

(a) $s \in K \implies \eta(s) = s$,
(b) $s \in \mathcal{U} \setminus K \implies (s, \eta(s))_* > 0$ and
(c) $s, t \in \mathcal{U} \implies (\eta(s), t)_* = (s, \eta(t))_*$.

Subject to these conditions, the above strategy leads to the sought interpolant s^* in finitely many steps when we use exact arithmetic, and in particular the familiar sequence

$$(7.10) \qquad\qquad \| s^* - s_j \|_*, \qquad j = 1, 2, 3, \ldots,$$

s_j denoting the approximation after $j - 1$ sweeps, decreases strictly monotonically as j increases until we reach s^*. We note immediately that conditions (a) and (b) imply the important nonsingularity statement that $\eta(s) = 0$ for some $s \in \mathcal{U}$ only if s vanishes. Indeed, if s is an element from K, our claim is trivially true. Otherwise, if s is not in the kernel K, then $s \neq 0$ and if $\eta(s)$ vanishes this contradicts (b).

Further, this nonsingularity statement is essential to the dimension of the space generated by (7.9). In fact, in the opposing case of singularity, the sequence (7.9) might not generate a whole j-dimensional subspace $\mathcal{U}_j$ and in particular may exclude the required solution s^* modulo an element of K, which would, of course, be a disaster. Condition (a) guarantees that the polynomial part of s^* from the kernel K will be recovered exactly by the method which is a natural and minimal requirement.

The coefficients of the iterates s_j with respect to the given canonical basis of each $\mathcal{U}_j$ are, however, never computed explicitly, because those bases are ill-conditioned. Instead, we begin an optimisation procedure based on the familiar conjugate gradients (e.g. Golub and Van Loan, 1989, see also the last section of this chapter) and compute for each iteration index $j = 1, 2, \ldots$

$$(7.11) \qquad s_{j+1} = s_j + \alpha_j d_j,$$

where d_j is a search direction

$$(7.12) \qquad d_j = \eta(s^* - s_j) + \beta_j d_{j-1},$$

which we wish to make orthogonal to d_{j-1} with respect to the native space semi-inner product. Moreover, $d_0 := 0$, $s_1 := 0$, and in particular $d_1 = \eta(s^*)$. No further directions have to be used in (7.12) on the right-hand side in order to obtain the required conjugacy. This is an important fact and it is a consequence of the self-adjointness condition (c).

We shall see that condition (b) is important for generating linearly independent search directions. The aforementioned orthogonality condition defines the β_j in (7.12), and the α_j is chosen so as to guarantee the monotonic decrease. The calculation ends if the residuals $|s_{j+1}(\xi) - f_\xi|$ are small enough uniformly in $\xi \in \Xi$, e.g. close to machine accuracy. In fact, as we shall note in the theorem below, full orthogonality

$$(7.13) \qquad (d_j, d_k)_* = 0, \qquad 1 \le j < k < k^*,$$

is automatically obtained with the aid of condition (c), k^* being the index of the final iteration. This fact makes this algorithm a conjugate gradient method, the search directions being conjugate with respect to our semi-inner product. However, the polynomial terms which belong to the interpolant are computed on every iteration, while a genuine conjugate gradient method only works on the preconditioned positive definite matrix and always modulo the kernel K of the semi-inner product. The necessary correction term from K is then added at the *end* of the conjugate gradient process.

The parameter β_j from (7.12) can be specified. It is

$$\beta_j = -\frac{\left(\eta(s^* - s_j), d_{j-1}\right)_*}{\|d_{j-1}\|_*^2}.$$

The minimising parameter in (7.11) is

$$\alpha_j = \frac{(s^* - s_j, d_j)_*}{\|d_j\|_*^2}.$$

It guarantees that $\|s^* - s_j\|_*$ is minimal amongst all $s_j \in \mathcal{U}_{j-1}$ through the orthogonality property

$$(s - s_j, g)_* = 0, \qquad \forall g \in \mathcal{U}_{j-1}.$$

We note that if the required interpolant is already in the kernel of the semi-inner product, then (a) implies that $d_1 = \eta(s^*) = s^*$ (recalling that s_1 is zero) and the choice $\alpha_1 = 1$ gives $s_2 = s^*$ as required. Otherwise, s_1 being zero, s_2 is computed as $\alpha_1 d_1$, where $d_1 = \eta(s^*)$ and α_1 minimises $\|s^* - \alpha_1 d_1\|_*$. Since $s^* \notin K$, (a) implies $d_1 \notin K$. Hence $(d_1, d_1)_*$ is positive and $\alpha_1 = (s^*, d_1)_*/(d_1, d_1)_*$ because of (b). We get, as required, $\|s^* - s_2\|_* < \|s^* - s_1\|_*$.

Thus, in Faul and Powell (1999b), the following theorem is proved by induction on the iteration index.

Theorem 7.3. *For $j > 1$, the Krylov subspace method with an operator η that fulfils conditions (a)–(c) leads to iterates s_j with uniquely defined search directions (7.12) that fulfil (7.13) and lead to positive α_j and strictly monotonically decreasing (7.10) until termination. The method stops in exact arithmetic in $k^* - 1$ steps (which is at most $m - \dim K$), where in the case that $\|s_j - s^*\|_*$ vanishes during the computation, the choice $\alpha_j = 1$ gives immediate termination.*

After the general description of the proposed Krylov subspace method we have to outline its practical implementation for the required radial basis function interpolants. To this end, the definition of the operator η is central. If the kernel K is trivial, i.e. the radial basis function is conditionally positive definite of order zero – subject to a sign-change if necessary – we use the operator

$$(7.14) \qquad \eta \colon s \mapsto \sum_{k=1}^{m} \frac{(L_k^{\mathrm{loc}}, s)_* L_k^{\mathrm{loc}}}{\lambda_{k\xi_k}},$$

where the local Lagrange functions are the same as in the BFGP method defined in the subsection before last for the sets $\mathcal{L}_k$ except that for all $k > m - q^*$ we define additionally $\mathcal{L}_k = \{\xi_k, \xi_{k+1}, \ldots, \xi_m\}$ and the associated local Lagrange functions. The semi-inner product $(\cdot\,,\,\cdot)_*$ is still the same as we have used before. This suits for instance the inverse multiquadric radial basis function.

The positive definiteness is a particularly simple case. In fact, the conditions (a)–(c) which the operator η has to meet follow immediately. Indeed, there is no polynomial reproduction to prove in this case. Condition (b) follows from the fact that the local Lagrange functions L_k^{loc} are linearly independent, as a consequence of the positivity of their 'leading (first nonzero) coefficient' $\lambda_{k\xi_k}$, and because the translates of the radial basis function are linearly independent due to the (stronger) fact of nonsingularity of the interpolation matrix for distinct centres. Property (c) is true for reasons of symmetry of the inner product and due to the definition (7.14).

Taking the approximate Lagrange functions for this scheme is justified by the following observations. Firstly, the theoretical choice of the *identity operator* as η would clearly lead to the sought solution in one iteration (with $s_2 = \alpha_1 s^*$ and taking $\alpha_1 = 1$), so it is reasonable to take some approximation of that. Secondly, we claim that this theoretic choice is equivalent to taking mutually orthogonal basis functions in (7.14). The orthogonality is, of course, in all cases with respect to $(\cdot\,,\,\cdot)_*$. This claim we establish as follows.

In fact, it is straightforward to see that the η is the identity operator if and only if the Lagrange functions therein are the *full* Lagrange functions, i.e. $q = |\Xi|$ – see also the subsection before last – and therefore the approximate Lagrange functions are useful. Further, we can prove the equivalence of orthogonality and fulfilment of the standard Lagrange conditions. Recalling the reproducing kernel property of the radial basis function with respect to the (semi-)inner product and Lemma 7.1, we observe that the orthogonality is a consequence of the assumption that the basis functions satisfy the global Lagrange conditions

$$(7.15) \qquad L_k^{\mathrm{loc}}(\xi_\ell) = 0, \qquad \ell > k,$$

if L_k^{loc} have the form (7.1), because condition (7.15) and the reproduction property in Lemma 7.1 imply the orthogonality

$$(7.16) \qquad (L_k^{\mathrm{loc}}, L_j^{\mathrm{loc}})_* = \sum_{\xi \in \mathcal{L}_j} \lambda_{j\xi} L_k^{\mathrm{loc}}(\xi) = 0$$

whenever $j > k$. This is indeed the aforementioned orthogonality. The converse – orthogonality implying Lagrange conditions – is also true as a consequence of (7.16) and the positivity of the coefficients $\lambda_{k\xi_k}$.

Property (7.2) is a good approximation to (7.15). For the same reasons as we have stated in the context of the BFGP method, fulfilling the complete Lagrange conditions is not suitable if we want to have an efficient iterative method because this would amount to solving the full linear system in advance, and therefore we choose to employ those approximate Lagrange functions. This requires at most $O(mq^3)$ operations.

When the kernel K is nontrivial, the following choice of η is fitting. We let η be the operator

$$(7.17) \qquad \eta: s \mapsto \sum_{k=1}^{m-q^*} \frac{(L_k^{\text{loc}}, s)_* L_k^{\text{loc}}}{\lambda_{k,\xi_k}} + \bar{\eta}(s),$$

where the local Lagrange functions are the same as before. Here $\bar{\eta}(s)$ is the interpolant from T^* which satisfies the interpolation conditions

$$\bar{\eta}(s)(\xi) = s(\xi), \quad \xi \in \mathcal{L}_{m-q^*+1}.$$

The fact that $\bar{\eta}(s)$ agrees with s for all s which are a linear combination of radial basis function translates $\phi(\|x - \xi\|)$, $\xi \in \mathcal{L}_{m-q^*+1}$, and possibly an element from K, by the uniqueness of the interpolants – it is a projection – immediately leads to the fulfilment of conditions (a)–(c).

As far as implementation is concerned, it is much easier, and an essential ingredient of the algorithm, to work as much as possible on the *coefficients* of the various linear combinations of radial basis functions and polynomials involved and *not* to work with the functions themselves (see Faul and Powell, 1999b). At the start of each iteration, the coefficient vectors $\lambda(s_j) = (\lambda(s_j)_i)_{i=1}^m$ and $\gamma(s_j) = (\gamma(s_j)_i)_{i=1}^\ell$ are available for the current approximation, by which we mean the real coefficients of the translates of the radial basis functions $\phi(\| \cdot - \xi_i \|)$ and of the monomials that span K, respectively. We shall also use in the same vein the notations $\lambda(d_j)$, $\lambda(d_{j-1})$ etc. for the appropriate coefficients of d_j, d_{j-1} and other expressions which have expansions, as always, in the translates of the radial basis functions $\phi(\| \cdot - \xi_i \|)$ plus a polynomial.

Further, we know the residuals

$$r_\xi^j = f_\xi - s_j(\xi), \qquad \xi \in \Xi,$$

and the values of the search direction at the ξs. Further, $\lambda(d_{j-1})$ and $\gamma(d_{j-1})$ are also stored, as are all $d_{j-1}(\xi_i)$, $i = 1, 2, \ldots, m$. The first step now is the computation of the corresponding coefficients of $\eta(s^* - s)$ in the new search direction which we do by putting $s = s^* - s_j$ in the definition of η and evaluating $(L_j^{\text{loc}}, s^* - s_j)_*$ as $\sum_{\xi \in \mathcal{L}_j} \lambda_{j\xi} r_\xi^j$ according to Lemma 7.1 and the above display. When $j = 1$ at the beginning of the process, the coefficients $\lambda(d_j)$ and $\gamma(d_j)$ are precisely the coefficients of $\eta(s^* - s)$, otherwise we set

$$\tilde{\beta}_j = -\frac{\sum_{i=1}^m \lambda\Big(\eta(s^* - s)\Big)_i d_{j-1}(\xi_i)}{\sum_{i=1}^m \lambda(d_{j-1})_i d_{j-1}(\xi_i)}.$$

Table 7.2

Centres	$q =$	Iter. Kryl. $\phi(r) = r$	Iter. Kryl. tps	It. BFGP tps
400	10	13	33	29
400	30	6	8	5
400	50	5	7	4
900	10	14	42	36
900	30	7	10	6
900	50	4	8	5

Thus $\lambda(d_j) = \lambda\left(\eta(s^* - s)\right) + \tilde{\beta}_j \lambda(d_{j-1})$ and $\gamma(d_j) = \gamma\left(\eta(s^* - s)\right) + \tilde{\beta}_j \gamma(d_{j-1})$. Further

$$\tilde{\alpha}_j = \frac{\sum_{i=1}^{m} \lambda(d_j)_i r_{\xi_i}^j}{\sum_{i=1}^{m} \lambda(d_j)_i d_j(\xi_i)}.$$

Finally, we set $\lambda(s_{j+1}) = \lambda(s_j) + \tilde{\alpha}_j \lambda(d_j)$ and further $\gamma(s_{j+1}) = \gamma(s_j) + \tilde{\alpha}_j \gamma(d_j)$.

Table 7.2 gives the iteration numbers needed in order to obtain accuracy $\varepsilon = 10^{-8}$ with a selection of scattered centres and thin-plate splines and $\phi(r) = r$ using the Krylov subspace method, and using thin-plate splines and the BFGP method, respectively. They are taken from Faul and Powell (1999a, 1999b).

7.3 The fast multipole algorithm

There is an intimate relation between the task of evaluating radial basis function approximants and performing particle simulations or the so-called N-body problem. One of the most successful methods for the numerical solution of the particle simulations is the fast multipole method.

The basic idea of the so-called fast multipole method is to distinguish, for each evaluation of the linear combination $s(x)$ of thin-plate spline terms at x, say, between the 'near field' of points in Ξ close to x and the 'far field' of points in Ξ far away from x. All thin-plate spline terms of the near field are computed explicitly whereas collective approximations are used to generate the far field contributions. It is important that all the dependencies on the centres of the far field expansions are contained in their coefficients and appear no longer in the form of translates of a basis function, because this saves the costly individual evaluations of the radial basis functions. Hierarchical structures on Ξ are generated and used to track down which assemblies of points are far away from others. This is quite comparable with the work in the previous section, where the set of centres was decomposed into small (though overlapping) 'clouds' of

points and where corresponding local Lagrange functions were computed. The
next step, however, is different.

In this method, a so-called multipole or Laurent series is used to approximate
the contribution of each whole cloud of points within that hierarchical structure
as a far field. A final step can be added by approximating several Laurent series
simultaneously by one finite Taylor expansion. To begin with we wish to be
explicit about the aforementioned Laurent expansions. We restrict this section
to thin-plate splines in two dimensions and are generally brief, the full details
are given in the introductory articles by Beatson and Greengard (1997) and
Greengard and Rokhlin (1997).

A typical Laurent series expansion to thin-plate spline terms – which we
still use as a paradigm for algorithms with more general classes of radial basis
functions – is as stated by Beatson and Newsam (1992) as given below. We take
the bivariate case which may be viewed alternatively as a case in one complex
variable.

Lemma 7.4.   *Let z and t be complex numbers, and define*

$$\phi_t(z): = \|t - z\|^2 \log \|t - z\|.$$

Then, for all $\|z\| > \|t\|$,

$$\phi_t(z) = \Re\left\{(\bar{z} - \bar{t})(z - t)\left(\log z - \sum_{k=1}^{\infty} \frac{1}{k}\left(\frac{t}{z}\right)^k\right)\right\}$$

which is the same as

$$(\|z\|^2 - 2\Re(\bar{t}z) + \|t\|^2)\log\|z\| + \Re\left\{\sum_{k=0}^{\infty}(a_k\bar{z} + b_k)z^{-k}\right\},$$

*where we are denoting the real part of a complex number $z \in \mathbb{C}$ by $\Re z$. Here
$b_k = -\bar{t}a_k$ and $a_0 = -t$ and $a_k = t^{k+1}/[k(k + 1)]$ for positive k. Moreover,
if the above series is truncated after $p + 1$ terms, the remainder is bounded
above by*

$$\frac{\|t\|^2}{(p + 1)(p + 2)}\frac{c + 1}{c - 1}\left(\frac{1}{c}\right)^p$$

with $c = \|z/t\|$.

One can clearly see from the lemma how expansions may be used to approximate the thin-plate spline radial basis functions for large argument and how
the point of truncating the series, i.e. the length of the remaining sum, influences the accuracy. A typical length of the truncated series in two-dimensional
applications using thin-plate spline is 20. There will be a multiple of $\log|\Xi|$
expansions needed in the implementation of the multipole algorithm.

It transpires that an important step in the algorithm is again the set-up for structuring the Ξ in a suitable way. Then it is required to compute suitable series expansions for the radial basis function at far-away points, that is $\phi(\|x\|)$ for large $\|x\|$. For the set-up we need to decide first on the desired acceptable accuracy. This determines where the infinite expansions are to be truncated. The next step is the hierarchical subdivision of the domain into panels, and finally the far field expansions for each panel are computed.

The set-up stage not only constructs the data structure but also prepares the series coefficients that are required later on in the evaluation stage. This is again very similar to the work in the previous section, where the coefficients of the local Lagrange functions were calculated in advance. The data are cast into a tree structure: If the data are fairly uniformly distributed within a square in two dimensions which we assume to be a typical case, then the whole square is the initial parent set which is then repeatedly subdivided into four – in some implementations two – equally sized subsquare child *panels*. This subdivision is then repeated iteratively until some fixed level. Here, we still remain with the two-dimensional setting, but a three-dimensional version is given in Greengard and Rokhlin, 1997, where an octtree rather than a quadtree is generated.

Next, for each childless panel Q we associate an 'evaluation list' for that panel. This works as follows. Every panel that is at the same or a less re-fined level, whose points are all far away from the panel Q and whose parent panel is *not* far away from a point in Q, goes into the evaluation list of Q. Thus, the far field for any point in Q is the collection of points in the panels in the evaluation list of Q. The near field of a panel Q contains all the re-maining points. For each panel R in the so-called evaluation list, the method proceeds by computing a Laurent expansion for all radial basis functions with centres in R, and the expansion is about the mid-point of R. We always use the highest level (largest possible) panels in the evaluation list for the far field expansions to get the maximum savings in operational cost by evaluating as many terms as possible at the same time. By taking the length of the expan-sion suitably (according to the expansions and remainders given for instance in Lemma 7.4), any accuracy can be achieved by this approach at the price of greater computational expense. Finally, the method calculates the Laurent expansions for higher level panels R by translating the centres of the Laurent ex-pansions of the children of the panel R to the centre of R and combining them. For this there is a mathematical analysis available in the paper by Beatson and Newsam (1992) cited in the bibliography whose work we summarise here and which shows how the series alter when their points of expansion are translated.

A typical expansion that approximates an interpolant without a polynomial added

$$s(x) = \sum_{\xi \in \Xi} \lambda_\xi \phi(\|x - \xi\|)$$

is then of the form

$$s(x) \approx \sum_{k=1}^{p} \mu_k \phi_k(x),$$

with the remainder

$$\sum_{\xi \in \Xi} \lambda_\xi \hat{R}(x, \xi),$$

so that remainder plus approximation gives the exact $s(x)$. Here, we have used an approximation of the form

$$\sum_{k=1}^{p} \phi_k(x) \gamma_k(\xi) \approx \phi(\|x - \xi\|)$$

whose remainder is, in turn, $\hat{R}(x, \xi)$, and that gives rise to the remainder in the previous display.

The algorithm can be further improved by re-expanding and combining the Laurent expansions as local Taylor expansions. This is possible because the Laurent series are infinitely smooth away from their centres, so can be approximated well by polynomials. Working from parents down to children, coefficients of the Taylor expansion centred on each panel R can be found efficiently by re-centring (see the paragraph before last) the expansion from the parent level to other panels and adding contributions from the other panels on the evaluation list of the child that are not on the evaluation list of the parent. The result is that all the contribution of a far field is contained in a single Taylor series instead of several Laurent series which come from the various panels in the evaluation list of a given panel.

The Laurent series may be computed also for multiquadrics for example, i.e. not only for thin-plate splines, and have the general form for a panel R

$$\phi_R(x) = \sum_{j=0}^{\infty} \frac{P_j(x)}{\|x\|^{2j-1}}, \qquad \|x\| \gg 0,$$

for homogeneous polynomials P_j of degree j. Then one needs also the local Taylor expansions of truncated Laurent series.

In summary, the principal steps of the whole algorithm as designed by Beatson *et al.* may be listed as follows.

Set-up

(1) Perform as described above the repeated subdivision of the square down to $O(|\log \Xi|)$ levels and sort the elements of Ξ into the finest level panels.

(2) Form the Laurent series expansions (i.e. compute and store their coefficients) for all fine level panels R.

(3) Translate centres of expansions and, by working up the tree towards coarser levels, form analogous Laurent expansions for all less refined levels.

(4) Working down the tree from the coarsest level to the finest, compute Taylor expansions of the whole far field for each panel Q.

Evaluation at x

(1) Locate the finest level panel Q containing the evaluation point x.

(2) Then evaluate the interpolant $s(x)$ by computing near field contributions explicitly with a direct linear equation solver for all centres near to x, and by using Taylor approximations of the far field. For the far field, we use all panels R that are far away from x and are not subsets of any coarser panels already considered.

The computational cost of the multipole method without set-up is a large multiple of $\log|\Xi|$ because of the hierarchical structure. The set-up cost is $O(|\Xi|\log|\Xi|)$, but the constant contained in this estimate may be large due to the computation of the various expansions and the complicated design of the tree structure. This is so although in principle each expansion is an order $O(1)$ procedure. In practice it turns out that the method is superior to direct computations if m is at least of the order of 200 points.

In what way is this algorithm now related to the computation of interpolation coefficients? It is related in one way because the efficient evaluation of the linear combination of thin-plate spline translates is required in our first algorithm presented in this chapter. There the residuals $f_\xi - s(\xi)$ played an important rôle, $s(\xi)$ being the current linear combination of thin-plate splines and the current approximation to the solution we require. Therefore, to make the BFGP algorithm efficient, fast evaluation of s is needed.

However, the importance of fast availability of residuals at the centres is not restricted to the BFGP method. Other iterative methods for computing radial basis function interpolants such as conjugate gradient methods are also in need of these residuals. In order to apply those conjugate gradient methods, however, usually a preconditioning method is needed, and this will be discussed in Section 7.4.

There are various possible improvements to the algorithm of this section. For instance, we can use an adaptive method for subdividing into the hierarchical structure of panels, so that the panels may be of different sizes, but always contain about the same number of centres. This is particularly advantageous when the data are highly nonuniformly distributed.

7.4 Preconditioning techniques

As we have already seen, the radial basis function interpolation matrix $\mathbf{A}$ is usually ill-conditioned when Ξ is a large set and when ϕ is, for example, the multiquadric function (the conditioning is bad for any value of the parameter, but the situation is getting worse for larger parameters c as we have noted in Chapter 4), its reciprocal or the thin-plate spline. In this section, we want to indicate how the matrix's condition number may be improved before beginning the computation of the interpolation coefficients. This is the standard approach when large linear systems of whatever origin become numerically intractable due to large condition numbers and ensuing serious numerical problems with rounding errors. It is standard also with spline interpolation and finite element methods, for instance, and indeed less related to the specific properties of the radial basis function than the methods in the two sections above. Nevertheless there is, of course, significant influence of the form of ϕ on the preconditioning technique we choose.

We begin by assuming that $\mathbf{A}$ is positive definite, as is the case when ϕ is the reciprocal multiquadric function. Thus, in principle, a standard conjugate gradient method can be applied (Golub and Van Loan, 1989) or even a direct method such as a Cholesky factorisation. The convergence speed of conjugate gradients, however, severely depends on the condition of the matrix of the linear system which is being solved. Indeed, if $\mathbf{A}$ is positive definite and we wish to solve (5.25) with the conjugate gradient method, then the error $\|\lambda_j - \lambda\|_A$ of the iterate λ_j in the 'A-norm' $\|y\|_{\mathbf{A}} := \sqrt{y^T \mathbf{A} y}$ is bounded by a multiple of

$$\left(\frac{1 - \sqrt{\mathrm{cond}_2(\mathbf{A})}}{1 + \sqrt{\mathrm{cond}_2(\mathbf{A})}} \right)^{2j}.$$

Convergence under the above conditions may therefore be slow or, indeed, may fail completely in practice due to rounding errors. This outcome is much more likely when we are faced with the presence of very large linear interpolation systems.

In order to improve the ℓ^2-condition-number of the positive definite matrix **A**, we solve the system

(7.18) $$\mathbf{PAP}\mu = \mathbf{Pf}$$

instead of (5.25), where **P** is a preconditioning matrix, nonsingular and usually symmetric – if it is nonsymmetric, the left-multiplications in (7.18) have to be by $\mathbf{P}^T$. **P** is chosen (e.g. a banded matrix) such that it is not too expensive to compute the matrix product on the left-hand side of (7.18) and such that the product is positive definite. We shall come to methods to achieve these properties below. If **P** is symmetric and $\mathbf{P}^2 = \mathbf{C}$, then the matrix product **CA** should ideally have a spectrum on the positive real half-axis consisting of a small number of clusters, one of them near one, because the conjugate gradient algorithm can deal particularly well with such situations and because $\mathbf{P}^2\mathbf{A} \approx \mathbf{I}$ is equivalent to $\mathbf{PAP} \approx \mathbf{I}$. Of course, the theoretic choice of $\mathbf{C} = \mathbf{A}^{-1}$ would be optimal, **CA** being the identity matrix. If **P** is chosen suitably, then the condition number of the preconditioned matrix is usually also small. There are always several choices of **P** possible and having made one choice, the desired coefficient vector is $\boldsymbol{\lambda} = \mathbf{P}\mu$ which is evaluated at the end.

Even if **A** is not itself positive definite because ϕ is a strictly conditionally positive definite function of nonzero order k, substantial savings in computational cost can often be achieved. A good example for this event is, as always, the choice of radial basis functions (4.4). We perform the preconditioning by choosing a **P** as above, except that it will be nonsymmetric and now annihilates a vector $\{p(\xi)\}_{\xi \in \Xi}$ for any polynomial p of total order less than k by premultiplication. Equivalently

$$\mathbf{P}K \mid_{\Xi} = 0,$$

where the left-hand side is a short notation for **P** applied to all elements of the null-space $K = \mathbb{P}_n^{k-1}$ evaluated only on Ξ.

When ϕ is a strictly conditionally positive definite function of nonzero order k, some simple additional equations have to be solved in order to identify the polynomial p in

$$s(x) = \sum_{\xi \in \Xi} \lambda_\xi \phi(\|x - \xi\|) + p(x).$$

If μ solves (7.18) and $\boldsymbol{\lambda} = \mathbf{P}\mu$, then the coefficient vector γ of p written as a linear combination of a basis $\{p_j(x)\}_{j=1}^{\ell}$ of K may be found by solving any nonredundant

$$\ell = \dim K = \dim \mathbb{P}_n^{k-1} = \binom{k+n-1}{n}$$

equations from the residual

$$\text{(7.19)}\qquad\qquad\qquad \mathbf{Q}\gamma = \mathbf{f} - \mathbf{A}\lambda,$$

where $\mathbf{Q} = \{p_j(\xi)\}_{\xi \in \Xi, j=1}^{\ell}$ and $p(x) = \gamma^T \{p_j(x)\}_{j=1}^{\ell}$. Of course, the side-conditions on the coefficients mean that λ must be in the null-space of $\mathbf{Q}^T$.

We take the radial basis functions (4.4) as examples with $K = \mathbb{P}_n^{k-1}$ now, and in that case, $\mathbf{P}$ is a rectangular $|\Xi| \times (|\Xi| - \ell)$ matrix whose columns are a basis for the null-space of $\mathbf{Q}^T$. Thus $\mathbf{P}^T \mathbf{A} \mathbf{P}$ is a $(|\Xi| - \ell) \times (|\Xi| - \ell)$ square matrix.

We remark that $\mathbf{P}^T \mathbf{A} \mathbf{P}$ is positive definite provided that $\mathbf{P}$ has precisely K as null-space. If there were a larger null-space, then the product would be positive semi-definite; indeed, a larger kernel of the matrix $\mathbf{P}$ would lead to zero eigenvalues for $\mathbf{P}^T \mathbf{A} \mathbf{P}$ (see also Sibson and Stone, 1991). There may be a sign change needed in the radial basis function for this, e.g. we recall that the multiquadric function is conditionally positive definite of order one when augmented with a negative sign. We are left with a great deal of freedom in our choice of $\mathbf{P}$ now. We use that to make as good an improvement to the condition number of $\mathbf{P}^T \mathbf{A} \mathbf{P}$ as possible. The fact that $\lambda = \mathbf{P}\mu$ guarantees that λ is indeed in the null-space of $\mathbf{Q}^T$, because $\mathbf{Q}^T \mathbf{P} = 0$.

The matrix thus preconditioned is symmetric and positive definite and therefore a Cholesky (direct) method may be used for the solution of the interpolation linear system. The advantage of $\mathbf{P}^T \mathbf{A} \mathbf{P}$ being positive definite is that the size of the off-diagonal elements is thus restricted while the elements of the $\mathbf{A}$ otherwise are growing off the diagonal for multiquadrics or (4.4). The Cholesky method decomposes the preconditioned interpolation matrix into a product $\mathbf{L}\mathbf{L}^T$, where $\mathbf{L}$ is lower-triangular. Therefore the given linear system

$$\mathbf{L}\mathbf{L}^T \mu = \mathbf{P}^T \mathbf{f}, \qquad \mathbf{P}\mu = \lambda,$$

can be solved by backward substitution in a straightforward manner. Given, however, that we address *large* sets of centres in this chapter, it is usually preferable to use iterative methods instead, such as conjugate gradients.

When using conjugate gradients, the number of iterations for solving the system numerically may in the worst case be of $O(|\Xi|)$, but, as we have pointed out before, the work on the computer can be made more efficient by using the results of Section 7.3. Specifically, if $\tilde{\mu}$ is the current approximation to μ of (7.18), the main work for getting the next estimate is always the calculation of the matrix (times vector) product $\mathbf{A}\tilde{v}$, where $\tilde{v} = \mathbf{P}\tilde{\mu}$, and the latter matrix product can be made cheap – $\mathbf{P}$ should be designed such that this is the case. The product $\mathbf{A}\tilde{v}$, however, is nothing other than a linear combination of radial basis function terms with coefficients from $\tilde{\mu}$, evaluated at the known centres which we have already indicated at the end of the description of the BFGP method. We note that this numerical value is exactly what can be evaluated fast by the algorithm in Section 7.3, in $O(|\Xi| \log |\Xi|)$ operations.

The preconditioning matrix $\mathbf{P}$ can be chosen conveniently by using our favourite method to generate orthogonal matrices, by Householder transformations or Givens rotations (Golub and Van Loan, 1989) for instance. In order to obtain a positive definite matrix $\mathbf{P}^T \mathbf{AP}$, we follow Powell (1996) and select the Householder method to give a suitable preconditioning matrix. Let as above

$$\mathbf{Q} = \{p_j(\xi)\}_{\xi \in \Xi, j=1,2,\dots,\ell},$$

the $p_1, p_2, \dots, p_\ell$ still being a basis of the polynomial space K which has to be annihiliated to make $\mathbf{P}^T \mathbf{AP}$ positive definite. Thus for this ℓ $\mathbf{Q}$ is a $|\Xi| \times \ell$ matrix and the resulting positive definite matrix $\mathbf{P}^T \mathbf{AP}$ will be $(|\Xi| - \ell) \times (|\Xi| - \ell)$. Now let $\mathbf{P}_1 = \mathbf{Q}$ and then, for $j = 1, 2, \dots, \ell$, compute the orthogonal transformations according to Householder,

$$\mathbf{P}_{j+1} = \left(I - \frac{2u_j u_j^T}{u_j^T u_j} \right) \mathbf{P}_j,$$

where the factors in parentheses are always $|\Xi| \times |\Xi|$ matrices and where u_j is chosen such that $\mathbf{P}_{j+1}$ has zeros below its diagonal in the first j columns. Here as always, I denotes the identity matrix. Thus the reverse product

$$\left(\prod_{j=\ell}^{1} \left(I - \frac{2u_j u_j^T}{u_j^T u_j} \right) \right) \mathbf{Q}$$

is an upper-triangular matrix whose last rows are zero by the choice of the u_j.

Now let $\mathbf{P}$ be the last $|\Xi| - \ell$ columns of

$$\prod_{j=1}^{\ell} \left(I - \frac{2u_j u_j^T}{u_j^T u_j} \right).$$

Therefore $\mathbf{P}^T \mathbf{AP}$ is a $(|\Xi| - \ell) \times (|\Xi| - \ell)$ square positive definite matrix and $\mathbf{Q}^T \mathbf{P} = 0$. There are at most $O(|\Xi|^2)$ operations needed for that procedure, since only $\ell \ll |\Xi|$ Householder transformations are required.

We can now apply a conjugate gradient scheme to the resulting matrix $\mathbf{P}^T \mathbf{AP}$. The total work for the above process is $O(|\Xi|)$ operations. The only task of an iteration of that method which may require more than $O(|\Xi|)$ operations is the multiplication of a vector $v = \mathbf{P}\mu$ by $\mathbf{P}^T \mathbf{A}$. As pointed out already above, fortunately, the scheme of Section 7.3 admits an $O(|\Xi| \log |\Xi|)$ procedure for that.

We mention an alternative preconditioning scheme suggested by Faul (2001). It is noted in her thesis that a preconditioning matrix can be used with the coefficients $\lambda_{k\xi_j}$ of the local Lagrange functions. In this case, the preconditioning is by a matrix defined through

$$\mathbf{P} = \left\{ \frac{\lambda_{k\xi_j}}{\sqrt{\lambda_{k\xi_k}}} \right\}_{j,k}.$$

The matrix $\mathbf{P}^T\mathbf{AP}$ is then positive definite and can be used within a standard conjugate gradient method. A disadvantage of this is that the Lagrange coefficients have to be computed first, although this can be done, according to the work of Section 7.2.

An example of an implementation of the conjugate gradient algorithm for the application in this context of radial basis function interpolation is as follows. Let μ_k be the value of μ at the beginning of the kth iteration of the algorithm. Let r^k be the residual of (7.18)

$$r^k = \mathbf{P}^T\mathbf{f} - \mathbf{P}^T\mathbf{AP}\mu_k.$$

We begin with a start-vector μ_1 whose value is arbitrary, often just 0. Of course, the iteration ends as soon as this residual is close to machine accuracy or any prechosen desired accuracy componentwise. Otherwise we continue with the choosing of a search direction vector d_k and set $\mu_{k+1} = \mu_k + \alpha_k d_k$, where α_k is chosen such that r^{k+1} is orthogonal to the search direction vector d_k. Specifically, the choice

$$\alpha_k = -\frac{d_k^T g_k}{d_k^T \mathbf{P}^T\mathbf{AP}d_k}$$

is good, where $g_1 = -d_1$ and $g_{k+1} = g_k + \alpha_k \mathbf{P}^T\mathbf{AP}d_k$. The search directions are then

$$d_k = -g_k + \frac{d_{k-1}^T \mathbf{P}^T\mathbf{AP}g_k}{d_{k-1}^T \mathbf{P}^T\mathbf{AP}d_{k-1}}d_{k-1}.$$

We begin with $d_1 = r^1$ and let the search direction d_k be as above such that the conjugacy condition

$$d_k^T \mathbf{P}^T\mathbf{AP}d_{k-1} = 0$$

holds. If there were no rounding errors in our computation, the residual vectors would all be mutually orthogonal and the algorithm would terminate with the exact solution in finite time.

In their seminal paper of 1986, Dyn, Levin and Rippa construct preconditioning matrices to radial basis function approximants using radial basis functions (4.4) in two dimensions especially for thin-plate splines by discretising the two-dimensional iterated Laplace operator Δ^k on a triangulation. The approach is summarised as follows in Powell (1996). The goal is once again as in the paragraph containing (7.18) to take $\mathbf{P}$ as the 'symmetric square root' of an approximation $\mathbf{C} \approx \mathbf{A}^{-1}$. This matrix $\mathbf{C}$ retains the positive semi-definiteness of $\mathbf{A}^{-1}$ with exactly ℓ of its eigenvalues being zero, the rest being positive. The number ℓ is zero if $k = 0$. Finally, as in the paragraph containing (7.19), $\mathbf{Q}^T \mathbf{C} = 0$ is required. If $\mathbf{P}^2 = \mathbf{C}$, then $\mathbf{Q}^T \mathbf{P} = 0$ and the columns of $\mathbf{P}$ span the null-space of $\mathbf{Q}^T$. Therefore, again, $\boldsymbol{\lambda} = \mathbf{P}\mu$ and we may solve (7.18) instead of the original interpolation equation. To this, again conjugate gradients may be applied.

The two highly important questions which remain are how to choose $\mathbf{C}$ and whether we may work with $\mathbf{C}$ instead of its square root $\mathbf{P}$ in the implementation of the conjugate gradient method. The second question is answered in the affirmative when we work with $\boldsymbol{\lambda}_k = \mathbf{P}\mu_k$, $c_k = \mathbf{P}d_k$ and $b_k = \mathbf{P}g_k$ instead of μ_k, d_k and g_k, respectively. This leads to the formulae

$$c_k = -b_k + \frac{c_{k-1}^T \mathbf{A} b_k}{c_{k-1}^T \mathbf{A} c_{k-1}} c_{k-1}$$

and $\boldsymbol{\lambda}_{k+1} = \boldsymbol{\lambda}_k + \alpha_k c_k$, $b_{k+1} = b_k + \alpha_k \mathbf{C} \mathbf{A} c_k$. Here, the step-length is

$$\alpha_k = \frac{c_k^T \mathbf{f} - c_k^T \mathbf{A} \boldsymbol{\lambda}_k}{c_k^T \mathbf{A} c_k},$$

recalling $(r^{k+1})^T d_k = 0$ and so $\alpha_k = d_k^T r^k / c_k^T \mathbf{A} c_k$ for this scheme.

Finding $\mathbf{C}$ is achieved by estimating $\|s\|_\phi^2$ by a quadrature rule with positive coefficients that will be the entries of $\mathbf{C}$. For this, a triangulation of the set Ξ is used and then the kth partial derivatives which occur, in $\|s\|_\phi^2$, when written out as an integral over partial derivatives are estimated by using finite difference approximations in two dimensions,

$$\frac{\delta^k}{\delta x^i \delta y^{k-i}}, \qquad i = 0, 1, \ldots, k,$$

to the kth partial derivatives, using the known values of s on Ξ only. Thus $\|s\|_\phi^2 \approx \text{const. } \mathbf{f}^T \mathbf{C} \mathbf{f}$ and indeed we recall that

$$\|s\|_\phi^2 = \text{const. } \mathbf{f}^T \mathbf{A}^{-1} \mathbf{f}$$

due to the reproducing kernel properties and the definition of $\boldsymbol{\lambda}$ and its use in s. Since kth derivatives annihilate $(k-1)$st degree polynomials, $\mathbf{Q}^T \mathbf{C}$ must be zero (see the last but one paragraph). Since the approximations to the derivatives are always taken from clusters of nearby points from Ξ, the matrix $\mathbf{C}$ is usually sparse, which is important for the success of the method.

This success can be demonstrated by numerical examples as follows. In numerical examples, the condition numbers of the thin-plate spline interpolation matrices for three different distributions of 121 centres are reduced from 6764, 12 633 and 16 107 to 5.1, 3.9 and 116.7, respectively, by this method. The discretisation of the above iterated Laplacian, however, becomes much simpler if the data lie on a grid.

In fact, in that event, one particularly simple way of preconditioning can be outlined as follows. Suppose that we have finitely many gridded data and $\mathbf{A} = \{\phi(\|i-j\|)\}_{i,j\in[-N,N]^n\cap\mathbb{Z}^n}$. If ϕ is, still for simplicity of exposition, such that $\mathbf{A}$ is positive definite and its symbol (as in Chapter 4) is well-defined and has no zero, then one may precondition $\mathbf{A}$ in the following way. We take as before $\sigma(\vartheta)$ to be the symbol corresponding to the usual bi-infinite interpolation matrix. Its reciprocal expands in an absolutely convergent series according to Wiener's lemma. Let the coefficients of this series be $\tilde{c}_k$. Under these circumstances we take as a preconditioner the finite *Toeplitz* matrix (that is, as we recall, one whose elements are constant along diagonals) $\mathbf{C} = \{c_{ij}\}$ with entries $c_{ij} = \tilde{c}_{i-j}$ if $\|i-j\|_\infty \leq M \ll N$, for a suitable M, otherwise $c_{ij} = 0$. So $\mathbf{C}$ is a banded Toeplitz matrix. We let $\tilde{\mathbf{C}}$ be the inverse of the full bi-infinite matrix $\{\phi(\|i-j\|)\}_{i,j\in\mathbb{Z}^n}$. Then the symbol of the latter multiplied by the symbol of the former is

$$\sigma(\vartheta) \times \sum_{k\in\mathbb{Z}^n} \tilde{c}_k e^{-i\vartheta\cdot k} = 1, \qquad \vartheta \in \mathbb{T}^n,$$

that is the symbol of the identity matrix. Therefore we are entitled to expect that $\mathbf{CA}$ has the desired properties, because the symbol of the matrix product $\mathbf{CA}$ is the product of the symbols

$$\sum_{j\in[-N,N]^n} \phi(\|j\|)e^{-i\vartheta\cdot j} \times \sum_{k\in[-M,M]^n} \tilde{c}_k e^{-i\vartheta\cdot k} \approx 1,$$

see also Lemma 4.18 which justifies the approximation of the solution of the full infinite interpolation problem on a cardinal grid by using a finite section of the bi-infinite interpolation matrix. Baxter (1992a) shows an example for the case when the radial basis function is a Gaussian where $\mathbf{C}$ is indeed a positive definite preconditioning matrix. The method is also used in a similar fashion in Buhmann and Powell (1990) where specifically the Lagrange functions for various radial basis functions in two dimensions (thin-plate splines, multiquadrics, linear) are computed and displayed.

One small further step has to be taken to obtain a suitable preconditioner: We recall from the fourth chapter that the decay of linear combinations of translates of a basis function and moment conditions on their coefficients are intricately linked. Therefore, in order that decay properties of linear combinations of translates (i.e. the preconditioned entries of the interpolation matrix) are satisfied, we may have to modify the 'truncated' series of coefficients $\tilde{c}_k$, so that they satisfy the moment conditions

$$\sum_{k \in [-M,M]^n} \tilde{c}_k p(k) = 0$$

for all p from a suitable class of polynomials of maximal total degree. Usually, this class is the kernel K of the semi-inner product. Baxter (1992a) has shown that this is a useful way of preconditioning the interpolation matrix if we have finitely many centres on a grid. He described this both for the Gaussian which gives rise to positive definite interpolation matrices and for multiquadrics. In the latter case the conjugate gradient method solves a simple linearly constrained minimisation problem. The conjugate gradient method here is in fact related to the BFGP method, in that finite parts of Lagrange functions are used, except that here no Lagrange conditions are satisfied at all, but the coefficients of the full Lagrange functions are truncated and the aforementioned moment conditions are required.

8

Least Squares Methods

In this chapter we shall summarise and explain a few results about the orders of convergence of least squares methods. These approximants are computed by minimising the sum of squares of the error on the Euclidean space over all choices of elements from a radial basis function space. The main differences in the various approaches presented here lie in the way in which 'sum of squares of the error' is precisely defined, i.e. whether the error is computed continuously over an interval – or the whole space – by an integral, or whether sums over measurements over discrete point sets are taken. In the event, it will be seen that, unsurprisingly, the same approximation orders are obtained as with interpolation, but an additional use of the results below is that *orthogonal bases* of radial basis function spaces are studied which are useful for implementations and are also in very close connection to work of the next chapter about wavelets.

8.1 Introduction to least squares

Interpolation was the method of choice so far in this book for approximation. This, however, is by no means the only approximation technique which is known and used in applications. Especially least squares techniques are highly important in practical usage. There is a variety of reasons for this fact. For one, *data smoothing* rather than interpolating is very frequently needed. This is because data often are inaccurate, contain noise or – as happens sometimes in practical applications – are too plentiful and cannot and need not be reasonably all interpolated at once. An additional case when least squares methods are required is whenever a 'dimension reduction' is needed, i.e. when we know from theoretical considerations that data come actually from a lower-dimensional linear space than – falsely – indicated by the number of data provided.

196

Moreover, smoothing ('regularisation') is almost always required as long as problems are ill-posed, which means that their solution depends – for theoretical reasons and not just because we are using a bad method – extremely sensitively on even the smallest changes in the input data. When in that event the so-called 'inverse problem' (inverting an ill-conditioned matrix is a simple example for this) is to be solved numerically, smoothing of the output data or data at intermediate stages of the solution process is necessary to dampen inaccuracies, or noise or rounding errors, which would otherwise be inflated through the ill-posedness of the problem and dominate the – thereby false – result.

We have already met the problem of ill-conditioned bases and the consequences of applying optimisation methods in our Subsection 7.2.4 on Krylov subspace methods.

An illuminating example is the computation of derivatives of a sufficiently differentiable function when it is given only at finitely many points that are close to each other. If only few derivatives are needed, using divided differences as approximations to the derivatives is fine so long as we know the function values at discrete points which are sufficiently close. However, if, say, the first seven or eight derivatives are required, a sufficiently high order spline, say, may be used to approximate the function in the least squares sense initially on the basis of the points where we need it. Then the spline can be differentiated instead, recursively and stably by a standard formula (cf., e.g. Powell, 1981, also as an excellent reference for divided differences). Further smoothing of the spline's derivatives is usually required as well along the process.

Now, interpolation is not suitable for this purpose because, if there are small rounding errors in the initially given function values, the approximant's derivatives will soon become very 'wiggly' indeed even if we have started with an actually very smooth function. In engineering and scientific applications, measurements can almost never be assumed to be exact but have small errors or noise. Therefore, smoothing of the initial information is needed, not interpolation. Incidentally, in the terminology of the experts in those 'inverse problems', numerical differentiation is only a weakly ill-posed problem, and there are far worse cases.

Until now, much of the motivation for our analysis and the use of radial basis functions came from interpolation and its highly favourable uniform convergence behaviour. That is why we have concentrated on describing interpolation (but also quasi-interpolation which can be used for smoothing because the basis functions ψ whose translates form the quasi-interpolants usually do not fulfil Lagrange conditions) with our examples and radial basis functions. We now turn to the least squares *ansatz*. We will continue to use approximants from the

radial basis function spaces, because the *spaces* as such have been established
to be highly suitable for approximation.

An important reason for still using radial function spaces is, incidentally,
that we have also in mind to integrate radial basis function approximants into
other numerical tools, e.g. to solve partial differential equations or nonlinear
optimisation problems (Gutmann, 1999). There, interpolation is normally not
demanded, because the algorithms themselves provide only approximate numerical solutions to a certain order, i.e. a power of the step size, for instance.
Thus one can use quasi-interpolation, or least squares approximations, of at
least the same approximation order as the main algorithm that uses the radial
basis approximant for its specific purposes. Of course the quasi-interpolation
is not *per se* a least squares approximation but it can nonetheless be used for
smoothing; it may therefore be highly applicable when the given data are noisy.

It is important to recall for the work in this chapter that, given a linear space
S with an inner product $(\cdot, \cdot) : S \times S \to \mathbb{R}$ (sometimes a semi-inner product
with a nontrivial kernel as in Chapter 5), seeking the least squares approximant
s^* to $f \in S$ from a linear subspace $\mathcal{U}$ means minimising

$$(f - s, f - s), \qquad s \in \mathcal{U}.$$

The least squares approximation s^* is achieved when the error $f - s^*$ is orthogonal to all of the space $\mathcal{U}$ with respect to the (semi-)inner product, i.e.

$$(f - s^*, s) = 0, \qquad s \in \mathcal{U}.$$

In our setting, $\mathcal{U}$ is a space spanned by suitable translates of a radial basis
function which sometimes has to be scaled as well, and S is a suitable smoothness space, usually a space of square-integrable functions or a Sobolev space
of distributions (or functions) whose derivatives of certain orders are square-
integrable. It follows from the expression in the last display that s^* is especially
easy to compute if we are equipped with an orthonormal basis of $\mathcal{U}$, because
then we may expand the solution in that basis with simple coefficients.

In all cases, the solution of the least squares problem is determined by the
linear equation

$$G\lambda = F,$$

where F is the vector of inner products of the approximand f with the basis
functions of $\mathcal{U}$ and G is the auto-correlation ('Gram'-) matrix of all inner products (M_i, M_j) of the basis functions M_i of $\mathcal{U}$. This matrix is always nonnegative
definite, and it is positive definite if the M_i are indeed linearly independent. If

they form an orthonormal basis, then the Gram-matrix is the identity matrix, which makes the computation of the coefficients of the expansion trivial.

An especially interesting aspect of least squares approximations with radial basis functions is the choice of centres. There are normally far fewer centres than the points where we are given data, for the reasons mentioned above. Letting the centres vary within the least squares problem makes the approach a nonlinear one, much like spline approximations with free knots. One interesting observation about this approach is that, when the approximand is very smooth such as a polynomial or an analytic function, the centres of the approximants will often tend to the boundary of the domain where we approximate due to the polynomial reproduction properties discussed in Chapter 4. This is because the polynomial reproduction takes place asymptotically through certain linear combinations of the shifts of the radial basis functions (4.4), for instance, where the coefficients of the linear combinations satisfy moment conditions such as (4.17)–(4.18) and, in particular, sum to zero. So far this behaviour was only observed in computational experiments and there are few theoretical results. We will not discuss this any further in the present book, but point out that it is a highly relevant, interesting field of research, as mentioned also in our final chapter, Chapter 10.

8.2 Approximation order results

We start with some pertinent remarks about the approximation power of the radial function spaces we studied already but now with respect to the standard $L^2(\mathbb{R}^n)$ norm. So here $\mathcal{S} = L^2(\mathbb{R}^n)$. Of course, the norm here is the canonical norm induced by the Euclidean inner product between functions

$$(8.1) \qquad (f, g) = \int_{-\infty}^{\infty} f(x)\overline{g}(x)\,dx, \quad f, g \in L^2(\mathbb{R}^n),$$

that is the norm is the standard Euclidean norm $\|f\|_2 = \sqrt{(f, f)}$. As we see immediately, we will need special considerations for our radial basis functions which themselves are usually unbounded and not at all square-integrable, especially if we wish to find the explicit form of best approximants to the approximand. In other words, we shall need different basis functions for the spaces we consider, namely closures of the spaces spanned by the integer translates of $\phi(\|h^{-1} \cdot -j\|)$, $j \in \mathbb{Z}^n$. An exception to this remark is of course the use of our positive definite radial basis functions with compact support of Chapter 6 which are all square-integrable due to their continuity and compact support.

Since the least squares approach is far more complicated when general, scattered data are admitted, this chapter treats only gridded data $\Xi = h\mathbb{Z}^n$. We have two types of results.

The first result in this chapter about least squares methods which we present gives abstract estimates on the approximation power, measured in the least squares norm, of radial basis function spaces without explaining exactly what the approximants will look like. What is measured here is thus the 'distance' between two spaces: the space of approximants, a radial basis function space spanned by translates of radial basis functions scaled by h, and the larger space of approximands $\mathcal{S}$. Therefore we actually need not yet worry about the integrability of our radial basis functions: the only expressions that have to be square-integrable are the *errors* of the approximations to f from the aforementioned L^2-closure $\mathcal{U}$ of all finite linear combinations of translates $\phi(\|h^{-1} \cdot - j\|)$, j a multiinteger, so that the errors can be measured in least squares norm. In fact, it even suffices that the error times s from $\mathcal{U}$ is square-integrable in order that the distance between these spaces may be estimated. In particular cases, the saturation orders are given, i.e. the best obtainable least squares approximation orders to sufficiently smooth but nontrivial functions – nontrivial in the sense that they are not already in the approximation space or identically zero, for instance. Recall the definition of the *nonhomogeneous* Sobolev space $W_2^k(\mathbb{R}^n)$ from Chapter 4. Ron (1992) proves for n-variate functions $\phi(\| \cdot \|): \mathbb{R}^n \to \mathbb{R}$ the following theorem.

Theorem 8.1. *If $f \in W_2^k(\mathbb{R}^n)$ and $\phi(\| \cdot \|): \mathbb{R}^n \to \mathbb{R}$ has a generalised Fourier transform $\hat{\phi}(\| \cdot \|)$ such that*

$$(8.2) \qquad \sum_{j \in \mathbb{Z}^n \setminus \{0\}} \hat{\phi}(\| \cdot + 2\pi j\|)^2$$

is bounded almost everywhere in a neighbourhood Ω of 0, then the distance

$$\mathrm{dist}_{L^2(\mathbb{R}^n)}(f, \mathcal{U}) := \inf_{g \in \mathcal{U}} \|f - g\|_2,$$

where we use the notation

$$\mathcal{U} = \overline{\mathrm{span}} \{\phi(\|h^{-1} \cdot - j\|) \mid j \in \mathbb{Z}^n\},$$

is bounded above by a constant multiple of

$$(8.3) \qquad \left\| \frac{\hat{f}}{\hat{\phi}(\|h \cdot \|)} \right\|_{2, \Omega h} + o(h^k) \|f\|_k, \ h \to 0.$$

If, moreover, $\hat{\phi}(\|t\|)$ becomes unbounded as $\|t\| \to 0$ and (8.2) is bounded below in Ω, then the bound in (8.3) can be attained (that is, it provides the saturation *order) and thus the approximation error is* not $o(h^k)$ *for any general, nontrivial class of arbitrarily smooth approximands.*

We do not prove this result here – a related theorem will be established in the next section – but nonetheless we wish to explain the result somewhat further through examples. By 'span' in (8.3) we mean the set of all finite linear combinations (of arbitrary length, though) of the stated translates $\phi(\|h^{-1} \cdot -j\|)$, and the closure is taken within $L^2(\mathbb{R}^n)$. The expression $\| \cdot \|_{2,\Omega h}$ denotes the Euclidean norm with the range of integration restricted to the set $\{xh \mid x \in \Omega\}$.

Examples are easily derived from (4.4) because there, the distributional Fourier transform $\hat{\phi}(\|x\|)$ is always a constant multiple of some negative power of $\|x\|$. In other words, if $\hat{\phi}(\|x\|) = \|x\|^{-\mu}$, then (8.3) provides as a dominant term a constant multiple of $h^\mu \| \| \cdot \|^\mu \hat{f} \|_2$, i.e. we achieve approximation order h^μ for sufficiently smooth approximands. The expression (8.3) also gives an easy bound if $\hat{\phi}(\|x\|)$ has an expansion at zero that begins with a constant multiple of $\|x\|^{-\mu}$, such as the multiquadric for $\mu = n + 1$. In both cases the bound (8.3) is attained, i.e. it is the best possible and the saturation order according to Theorem 8.1. Explicit forms of approximants, however, left unstated here, we address the question of explicit solutions in the following section.

8.3 Discrete least squares

In the preceding theorem, the usual continuous least squares error estimates are made and the resulting errors are estimated in the form of the distance between approximation spaces and approximands. In real life, however, explicit forms of approximants are important and, moreover, discrete norms are much more appropriate because they can be evaluated on a computer and suit practical applications better. Hence, a very explicit approach is taken now in the second theorem we present, where for discrete ℓ^2-approximations, orthogonal functions are constructed to represent the best approximations explicitly in expansions and estimate their least squares error. We recall from the standard least squares approximation problem that its solution can very easily be expressed once we know orthogonal generators or even orthonormal bases for the approximation space.

In our discrete approach, we use two grids, namely one grid whereby the inner products and error estimates are formed (a finer grid) and then another, coarser grid whose points are used to serve as the centres of the radial basis functions. This represents a very reasonable model for the practical questions that arise: measurements are usually frequent (i.e. here on a fine grid) and centres for the radial basis functions that are used for the approximation are sparse and usually come from a subset of the measurement points. For convenience, the centres form a subgrid of the fine grid, i.e. $h' = H^{-1}$ with $H \in \mathbb{N}$, and $h' h \mathbb{Z}^n$ are the points of measurements, $h \mathbb{Z}^n$ are the centres. Thus, for any $h > 0$, $h \mathbb{Z}^n \subset h' h \mathbb{Z}^n$. Of course we wish to study the case $h \to 0$. This nestedness is, incidentally, not necessary for the theory, but simplifies greatly the expressions we shall have to deal with.

The inner product which we use is therefore discrete and has the form with a suitable scaling by h'^n

$$(8.4) \qquad (f, g)_{\text{discr}} := h'^n \sum_{j \in \mathbb{Z}^n} f(jhh') \, \overline{g}(jhh').$$

These ideas admit application of our earlier approaches of interpolation with square cardinal grids, with cardinal interpolants and quasi-interpolations which are able to reproduce polynomials and provide approximation orders that we were able to identify in Chapter 4. They especially admit useful applications of Fourier transform methods such as the Poisson summation formula and the like.

Now, in this chapter, we look for functions whose equally spaced translates with respect to the coarsely spaced grid are orthonormal with respect to the inner product (8.4). They are of the following form very similar to Lagrange functions:

$$(8.5) \qquad M_{h'}(x) = \sum_{m \in \mathbb{Z}^n} c_m^{h'} \, \phi(\|x - m\|), \quad x \in \mathbb{R}^n.$$

The superscript h' in the $c_m^{h'}$ indicates the coefficient's dependence on h'. Then, we study approximations or, rather, straight orthogonal projections on the space spanned by the translates of functions (8.5). They have the form

$$(8.6) \qquad s_h(x) = \sum_{k \in \mathbb{Z}^n} \left(f, M_{h'} \left(\frac{\cdot}{h} - k \right) \right)_{\text{discr}} M_{h'} \left(\frac{x}{h} - k \right), \quad x \in \mathbb{R}^n.$$

The inner products that appear in (8.6) are still the same discrete inner product as defined above, and it is the orthonormality of the translates of the functions (8.5) that allows s_h to have such a simple form. Many of the properties of (8.6), including the least squares approximation orders generated by it, are fairly simple consequences of the properties of (8.5). We have to be concerned with the existence of the latter first.

Indeed, the functions (8.5) exist and can, as we will show below in Theorem 8.2, be defined by their Fourier transforms $\hat{M}_{h'}(x)$, in a form that is strongly reminiscent of our Lagrange functions' Fourier transform. It is

$$(8.7) \qquad \hat{M}_{h'}(x) = \frac{\left(h'\right)^{-n/2} \hat{\phi}(\|x\|)}{\sqrt{\displaystyle\sum_{\ell \in \mathbb{Z}^n \cap [0,H)^n} \left| \sum_{k \in \mathbb{Z}^n} e^{2\pi i \ell \cdot k / H} \hat{\phi}(\|x + 2\pi k\|) \right|^2}}.$$

The validity of this form will be shown below, but we note in passing that (8.7) and the coefficients in (8.5), namely the $c_m^{h'}$, are related in much the same way as the Lagrange functions and their coefficients of Chapter 4 are related. Specifically, $c_m^{h'}$ is the multivariate Fourier coefficient

$$(8.8) \qquad \int_{\mathbb{T}^n} \frac{\left(2\pi \sqrt{h'}\right)^{-n} e^{im \cdot t}\, dt}{\sqrt{\displaystyle\sum_{\ell \in \mathbb{Z}^n \cap [0,H)^n} \left| \sum_{k \in \mathbb{Z}^n} e^{2\pi i \ell \cdot k / H} \hat{\phi}(\|t + 2\pi k\|) \right|^2}},$$

similar to the coefficients of the Lagrange functions in Section 4.1.

Theorem 8.2. *Let conditions (A1), (A2a), (A3a) on the radial basis function* $\phi : \mathbb{R}_+ \to \mathbb{R}$ *of Chapter 4 hold. Then the functions defined through (8.5) and (8.8) are continuous, satisfy*

$$(8.9) \qquad |M_{h'}(x)| = O\left((1 + \|x\|)^{-n-\mu}\right),$$

and are thus integrable with a Fourier transform (8.7). Their multiinteger translates are orthonormal with respect to the discrete inner product (8.4). Finally, if f is continuous, satisfies $|f(x)| = O\left((1 + \|x\|)^{-\frac{n}{2}-\varepsilon}\right)$ and the smoothness condition $|\hat{f}(t)\|t\|^\mu| = O\left((1 + \|t\|)^{-\frac{n}{2}-\varepsilon}\right)$ for a positive ε, (8.6) provides the least squares error

$$(8.10) \qquad \|f - s_h\|_2 = O(h^\mu), \quad h \to 0.$$

We have called the condition on f's Fourier transform a smoothness condition in the statement of the theorem, because asymptotic decay of the Fourier transform at infinity, together with some additional conditions, leads to higher differentiability of the function itself, cf. the Appendix.

Proof of Theorem 8.2: We can establish through conditions (A1), (A2a), (A3a) and (8.7) and (8.8) that the decay estimates

$$|c_m^{h'}| = O\left((1 + \|m\|)^{-n-\mu}\right)$$

and (8.9), the one claimed for $M_{h'}$, hold. This is done precisely in the same way as in the proofs of Theorems 4.2 and 4.3. Thus, according to the work in

the fourth chapter, M_h' can alternatively be defined through (8.7) or its inverse
Fourier transform (note that (8.7) is absolutely integrable and square-integrable
by the properties of $\hat{\phi}$ through (A1), (A2a), (A3a)), or through (8.5) and (8.8),
the series (8.5) being absolutely convergent. This leaves us to show that (8.5)
does indeed satisfy the orthogonality conditions with respect to (8.4) and the
convergence result at the end of the statement of the theorem.

We demonstrate the orthogonality here, because it is very important in the
context of this chapter and shows where the explicit form of the orthogonal functions $M_{h'}$ comes from. By a discrete Fourier transform applied to the required
orthogonality conditions (setting $h = 1$ without loss of generality)

$$\left(M_{h'}, \, M_{h'}(\cdot - k) \right)_{\text{discr}} = \delta_{0k}, \quad k \in \mathbb{Z}^n, \quad h' > 0,$$

we get the requirement expressed alternatively by the identity

$$(8.11) \qquad \sum_{j \in \mathbb{Z}^n} \sum_{k \in \mathbb{Z}^n} e^{-i\vartheta \cdot k} M_{h'}(jh') \, M_{h'}(jh' - k) = \left(h' \right)^{-n}, \quad \vartheta \in \mathbb{T}^n,$$

because $M_{h'}$ are real-valued and so there is no complex conjugate in the inner
product.

By the Poisson summation formula applied with respect to the summation
over the index k, we get from this the alternative form

$$\sum_{j \in \mathbb{Z}^n} \sum_{k \in \mathbb{Z}^n} e^{-ih'(\vartheta + 2\pi k) \cdot j} M_{h'}(jh') \, \overline{\hat{M}_{h'} (\vartheta + 2\pi k)} = \left(h' \right)^{-n},$$

which is tantamount to

$$\sum_{\ell \in \mathbb{Z}^n \cap [0, H)^n} \sum_{j \in \mathbb{Z}^n} \sum_{k \in \mathbb{Z}^n} e^{-ij \cdot \vartheta - i\ell \cdot (\vartheta + 2\pi k)/H} M_{h'}(j + \ell/H)$$

$$\times \overline{\hat{M}_{h'}(\vartheta + 2\pi k)} = \left(h' \right)^{-n}, \quad \vartheta \in \mathbb{T}^n,$$

recalling $h' = H^{-1}$.

Another application of the Poisson summation formula leads to the equivalent
expression

$$\sum_{\ell \in \mathbb{Z}^n \cap [0, H)^n} \sum_{j \in \mathbb{Z}^n} \sum_{k \in \mathbb{Z}^n} e^{i\ell \cdot (\vartheta + 2\pi j)/H - i\ell \cdot (\vartheta + 2\pi k)/H}$$

$$\times \hat{M}_{h'}(\vartheta + 2\pi j) \overline{\hat{M}_{h'}(\vartheta + 2\pi k)} = \left(h' \right)^{-n}.$$

Finally, the last line simplifies to

$$\sum_{\ell \in \mathbb{Z}^n \cap [0,H)^n} \sum_{j \in \mathbb{Z}^n} \sum_{k \in \mathbb{Z}^n} e^{i2\pi \ell \cdot (j-k)/H} \hat{M}_{h'}(\vartheta + 2\pi j) \overline{\hat{M}_{h'}(\vartheta + 2\pi k)}$$

which should equal $(h')^{-n}$. Next we insert (8.7) into the above display twice which confirms the required identity for $M_{h'}$'s Fourier transform by inspection.

We now restrict the support of f's Fourier transform to the cube $[-\pi/h, \pi/h]^n$. Such an f differs from any function f which satisfies the assumptions of the theorem by $O(h^\mu)$ in the least squares norm. This is because we may estimate

$$\int_{\|x\|_\infty > \pi/h} |\hat{f}(\vartheta)|^2 \, d\vartheta \leq \int_{\|x\|_\infty > \pi/h} \|\vartheta\|^{-2\mu} (1 + \|\vartheta\|)^{-n-2\varepsilon} \, d\vartheta = O(h^{2\mu}).$$

Therefore the restriction of $\hat{f}$'s support means no loss of generality.

Hence, analogously to our manipulations of (8.11) we get by inserting the definition of the discrete inner product and using the Poisson summation formula that the square of the least squares error between f and s_h is

$$\int_{\mathbb{R}^n} \left| \hat{f}(\vartheta) - h^n \sum_{k \in \mathbb{Z}^n} \left(f, M_{h'}\left(\frac{\cdot}{h} - k \right) \right)_{\text{discr}} e^{-ih\vartheta \cdot k} \hat{M}_{h'}(h\vartheta) \right|^2 \, d\vartheta$$

$$= \int_{h^{-1}\mathbb{T}^n} \sum_{m \in \mathbb{Z}^n} \left| \hat{f}(\vartheta + h^{-1} 2\pi m) - \hat{M}_{h'}(h\vartheta + 2\pi m) \right.$$

$$\left. \times \sum_{k \in \mathbb{Z}^n} \sum_{\ell \in \mathbb{Z}^n} \hat{f}\left(\vartheta + \frac{2\pi \ell}{hh'} + \frac{2\pi k}{h} \right) \overline{\hat{M}_{h'}(h\vartheta + 2\pi k)} \right|^2 \, d\vartheta$$

where we have used periodisation of $\hat{f}$. This is, by the band-limitedness, the same as

$$\int_{h^{-1}\mathbb{T}^n} \sum_{m \in \mathbb{Z}^n} \left| \delta_{0m} \hat{f}(\vartheta) - \hat{M}_{h'}(h\vartheta + 2\pi m) \hat{f}(\vartheta) \right.$$

$$\left. \times \sum_{k \in \mathbb{Z}^n} \overline{\hat{M}_{h'}\left(h\vartheta + \frac{2\pi k}{h'} \right)} \right|^2 \, d\vartheta,$$

The last display may be bounded above by a fixed multiple of $h^{2\mu}$. This we shall establish as follows. The integrand of the integral above consists of a product of $|f(\vartheta)|^2$ times the following expression that we get by rearranging terms:

$$I(h\vartheta) = 1 - 2 \, \Re \, \hat{M}_{h'}(h\vartheta) \sum_{k \in \mathbb{Z}^n} \hat{M}_{h'}\left(h\vartheta + \frac{2\pi k}{h'}\right)$$

$$+ \sum_{m \in \mathbb{Z}^n} \left| \hat{M}_{h'}(h\vartheta + 2\pi m) \sum_{k \in \mathbb{Z}^n} \hat{M}_{h'}\left(h\vartheta + \frac{2\pi k}{h'}\right) \right|^2$$

$$= \sum_{m \in \mathbb{Z}^n} \left| \hat{M}_{h'}(h\vartheta + 2\pi m) \right|^2 \cdot \left| \sum_{k \in \mathbb{Z}^n} \hat{M}_{h'}\left(h\vartheta + \frac{2\pi k}{h'}\right) \right|^2$$

$$+ 1 - 2 \, \Re \, \hat{M}_{h'}(h\vartheta) \sum_{k \in \mathbb{Z}^n} \hat{M}_{h'}\left(h\vartheta + \frac{2\pi k}{h'}\right).$$

By virtue of (8.7) and condition (A3a),

$$(8.14) \quad I(h\vartheta) = \left(1 + O(\|h\vartheta\|^{2\mu})\right)^2 + 1 - 2\left(1 + O(\|h\vartheta\|^{2\mu})\right) = O(\|h\vartheta\|^{2\mu}),$$

because for small h

$$\left| \sum_{j \in \mathbb{Z}^n} \hat{\phi}(\|h\vartheta + 2\pi j\|) \right|^{-2} = C\|h\vartheta\|^{2\mu} + O(\|h\vartheta\|^{2\mu + 2\varepsilon}).$$

Estimate (8.14) implies our desired for $h \to 0$ result because afterwards we may use that

$$\int_{h^{-1}\mathbb{T}^n} I(h\vartheta) |\hat{f}(\vartheta)|^2 \, d\vartheta \leq C \int_{h^{-1}\mathbb{T}^n} \|h\vartheta\|^{2\mu} |\hat{f}(\vartheta)|^2 \, d\vartheta$$

$$\leq Ch^{2\mu} \int_{\mathbb{R}^n} \|\vartheta\|^{2\mu} |\hat{f}(\vartheta)|^2 \, d\vartheta$$

$$= O(h^{2\mu}) \, .$$

The result now follows from the assumptions of the theorem.

As in Theorem 4.4 we can easily deduce from the Fourier transform (8.7) that the least squares approximation (8.6) recovers all polynomials exactly of order less than μ in total, where μ is the constant from conditions (A1), (A2a), (A3a) which, as we recall, are still assumed to hold.

To this end, we have to verify the conditions (4.16)–(4.18) in the same way as in Chapter 4 by making use in particular of the high order zeros of (8.7) at the 2π-multiples of multiintegers. In other words, if f is such a polynomial, then we have the polynomial reproduction property

$$\sum_{k \in \mathbb{Z}^n} \left(f, M_{h'}\left(\frac{\cdot}{h} - k\right)\right)_{\text{discr}} M_{h'}\left(\frac{x}{h} - k\right) = f(x), \quad x \in \mathbb{R}^n \, ,$$

the infinite sum being absolutely convergent by (8.9).

8.4 Implementations

We finish this chapter with a very few remarks about aspects of implementations of the least squares approach. The implementations of least squares methods using radial basis functions can be based on several different approaches. If we wish to implement the discrete least squares approach of the previous section, we may compute the orthonormal bases described therein by FFT methods because of the periodicity of the data. Concretely, the series

$$\sqrt{\sum_{\ell \in \mathbb{Z}^n \cap [0,H)^n} \left| \sum_{k \in \mathbb{Z}^n} e^{2\pi i \ell \cdot k / H} \, \hat{\phi}(\|x + 2\pi k\|) \right|^2}$$

that appears in the definition of the Fourier coefficients of the orthonormal bases contains a series of Fourier transforms of ϕ which converge fast, especially if the radial basis function is the multiquadric function. For this, we recall that $\hat{\phi}$ is in this case an exponentially decaying function. Thus only very few terms of the infinite series need be considered (summed up and then periodised) for a good approximation to the infinite series. The coefficients can then be computed by standard FFT implementations which work very fast. Once we have an orthonormal basis, the approximations can be expressed trivially with respect to that basis. Since the orthonormal basis functions decay quickly as we have asserted in (8.9), the infinite expansions (8.6) may be truncated with small loss of accuracy.

When radial basis functions are used for least squares approximations and the centres are *scattered*, (Gram-)matrices turn up which are nonsingular and can be analysed in the same fashion as at the end of Chapter 5, i.e. bounds on their condition numbers can be found which depend especially on the separation radius of the centres. This work has been done by Quak, Sivakumar and Ward (1991) and it strongly resembles the analysis of the last section of Chapter 5. One of the main differences is that the matrices are no longer collocation matrices, but they can be reformulated as collocation matrices where the collocation points and the centres differ, so we end up with nonsymmetric matrices. Their properties are much harder to analyse than those of our standard symmetric interpolation matrices.

The computation of the coefficients of the least squares solution is done in a standard way by a QR decomposition of the Gram-matrix, rather than solving the normal equations by a direct method (Powell, 1981, for example).

8.5 Neural network applications

Neural network applications can be seen as high-dimensional least squares problems (Broomhead, 1988). For a small number of centres ξ and a prescribed radial basis function ϕ, coefficients λ_ξ are sought such that the typical linear combination of translates matches as well as possible a given input (ξ, f_ξ) of many more trials than given centres. This is then treated usually as an overdetermined linear system and solved by a least squares method. A radial basis function approximation of the form

$$\sum_{\xi \in \Xi} \lambda_\xi \phi(\|x - \xi\|), \qquad x \in \mathbb{R}^n,$$

can be viewed as a 'single layer neural network with hidden units' in the language of neural network research. In neural network applications, a multiplicative term $\rho > 0$ is often inserted into the $\phi(\cdot)$. The questions thus arising are the same as in our radial basis function context: what classes of function can be approximated well, up to what order, what algorithms are available? Answers to the first question are provided by the research of Pinkus (1995–99), for the others see Evgeniou, Pontil and Poggio (2000), for instance.

On the other hand, the classical, so-called regularisation networks minimise the expressions

$$|\Xi|^{-1} \sum_{\xi \in \Xi} \left(s(\xi) - f_\xi \right)^2 + \lambda \|s\|_\phi^2$$

whose solution is a linear combination of translates of a radial basis function as we know it. This is also called a smoothing spline as it does not satisfy interpolation conditions, but smoothness is regulated by the above parameter λ. The solution to the above smoothing problem exists and is unique if the Ξ contain a unisolvent set for the polynomial kernel of $\| \cdot \|_\phi$ (Bezhaev and Vasilenko, 2001, for example). In general, a smoothing spline s from a real separable Hilbert space X exists that minimises

$$\|As - \mathbf{f}\|_Z^2 + \lambda \|\Phi s\|_Y^2$$

if Y and Z are real separable Hilbert spaces and $A : X \to Z$ and Φ are linear bounded operators with closed ranges, the null-space of $\Phi : X \to Y$ is finite-dimensional, and its intersection with the null-space of A is trivial. In our application, Z is the discrete ℓ_2-norm, A maps the argument to the vector of evaluations on Ξ and $\|\Phi s\|_Y^2 = \|s\|_\phi^2$, see also Wahba (1981) for many more details on spline smoothing.

9

Wavelet Methods with
Radial Basis Functions

9.1 Introduction to wavelets and prewavelets

Already in the previous chapter we have discussed in what cases L^2-approximants or other smoothing methods such as quasi-interpolation or smoothing splines with radial basis functions are needed and suitable for approximation in practice, in particular when data or functions f underlying the data are at the beginning not very smooth or must be smoothed further during the computation. The so-called wavelet analysis that we will introduce now is a further development in the general context of L^2-methods, and indeed everything we say here will concern L^2-functions, convergence in the L^2-norm etc. only. Many important books have been written on wavelets before, and since this is not at all a book on wavelets, we will be fairly short here. The reader who is interested in the specific theory of wavelets is directed to one of the excellent works on wavelets mentioned in the bibliography, for instance the books by Chui, Daubechies, Meyer and others. Here, our modest goal is to describe what wavelets may be considered as in the context of radial basis functions. The radial basis functions turn out to be useful additions to the theory of wavelets because of the versatility of the available radial basis functions.

Given a square-integrable function f on $\mathbb{R}$, say, the aim of wavelet analysis is to decompose it simultaneously into its time *and* its frequency components. Therefore, a wavelet decomposition is always a double series, which should be contrasted with the simple orthogonal decompositions that are usual in the L^2-theory of 2π-periodic functions, i.e. mainly Fourier analysis, where only *one* orthogonal series (that is, with one summation index) is employed to represent the function f. The wavelet expansion also uses basis functions that are orthogonal, like exponential functions in the L^2-theory of 2π-periodic functions with some suitable coefficients to multiply each exponential. One consequence of our goal of decomposing in time and frequency simultaneously is that we need to find basis functions which are

(i) mutually orthogonal or orthonormal (for example – as with our basis functions $M_{h'}$ for least squares approximation in the previous chapter – generated by shifting just one function, call it ω, whose integer translates are orthogonal),

(ii) in 'some way', which will be explained shortly, representatives of different frequency components of a signal to be decomposed,

(iii) spanning $L^2(\mathbb{R})$ so that every square-integrable function can be expanded in series of those functions.

While it is well-known in principle how to achieve (i) and (iii) by various function systems, such as those we have encountered in the previous chapter, we need to be much more concrete as to condition (ii). The 'wavelet way' to obtain property (ii) is to seek a univariate square-integrable function ω called a wavelet, whose scales ('dilates') by powers of two are mutually orthogonal:

$$(9.1) \qquad \omega(2^j \cdot -k) \perp \omega(2^{j'} \cdot -k'), \quad \forall j \neq j', \quad \forall k \neq k',$$

for j, j', k, k' from $\mathbb{Z}$. The orthogonality is always in this chapter with respect to the standard Euclidean inner product (8.1) of two square-integrable functions. Powers other than powers of two are possible in principle as well, but the standard is to use 2^j. If other integral powers or rational numbers are used, more than one wavelet ω will normally be required.

It is immediately clear why (9.1) is a reasonable condition: the scaling by powers of two (which may be negative or may be positive powers) stretches or compresses the function ω so that it is suitable to represent lower frequency or higher frequency oscillations of a function. If, on top of this, ω is a local function (e.g. compactly supported or quickly decaying for large argument), it will be able to represent different frequencies at different locations $2^{-j}k$ individually, which is precisely what we desire, the different locations being taken care of by translation. Moreover, the lengths of the translates are scaled accordingly: at level 2^j we translate through $2^{-j}k$ by taking $\omega(2^j \cdot -k) = \omega(2^j[\cdot - 2^{-j}k])$.

This should be contrasted with Fourier decompositions which are completely local in frequency – the exponentials with different arguments are linearly independent and each representing one frequency exactly – but not local at all in real space since the exponential functions with imaginary argument do not decay in modulus; they are constant ($|e^{ix}| = 1$) in modulus instead. Therefore all frequencies that occur in a signal can be recovered precisely with Fourier analysis but it will remain unknown *where* they occur. This is especially of little use for filtering techniques where it is usually not desirable to remove certain frequencies independently of time everywhere. Instead, local phenomena have to be taken into account which means that some frequencies must be kept or removed at one time or place, but not always.

We record therefore that, in particular, for higher frequencies, which means large j in the expression $\omega(2^j \cdot -k)$, the translates are shorter and for lower frequencies they are longer, which is suitable to grasp the fast and the slower oscillations of the approximand at different times, respectively.

By these means, wavelet decompositions can be extremely efficient for computation, because, in practice, almost all functions f (here also called signals) contain different frequencies at different times (e.g. music signals or speech) and are therefore not 'stationary' such as a single, pure tone represented by just one exponential. This is also the case for example for numerical solutions of partial differential equations that are expanded in orthonormal bases when spectral methods are used (Fornberg, 1999, for instance).

How are $\omega \in L^2(\mathbb{R})$ that satisfy (9.1) and span $L^2(\mathbb{R})$ computed (exactly: whose dilates and translates span $L^2(\mathbb{R})$ when arbitrary square-summable co-efficient sequences are admitted)? And, in our context, how are they identified from radial basis function spaces?

One simple instance for a wavelet is the famous Haar wavelet ω that is, in the one-dimensional setting, defined by

$$\omega(x) = \begin{cases} 1 & \text{if } 0 \leq x < \frac{1}{2}, \\ -1 & \text{if } \frac{1}{2} \leq x < 1 \text{ and} \\ 0 & \text{otherwise.} \end{cases}$$

If we scale and dilate this Haar wavelet by using $\omega(2^j \cdot -k)$, letting j and k vary over all integers, we obtain the required orthogonal decomposition of square-integrable functions f by double series

$$\sum_{j=-\infty}^{\infty} \sum_{k=-\infty}^{\infty} c_{jk}\omega(2^j \cdot -k),$$

the orthogonality

$$2^{j_1} \int_{-\infty}^{\infty} \omega(2^{j_1}x - k_1)\omega(2^{j_2}x - k_2)\,dx = \delta_{j_1 j_2}\delta_{k_1 k_2}$$

for all integers j_1, j_2, k_1, k_2 being trivial.

Before we embark further on those questions, we point out that what we develop here will, strictly speaking, be prewavelets, not wavelets, in that there is no orthogonality in (9.1) with respect to k required; we only demand that different *scales* of the ω are orthogonal whatever the k, k' are. It is usual to replace orthogonality in k on each frequency 2^j by a suitable (Riesz) stability condition, as we shall see. For the application of these functions, this is in most instances just as good, since it is the orthogonality between different frequencies ('levels') j or 2^j that is decisive especially for the existence of a fast algorithm for the computation of the decomposition. The latter is very important for the usefulness of the approach, since only in connection with a fast algorithm – here it is the so-called fast wavelet transform (FWT) – will a new method for the purpose of analysing and e.g. filtering functions and signals be acceptable and useful in practice. The prime example for this fact is the development of the fast Fourier transform (FFT) some 50 years ago which made the use of Fourier techniques in science and engineering a standard, highly useful tool.

Furthermore, we point out that we will deal now with prewavelets in n dimensions, $\mathbb{R}^n$, since this book is centred on the multivariate theory; however, all the concepts and much of the notation will remain intact as compared with the univariate explanations we have given up to now.

In the literature, especially in the books by Chui, Daubechies and Meyer we have included in the bibliography, mostly wavelets and prewavelets either from univariate spline spaces are studied (especially in the book by Chui) or compactly supported (orthogonal) wavelets that are defined recursively and have no simple analytic expression (Daubechies wavelets) are studied. In both cases, they differ substantially from our prewavelets here, because the spaces where they originate from are usually spanned by compactly supported functions, even if the wavelets themselves are sometimes not compactly supported. The standard approach there is *not* to start from the approximation spaces and develop the prewavelets from there as we do in this book, but to begin with certain requirements on the (pre)wavelets such as compact support, smoothness or polynomial moment conditions, and construct them and their underlying approximation spaces from there. We prefer the reverse way as our motivation begins in this book from the radial basis function spaces, their analysis and their uses. Also, the objects we call prewavelets here are often called semi-orthogonal wavelets in the literature, especially by Chui.

9.2 Basic definitions and constructions

Since radial basis functions are usually not square-integrable, we use some quasi-interpolating basis functions of the type introduced in our Chapter 4 to generate the prewavelets instead which are. As we recall, they are finite linear combinations of radial basis functions, so we are still dealing with radial basis function spaces when we form prewavelets from the quasi-interpolating functions ψ of Chapter 4.

If the radial basis function is such that a quasi-interpolating basis function ψ which is square-integrable exists, then we can make an *ansatz* for prewavelets from radial basis function spaces. We might, incidentally, equally well use the cardinal functions L of Chapter 4 and their translates instead which, in most cases, have superior localisation. However, this is not strictly needed for our prewavelet application here because everything in this chapter will be dealt with just in the context of square-integrable functions. Furthermore, the quasi-interpolating functions have coefficient sequences with finite support and are thereby much simpler to use.

Now, this *ansatz* works particularly simply for instance for thin-plate splines and more generally for the radial basis functions belonging to the class defined in (4.4), and we shall begin with a description of that specific choice and add remarks about multiquadrics below. The main reason for this is that the radial basis functions of the thin-plate spline type (4.4) – unlike, for example, the multiquadric radial basis function – have easy distributional Fourier transforms which are reciprocals of even order homogeneous polynomials that have no roots except at zero. Up to a nonzero constant whose value is immaterial here, $\hat{\phi}$ corresponding to (4.4) is $\| \cdot \|^{-2k}$ as we recall. The aforementioned transforms are therefore analytic in a small complex tube about the real axis except at the origin. This has three consequences that are very important to us in this chapter and that we shall consider here in sequence.

Firstly, one can construct decaying symmetric and finite differences ψ of these radial basis functions in an especially simple way. The reason why we prefer in this chapter a simpler approach to the one of Chapter 4 is that we wish to use for the construction of our prewavelets the simplest possible formulation of ψ. Symmetric differencing in the real domain amounts to multiplying the Fourier transform of the radial basis function by an even order multivariate trigonometric polynomial g with roots at zero, as we have seen already in the one-dimensional case in Chapter 2. This g can be expressed as a sum of sin functions, because, analogously to one dimension, multiplication of a function in the Fourier domain by

$$g(y) = \left(\sum_{s=1}^{n} \sin^2\left(\frac{1}{2} y_s\right) \right)^k,$$

where $y = (y_1, y_2, \ldots, y_n)^T \in \mathbb{R}^n$, is equivalent to applying k times symmetric differences of that function of order two in the tensor product way, i.e. separately in all n coordinate directions.

The trigonometric polynomial g resolves the singularity of $\hat{\phi}$ at zero if the differences are of high enough order. In this way the conditions (4.16)–(4.18) are met. The rate at which these differences ψ decay depends only on the order of contact of the trigonometric polynomial and $\| \cdot \|^{-2k}$ at the origin, i.e. on the smoothness of the product of the trigonometric polynomial g and the Fourier transform $\hat{\phi}(\| \cdot \|)$. It can therefore be arbitrarily high due to the aforementioned analyticity. As a result of our differencing, a multivariate difference ψ of $\phi(\| \cdot \|)$ can be defined by its Fourier transform as follows:

$$(9.2) \qquad \hat{\psi}(y) = g(y)\hat{\phi}(\|y\|) = \frac{\left(\sum_{s=1}^{n} \sin^2\left(\frac{1}{2}y_s\right)\right)^k}{\left\|\frac{1}{2}y\right\|^{2k}},$$

see also Rabut (1990). This straightforward form will also aid us as an example in connection with the introduction of a so-called multiresolution analysis below. Note especially that $\hat{\psi} \in C(\mathbb{R}^n)$ or, rather, it has a removable singularity and can be extended to a continuous function because of the well-known expansion of the sin function at the origin. If we apply the same methods to prove decay of a function ψ at infinity by properties of its Fourier transform $\hat{\psi}$ as we did in Chapter 4, we can show that this function ψ defined through (9.2) satisfies the decay estimate

$$|\psi(x)| = O\left((1 + \|x\|)^{-n-2}\right).$$

So, in particular, $\psi \in L^2(\mathbb{R}^n)$, a fact that incidentally also follows from the fact that (9.2) denotes a square-integrable function due to $2k > n$ and from the isometry property of the Fourier transform operator as a map $L^2(\mathbb{R}^n) \to L^2(\mathbb{R}^n)$ as mentioned in the Appendix.

As a second important consequence of the aforementioned qualities which the radial basis functions (4.4) enjoy, these differences are able to generate a so-called *multiresolution analysis*. We want to explain this point in detail in the next section, because it is central to the derivation and analysis of all prewavelets and wavelets. The third consequence will be discussed later on, in Subsection 9.3.4.

9.3 Multiresolution analysis and refinement

9.3.1 Definition of MRA and the prewavelets

A multiresolution analysis (abbreviated to MRA) is an infinite sequence of closed subspaces V_j of square-integrable functions

$$(\text{MRA1}) \qquad \cdots \subset V_{-1} \subset V_0 \subset V_1 \subset \cdots \subset L^2(\mathbb{R}^n)$$

whose union is dense in the square-integrable functions, that is

(MRA2) $$\overline{\bigcup_{j=-\infty}^{\infty} V_j} = L^2(\mathbb{R}^n),$$

and whose intersection contains only the zero function:

(MRA3) $$\bigcap_{j=-\infty}^{\infty} V_j = \{0\}.$$

It is an additional condition in the so-called stationary case that the spaces in the infinite sequence (MRA1) satisfy

(MRA1a) $\qquad\qquad f \in V_j \quad$ if and only if $\quad f(2\cdot) \in V_{j+1}.$

Finally, it is *always* required that V_j have a basis of translates of a single square-integrable function, call it ψ_j:

(MRA4) $\qquad\qquad V_j = \mathrm{span}\{\psi_j(2^j \cdot -i) \mid i \in \mathbb{Z}^n\},$

the span being the L^2-closure of the set generated by all finite linear combinations. We recall from Chapter 4 that V_0 is a *shift-invariant space*, in fact a 'principal' shift-invariant space, because multiinteger shifts by any $i \in \mathbb{Z}^n$ of any function stemming from that space are again elements of the space. The principal refers to the fact that it is only generated by one function. Often, the generators in (MRA4) are required to form a Riesz basis but this is not always required here. We will come back to this later.

In the case of a stationary MRA, ψ_j remains the same, single ψ_0: $\psi_j(2^j \cdot -i) = \psi(2^j \cdot -i)$. That this is so is an easy consequence of the condition (MRA1a).

We summarise our statements so far in the following important definition.

Definition 9.1. *We call a sequence of closed subspaces (MRA1) a* multiresolution analysis (MRA) *if (MRA2), (MRA3) and (MRA4) hold. It is called a* stationary MRA *when additionally (MRA1a) holds, otherwise it is a* nonstationary MRA.

In the stationary case, it follows from (MRA4) and from our remarks in the paragraph before the definition that the stationary multiresolution analysis has the form (MRA1), where the V_j are defined by

(9.3) $\qquad\qquad V_j := \mathrm{span}\{\psi(2^j \cdot -i) \mid i \in \mathbb{Z}^n\}$

for a single j-independent square-integrable function ψ.

We shall explain later several cases of sequences (MRA1) when conditions (MRA2) and (MRA3) are automatically fulfilled under the weak condition that the support of $\hat{\psi}_j$ which is, as we recall, *per definitionem* a closed set, be $\mathbb{R}^n$ for each j. This condition is fulfilled, for instance, by the functions defined by (9.2).

We now define for an MRA the *prewavelet spaces* W_j by the orthogonal and direct sum

$$V_{j+1} = W_j \oplus V_j,$$

so in particular $V_j \cap W_j = \{0\}$. This can be expressed in a rather informal but very useful way of writing

$$W_j = V_{j+1} \ominus V_j,$$

with orthogonality $W_j \perp V_j$ with respect to the standard Euclidean inner product for all integers j and, in particular, $W_j \subset V_{j+1}$. This decomposition is always possible because the spaces V_j which form the multiresolution analysis are closed: as a result, the elements of W_j can be defined by the set of all elements of V_{j+1} minus their least squares projection on V_j. By virtue of the conditions of the MRA it is an easy consequence of this definition that

$$L^2(\mathbb{R}^n) = \bigoplus_{j=-\infty}^{\infty} W_j.$$

For the sake of completeness we provide the standard proof of this fact. Indeed, given any positive ε and any square-integrable f, there exist integral j and $f_j \in V_j$ such that the error $\| f - f_j \|_2$ is less than $\frac{1}{2}\varepsilon$, say. This is due to condition (MRA2). Moreover, it is a consequence of our orthogonal decomposition of the spaces V_j and of (MRA3) that there are $i \in \mathbb{Z}$ and

$$g_{j-1} \in W_{j-1}, \quad g_{j-2} \in W_{j-2}, \quad \ldots, \quad g_{j-i} \in W_{j-i},$$

and $f_{j-i} \in V_{j-i}$ such that $\| f_{j-i} \|_2 < \frac{1}{2}\varepsilon$ and

$$f_j = g_{j-1} + g_{j-2} + \cdots + g_{j-i} + f_{j-i}$$
$$\in W_{j-1} + W_{j-2} + \cdots + W_{j-i} + V_{j-i}.$$

Since ε was chosen arbitrarily, the assertion is proved, by letting both j and i tend to infinity, the series in g_j converging in L^2, and by appealing to (MRA2) once again.

Our demanded decomposition is a consequence of the last displayed equation: If f is a square-integrable function, then there are $g_j \in W_j$ such that f is the infinite sum of the g_j, the sum being convergent in L^2, and each g_j is decomposable in the form

$$g_j = \sum_{\substack{i \in \mathbb{Z}^n \\ e \in E^*}} c_{j,i} \omega_{j,e}(2^j \cdot -i).$$

In this equation, the translates of the $\omega_{j,e}$ span W_j, where $e \in E^*$ and E^* is a finite suitable index set of which we have standard examples below. The functions $\omega_{j,e}$ that therefore must span the space

$$W_j = \text{span}\left\{\omega_{j,e}(2^j \cdot -i) \middle| i \in \mathbb{Z}^n, e \in E^*\right\}$$

will be called the *prewavelets*. We shall give a more formal definition in Definition 9.4.

9.3.2 The fast wavelet transform FWT

In fact, it is straightforward to derive the method of the so-called fast wavelet transform FWT (also known as the cascade algorithm or pyramid scheme) from the above description of decomposing any square-integrable f into its prewavelet parts, or reconstructing it from them. The FWT is a recursive method whose existence and efficiency are central to the usefulness of the wavelet theory in practice as we have pointed out in the introduction to this chapter. To explain the method, let ψ_{j+1}, that is the generator of V_{j+1}, be represented by infinite linear combinations

$$(9.4) \qquad \psi_{j+1}(2(\cdot - e)) = \sum_{i \in \mathbb{Z}^n} a_{i,e}^j \psi_j(\cdot - i) + \sum_{\substack{i \in \mathbb{Z}^n \\ f \in E^*}} b_{i,e}^{j,f} \omega_{j,f}(\cdot - i).$$

Here, $e \in E$, the set of corners of the unit cube $[0, 1/2]^n$ in $\mathbb{R}^n$. This decomposition is possible for suitable infinite coefficient sequences $a_{i,e}^j$ and $b_{i,e}^{j,f}$ because $V_{j+1} = V_j \oplus W_j$ and because $\mathbb{Z}^n = 2(\mathbb{Z}^n + E)$. The coefficient sequences are at least square-summable, but we will require them to be absolutely summable below. In the stationary setting, the dependencies of the coefficient sequences on j may be dropped. Then we note that, for each element f_{j+1} of V_{j+1}, we can split the expansion in the following way:

$$(9.5) \qquad f_{j+1} = \sum_{i \in \mathbb{Z}^n} c_i^{j+1} \psi_{j+1}(2^{j+1} \cdot -i)$$

$$= \sum_{e \in E} \sum_{m \in \mathbb{Z}^n} c_{2(m+e)}^{j+1} \psi_{j+1}\left(2^{j+1} \cdot -2(m + e)\right)$$

$$= \sum_{e \in E} \sum_{m \in \mathbb{Z}^n} c_{2(m+e)}^{j+1} \psi_{j+1}\left(2(2^j \cdot -(m + e))\right)$$

with square-summable coefficients c_i^{j+1}. Next it can be decomposed into its parts from V_j and W_j. For this, we insert and translate the decomposition (9.4) into the expansion (9.5) as follows (questions of convergence and interchanges of infinite series etc. will be settled below):

$$f_{j+1} = \sum_{e \in E} \sum_{m \in \mathbb{Z}^n} \sum_{i \in \mathbb{Z}^n} c_{2(m+e)}^{j+1} a_{i,e}^j \psi_j(2^j \cdot -m - i)$$

$$+ \sum_{\substack{e \in E \\ }} \sum_{m \in \mathbb{Z}^n} \sum_{\substack{i \in \mathbb{Z}^n \\ f \in E^*}} c_{2(m+e)}^{j+1} b_{i,e}^{j,f} \omega_{j,f}(2^j \cdot -m - i)$$

$$= \sum_{i \in \mathbb{Z}^n} \sum_{e \in E} \sum_{m \in \mathbb{Z}^n} c_{2(m+e)}^{j+1} a_{i-m,e}^j \psi_j(2^j \cdot -i)$$

$$+ \sum_{\substack{i \in \mathbb{Z}^n \\ f \in E^*}} \sum_{e \in E} \sum_{m \in \mathbb{Z}^n} c_{2(m+e)}^{j+1} b_{i-m,e}^{j,f} \omega_{j,f}(2^j \cdot -i).$$

Through the use of Cauchy's summation formula we see that the result can be identified by convolving the coefficient sequence $c^{j+1} = \{c_i^{j+1}\}_{i \in \mathbb{Z}^n}$ with $a^j = \{a_{i,e}^j\}_{i \in \mathbb{Z}^n, e \in E}$ and $b^{j,f} = \{b_{i,e}^{j,f}\}_{i \in \mathbb{Z}^n, e \in E}$, respectively. Call the resulting discrete convolutions and new coefficient sequences

$$c^j = a^j * c^{j+1} = \left\{ \sum_{\substack{m \in \mathbb{Z}^n \\ e \in E}} c_{2(m+e)}^{j+1} a_{\ell-m,e}^j \right\}_{\ell \in \mathbb{Z}^n}$$

and $d^{j,f} = b^{j,f} * c^{j+1}$. The first one of those, namely c^j, leads, if square-summable, to another function f_j in V_j with a decomposition into its parts in V_{j-1} and W_{j-1} etc. Therefore we get a recursive decomposition of any square-integrable function f which is initially to be approximated by $f_{j+1} \in V_{j+1}$ for a suitable j, and then decomposed in the fashion outlined above. The important initial approximation is possible to any accuracy due to condition (MRA2).

The decompositions are computed faster if the $a_{.,e}^j$ and $b_{.,e}^{j,f}$ sequences are local, e.g. exponentially decaying or even compactly supported with respect to the index, because then the convolutions, i.e. the infinite series involved, can be computed faster, although $c^j, d^{j,f} \in \ell^2(\mathbb{Z}^n)$ are already ensured by Young's inequality from the fifth chapter if the sequences $a_{.,e}^j, b_{.,e}^{j,f}$ are absolutely summable. This also implies validity of the above operations on the series. In summary the decomposition can be pictured by the sequence

$$f \approx f_{j+1} \to g_j \text{ and } f_j \to g_{j-1} \text{ and } f_{j-1} \to \cdots$$

or, with respect to its coefficients that are obtained by convolution as described above,

$$c^{j+1} \to d^{j,e} \text{ and } c^j \to d^{j-1,e} \text{ and } c^{j-1} \to \cdots.$$

Re-composing a function from its prewavelets parts is executed in the same, recursive way, albeit using different coefficient sequences for the convolution, see also Daubechies (1992).

It is relatively easy to analyse the complexity of the above FWT. At each stage the complexity only depends on the length of the expansions which use the coefficient sequences a^j and $b^{j,f}$. If they are compactly supported with respect to their index, this introduces a fixed constant into the complexity count. Otherwise, it is usual to truncate the infinite series, which is acceptable at least if they are exponentially decaying sequences. The number of stages we use in the decomposition or reconstruction depends on the resolution we wish to achieve because the remainder $f_{j-1} \in V_{j-1}$ that we get at each stage of the algorithm becomes a more and more 'blurred' part of the original function and will finally be omitted.

9.3.3 The refinement equation

It is now our task to find suitable bases or generators

$$\{\omega_{j,e}(2^j \cdot -i)\}_{i \in \mathbb{Z}^n, e \in E^*}$$

for these prewavelet spaces W_j. They are the prewavelets which we seek. There is a multitude of ways to find these generators and consequently there is a multitude of different prewavelets to find. It is standard, however, at least to normalise the prewavelets to Euclidean norm one. Our restricted aim in this chapter is to exhibit a few suitable choices of prewavelets involving radial basis functions and therefore we begin by constructing multiresolution analyses with radial basis functions.

There is a convenient and very common sufficient condition for the first property of multiresolution analysis (MRA1), that is the nestedness condition (MRA1) is always a consequence of the so-called *refinement equation* which is a standard requirement in the stationary setting and which we shall employ in connection with radial basis functions. We shall explain this notion as follows. We recall the notation $\ell^1(\mathbb{Z}^n)$ for the set of absolutely summable sequences indexed ones $\mathbb{Z}^n$.

Definition 9.2. *The square-integrable function ψ satisfies a stationary refinement equation if there exists a sequence $a \in \ell^1(\mathbb{Z}^n)$ such that*

$$\psi(x) = \sum_{j \in \mathbb{Z}^n} a_j \psi(2x - j), \qquad x \in \mathbb{R}^n,$$

where $a = \{a_j\}_{j \in \mathbb{Z}^n}$. If ψ and ψ_1 from $L^2(\mathbb{R}^n)$ satisfy for a sequence $a \in \ell^1(\mathbb{Z}^n)$

$$\psi(x) = \sum_{j \in \mathbb{Z}^n} a_j \psi_1(2x - j), \qquad x \in \mathbb{R}^n,$$

then they provide a nonstationary refinement equation.

This implies indeed in both cases that the spaces defined in (MRA4) are nested if $\psi \in V_0$, $\psi_1 \in V_1$ are the respective generators, that is $V_0 \subset V_1$, as desired for a stationary or nonstationary MRA.

That fact is again due to Young's inequality quoted in Chapter 5. We apply it for the stationary case: the inequality implies that, if we replace the shifts of $\psi(2\cdot)$ on the right-hand side of the refinement equation by a sum of such translates (with ℓ^2-coefficients, i.e. square-summable ones), then the result can be written as a sum of translates of ψ with ℓ^2-coefficients:

$$\sum_{k \in \mathbb{Z}^n} c_k \psi(\cdot - k) = \sum_{k \in \mathbb{Z}^n} c_k \sum_{j \in \mathbb{Z}^n} a_j \psi(2 \cdot -2k - j) = \sum_{j \in \mathbb{Z}^n} d_j \psi(2 \cdot -j).$$

Here $d_j = \sum_{k \in \mathbb{Z}^n} a_{j-2k} c_k$ for all multiintegers j. (Incidentally, $a \in \ell^1(\mathbb{Z}^n)$ is for this not a necessary condition but suffices. It is only necessary that $a \in \ell^2(\mathbb{Z}^n)$. We usually require summability because of the above application of Young's inequality which makes the analysis simple, and because then the associated trigonometric sum

$$a(x) := \sum_{j \in \mathbb{Z}^n} a_j e^{-ij \cdot x}$$

is a continuous function.)

In particular the above coefficients d_j are square-summable. In the nonstationary case, the argument remains essentially the same, where ψ has to be replaced by ψ_1 on the right-hand side of the refinement equation.

We continue with our example of radial basis functions (4.4) for the application of the above in order to show its usefulness. To this end, we shall show next that the refinement equation is always satisfied in our setting for the radial basis functions (4.4) or, rather, for the thereby generated differences ψ, due to the homogeneity of $\hat\phi$ when our radial basis functions (4.4) are used.

In our case of choosing quasi-interpolating radial basis functions as above, the a_j are a constant multiple of the Fourier coefficients of $g(2\cdot)/g$, where $g: \mathbb{T}^n \to \mathbb{R}$ is the numerator in (9.2). Indeed, if we take Fourier transforms in the first display in Definition 9.2 on both sides, then the refinement equation reads

$$\hat\psi(x) = \frac{1}{2^n} \sum_{j \in \mathbb{Z}^n} a_j e^{-ij \cdot x/2} \hat\psi\left(\frac{x}{2}\right),$$

from which we get for our example that

$$\left(\frac{1}{2}\right)^n a(x) = \frac{\hat\psi(2x)}{\hat\psi(x)} = \frac{g(2x)}{2^{2k} g(x)},$$

the single isolated zero in $\mathbb{T}^n$ of the denominator on the right-hand side cancelling against the same in the numerator because of the form of g in (9.2).

That the Fourier coefficients of the expansion on the left-hand side of the above display are absolutely summable is therefore a consequence of Wiener's lemma stated in Chapter 2, so indeed the trigonometric sum converges absolutely and is a continuous function. In order to finish the example, it is required further that the above ψ, and the V_j generated from its translates and dilates, generate a multiresolution analysis, i.e. our remaining assertion to be established is that the conditions (MRA2) and (MRA3) hold, (MRA1) being true anyhow by the construction and the refinement equation.

Indeed, the second condition of multiresolution analysis (MRA2) holds because the translates and dilates of ψ provide approximations in V_j to all continuous, at most linearly growing functions f, say, and those approximations converge uniformly for $j \to \infty$. We exemplify those approximants by using the simplest form of a quasi-interpolant – which we recognise from Chapter 4 – with spacing $h = 2^{-j}$,

$$Q_{2^{-j}} f(x) = \sum_{k \in \mathbb{Z}^n} f(k 2^{-j}) \psi(2^j x - k), \qquad x \in \mathbb{R}^n.$$

This quasi-interpolant is well-defined and exact *at least* on all linear polynomials and, additionally, converges uniformly to f on the whole n-dimensional Euclidean space as $j \to \infty$ for all functions f from a suitable class that is dense in $L^2(\mathbb{R}^n)$, for instance the class of twice continuously differentiable square-integrable functions with bounded derivatives, cf. for instance Theorem 4.5. We recall that this is only a simple example of a suitable approximation and much better approximation results can be found but this it sufficient in our present context of square-integrable functions, because we are satisfied if we can show density. Therefore condition (MRA2) of MRA is verified while (MRA3) requires further work which will be addressed in the next subsection.

9.3.4 The Riesz basis property

The condition (MRA3) of multiresolution analysis holds because the translates and dilates of ψ form Riesz bases of the V_j. This is, again, a most useful and commonly required property to fulfil the third property of multiresolution analysis (MRA3), which also has other pertinent consequences. We will define and explain Riesz bases now for stationary multiresolution analyses.

Definition 9.3. *Let j be an integer. The translates $\psi(2^j \cdot -k)$ for multiintegers k form a Riesz basis of V_j if there exist positive finite constants μ_j and M_j such that the 'Riesz stability' inequalities*

$$(9.6) \quad \mu_j \|c\| \le \left\| \sum_{k \in \mathbb{Z}^n} c_k \psi(2^j \cdot -k) \right\|_2 \le M_j \|c\|, \quad c = \{c_k\}_{k \in \mathbb{Z}^n} \in \ell^2(\mathbb{Z}^n),$$

*hold for all $j \in \mathbb{Z}$ independently of c and the ratio $M_j/\mu_j \geq 1$ is uniformly
bounded in j. Here $\| \cdot \|$ is the norm for the sequence space $\ell^2(\mathbb{Z}^n)$, while $\| \cdot \|_2$
denotes the standard $L^2(\mathbb{R}^n)$ norm.*

In Fourier transform form, that is, after an application of the Parseval–Plancherel
identity, this expression (9.6) reads for $j = 0$, using here the common and
convenient abbreviation $c(x)$ for the trigonometric series $\sum_{k \in \mathbb{Z}^n} c_k e^{-ix \cdot k}$,

$$(2\pi)^{n/2}\mu_0\|c\| \leq \|c(\cdot) \times \hat{\psi}(\cdot)\|_2 \leq M_0\|c\|(2\pi)^{n/2}.$$

It is useful to express this Riesz condition after periodisation in the alternative
formulation

$$(9.7) \qquad (2\pi)^{n/2}\mu_0\|c\| \leq \left\| c(\cdot) \sum_{k \in \mathbb{Z}^n} \hat{\psi}(\cdot + 2\pi k) \right\|_{2,\mathbb{T}^n} \leq (2\pi)^{n/2}M_0\|c\|,$$

with the new notation $\| \cdot \|_{2,\mathbb{T}^n}$ denoting the Euclidean norm now restricted to the
cube $\mathbb{T}^n$ in (9.7). We absorb constant multipliers such as $(2\pi)^{n/2}$ in (9.7) into the
constants μ_0 and M_0 from now on to keep our formulae as simple as possible.

We wish to come back to our example now and apply our theory to that
example. Thus we shall establish that the translates and scales of our choice of
ψ form a Riesz basis of each V_j, for a suitable g, by fulfilling (9.7). This is in fact
the *third important consequence* that we have already alluded to of the particular
shape of $\hat{\phi}$ when ϕ is from the class (4.4). Indeed, if $\hat{\psi}(\cdot) = g(\cdot) \times \hat{\phi}(\| \cdot \|)$,
then the last display (9.7) reads by g's periodicity

$$\mu_0\|c\| \leq \left\| c(\cdot)g(\cdot) \sum_{k \in \mathbb{Z}^n} \hat{\phi}(\| \cdot + 2\pi k \|) \right\|_{2,\mathbb{T}^n} \leq M_0\|c\|.$$

The reason why these inequalities hold for the ψ in (9.2) here is that $\hat{\phi}$ has no
zero and that our chosen g exactly matches $\hat{\phi}$'s singularity at zero by a high
order zero at the origin without having any further zeros elsewhere or higher
order zeros at 0. Still taking $j = 0$, the upper and the lower bounds in the
display above can be easily specified. They are the finite supremum and the
positive infimum of the square root of

$$\sum_{k \in \mathbb{Z}^n} |\hat{\psi}(t + 2\pi k)|^2 = |g(t)|^2 \sum_{k \in \mathbb{Z}^n} \hat{\phi}(\|t + 2\pi k\|)^2, \qquad t \in \mathbb{T}^n,$$

respectively, because of the well-known identity for Fourier series and square-
integrable sequences

$$(9.8) \qquad\qquad \|c(\cdot)\|_2 = (2\pi)^{n/2}\|c\|$$

from the Appendix and because of the following simple estimate. We get by appealing to Hölder's inequality and taking supremum

$$\left\| c(\cdot) g(\cdot) \sum_{k \in \mathbb{Z}^n} \hat{\phi}(\| \cdot + 2\pi k \|) \right\|_{2, \mathbb{T}^n} \leq \| c(\cdot) \|_2 \sup_{t \in \mathbb{T}^n} \sqrt{\sum_{k \in \mathbb{Z}^n} |\hat{\psi}(t + 2\pi k)|^2}.$$

Similarly we get an estimate for the *infimum* by reversing the inequality. Both of these quantities sup and inf are positive and finite due to the positivity of $|\hat{\psi}|^2$ and by the definition of g, the summability of the infinite series being guaranteed as usual because $2k > n$. For other integers j, (9.6) follows immediately from scaling of the so-called Riesz constants μ_0 and M_0 by $2^{jn/2}$, but it is sufficient to consider the decomposition $V_1 = V_0 + W_0$ instead of all the decompositions $V_{j+1} = V_j + W_j$ when establishing the Riesz property of bases. In particular, the ratios M_j/μ_j are constant here. It is usual to restrict the above condition (9.6) of a Riesz property to $j = 0$. We leave the examples (9.2) now and return to the general case.

Having verified the Riesz basis property, we can prove that the property (9.6) implies (MRA3) by the following argument. We get from (9.6) that for every $f \in V_0$ with expansion $f = \sum_{k \in \mathbb{Z}^n} c_k \psi(\cdot - k)$ by the Cauchy–Schwarz inequality

$$\| f \|_\infty = \sup \left| \sum_{k \in \mathbb{Z}^n} c_k \psi(\cdot - k) \right| \leq C \| c \| \leq C \mu_0^{-1} \| f \|_2.$$

Here, sup denotes the essential supremum, i.e. the supremum taken everywhere except on a set of measure zero.

Therefore, by a change of variables, we get for a specific $g \in V_{-j}, j \in \mathbb{Z}$, and a special f, namely that specified by $f := g(2^j \cdot)$,

$$\| g \|_\infty \leq C \mu_0^{-1} \| f \|_2 = 2^{-j/2} C \mu_0^{-1} \| g \|_2 \to 0, \qquad j \to \infty.$$

Thus it follows that $g = 0$ almost everywhere as soon as it lies in the intersection of all $V_{-j}, j \in \mathbb{Z}$, and (MRA3) is therefore established by this argument.

As we have already pointed out, this Riesz property is a condition that is sometimes incorporated into the requirements of stationary multiresolution analysis as well. It can be viewed as a suitable replacement for the orthonormality condition that is imposed on the translates of *wavelets* (as opposed to prewavelets) at each level j of scaling. In fact, if the translates are orthonormal, then $\mu_j \equiv M_j \equiv 1$ are the correct choices in (9.6) because the orthonormality of translates of square-integrable functions ψ is equivalent to

(9.9)
$$\sum_{j \in \mathbb{Z}^n} |\hat{\psi}(\cdot + 2\pi j)|^2 \equiv 1$$

almost everywhere and because of (9.8). The identity (9.9) is equivalent to orthonormality because the orthonormality conditions are, by application of the Parseval–Plancherel formula for square-integrable functions from the Appendix and by periodisation of ψ,

$$
\begin{aligned}
\delta_{0k} &= \int_{\mathbb{R}^n} \psi(x)\overline{\psi(x-k)}\,dx \\
&= \frac{1}{(2\pi)^n} \int_{\mathbb{R}^n} e^{ix\cdot k} |\hat{\psi}(x)|^2\,dx \\
&= \frac{1}{(2\pi)^n} \int_{\mathbb{T}^n} \sum_{j\in\mathbb{Z}^n} |\hat{\psi}(x+2\pi j)|^2 e^{ix\cdot k}\,dx.
\end{aligned}
$$

This equals δ_{0k} for all $k \in \mathbb{Z}^n$ precisely whenever (9.9) is true.

9.3.5 First constructions of prewavelets

It is time to make use of multiresolution analysis because it opens the door to many simple and useful constructions of prewavelets for us both in general settings and in our example which is relevant in the context of radial basis functions and in our book.

Definition 9.4. *Let a sequence of closed subspaces V_j of $L^2(\mathbb{R}^n)$ satisfy the conditions of a stationary multiresolution analysis. A set $\{\omega_e\}_{e\in E^*}$, E^* being an index set, of functions with L^2-norm one is a set of* prewavelets *if its elements are in V_1, are orthogonal to V_0, and if their translates together span $V_1 \ominus V_0$. In the nonstationary case, the functions $\omega_{j,e}$ must be in V_{j+1}, orthogonal to V_j and the $\omega_{j,e}(2^j\cdot - i)$, $i \in \mathbb{Z}^n$, $e \in E^*$, span $V_{j+1} \ominus V_j$ for all integral j in order that they be prewavelets.*

The simplest method to find prewavelets is to note that in the stationary case, where we remain for the time being,

$$
V_1 = \mathrm{span}\left\{ \psi_1(\cdot - e - j) \,\middle|\, j \in \mathbb{Z}^n,\ e \in E \right\},
$$

where $\psi_1 := \psi(2\cdot)$, and define

$$
\omega_e = \psi_1(\cdot - e) - P\psi_1(\cdot - e)
$$

for all $e \in E^* := E \setminus \{0\}$. Here $P\colon V_1 \to V_0$ is the least squares (orthogonal) projection. Then it is easy to see that those ω_e and their multiinteger translates generate W_0 due to the fact that $W_0 = V_1 \ominus V_0$ contains precisely the residuals of the least squares projection from V_1 onto V_0 and this is what we compute in the above display for V_1's generators. In particular, orthogonality of the prewavelets on V_0 is a trivial consequence of the projection we are using in the prewavelet's definition.

A description by Fourier transform of this ω_e defined in the previous display is

$$\hat{\omega}_e := \hat{\psi}_e - \frac{\sum_{j\in\mathbb{Z}^n} \hat{\psi}_e(\cdot+2\pi j)\overline{\hat{\psi}(\cdot+2\pi j)}}{\sum_{j\in\mathbb{Z}^n} |\hat{\psi}(\cdot+2\pi j)|^2} \times \hat{\psi},$$

if the denominator is positive, where ψ_e is defined by $\psi_e = \psi_1(\cdot-e)$ and where the $\hat{\psi}$ can be for example the function (9.2). This is so because the orthogonal (L^2)-projection $P: V_1 \to V_0$ can be characterised through the useful proposition below. We still assume the V_j to form an MRA.

Proposition 9.1. *Let $f \in V_1$ and ψ be as above. Then the orthogonal ('least squares') projection Pf of f with respect to the standard Euclidean inner product on square-integrable functions from $V_1 \to V_0$ is characterised by its Fourier transform*

$$\widehat{Pf} = \hat{\psi} \times \frac{\sum_{j\in\mathbb{Z}^n} \hat{f}(\cdot+2\pi j)\overline{\hat{\psi}(\cdot+2\pi j)}}{\sum_{j\in\mathbb{Z}^n} |\hat{\psi}(\cdot+2\pi j)|^2}$$

so long as the right-hand side of the display is in $L^2(\mathbb{R}^n)$.

Proof: The proof of this fact is easy and it works out as follows. As we have pointed out in the previous chapter, we just need to show the orthogonality of $f - Pf$, Pf being defined through its Fourier transform in the last display, to any element of V_0.

Recalling ψ's square-integrability and using the Parseval–Plancherel identity for square-integrable functions, again from the Appendix, this amounts to showing for all $k \in \mathbb{Z}^n$ the identity

$$\int_{\mathbb{R}^n} \left(\hat{f}(x) - \widehat{Pf}(x)\right)\bar{\hat{\psi}}(x)e^{ik\cdot x}\, dx = 0.$$

That is, by 2π-periodisation of the integrand, and inserting the $\widehat{Pf}(x)$ from the above definition,

$$\int_{\mathbb{T}^n} \left(\sum_{j\in\mathbb{Z}^n} \hat{f}(x+2\pi j)\overline{\hat{\psi}(x+2\pi j)} \sum_{j\in\mathbb{Z}^n} |\hat{\psi}(x+2\pi j)|^2 \right.$$
$$\left. - \sum_{j\in\mathbb{Z}^n} |\hat{\psi}(x+2\pi j)|^2 \sum_{j\in\mathbb{Z}^n} \hat{f}(x+2\pi j)\overline{\hat{\psi}(x+2\pi j)}\right)$$
$$\times \left(\sum_{j\in\mathbb{Z}^n} |\hat{\psi}(x+2\pi j)|^2\right)^{-1} e^{ik\cdot x} dx = 0.$$

This is true by construction for all k. $\qquad\square$

9.4 Special constructions

9.4.1 An example of a prewavelet construction

In this first set-up, the prewavelets are found that are described in our next theorem. Their construction is related to the construction that led to the orthonormalisation used in our Theorem 8.2 and to the work in Chui and Wang (1991) and Micchelli, Rabut and Utreras (1991).

Theorem 9.2. *Let ψ be as specified through (9.2), and let the MRA be generated by the V_j as above. Define the square-integrable ω_0 by its Fourier transform*

$$\hat{\omega}_0(y) = \|y\|^{2k} \frac{|\hat{\psi}(\frac{y}{2})|^2}{\sum_{\ell \in \mathbb{Z}^n} |\hat{\psi}(\frac{y}{2} + 2\pi\ell)|^2}, \qquad y \in \mathbb{R}^n,$$

where as before $2k > n$. Further define the translates along the set E^* of corners of the unit cube, except zero, by the identity

$$(9.10) \qquad \omega_e(y) = \omega_0(y - e), \qquad y \in \mathbb{R}^n, \ e \in E^*.$$

Let finally spaces be defined by

$$W_{j,e} := \operatorname{span}\{\omega_e(2^j \cdot -i) \mid i \in \mathbb{Z}^n\}, \ e \in E^*,$$

and their direct sum

$$W_j := \bigoplus_{e \in E^*} W_{j,e}, \qquad j \in \mathbb{Z}.$$

Then we have orthogonality $W_j \perp W_\ell$, and in particular $V_{j+1} = V_j \oplus W_j$, $W_j \perp V_j$, for all integers $j \neq \ell$, and we have the expansion for all square-integrable functions as an infinite direct sum

$$L^2(\mathbb{R}^n) = \bigoplus_{j=-\infty}^{\infty} W_j.$$

In particular, the normalised $\omega_e/\|\omega_e\|_2$, $e \in E^*$, form a set of prewavelets.

We note first that the functions defined in (9.10) are indeed in V_1. If we take the Fourier transform of (9.10), we get

$$e^{-iy\cdot e} \|y\|^{2k} \frac{|\hat{\psi}(\frac{y}{2})|^2}{\sum_{\ell \in \mathbb{Z}^n} |\hat{\psi}(\frac{y}{2} + 2\pi\ell)|^2} = h(y)|\hat{\psi}(\tfrac{y}{2})|^2 \hat{\phi}(\|y\|)^{-1},$$

where h is a square-integrable 4π-periodic function. Moreover,

$$\left|\hat{\psi}\left(\frac{y}{2}\right)\right|^2 \hat{\phi}(\|y\|)^{-1}$$

is a constant multiple of $\bar{g}(\frac{1}{2}y)\hat{\psi}(\frac{1}{2}y)$, so in total we have an expression of the form $\tilde{h}(y)\hat{\psi}(\frac{1}{2}y)$, where $\tilde{h}$ is 4π-periodic and square-integrable. Hence, by reverting from the Fourier domain to the real domain and recalling that V_1 is spanned by translates $\psi(2 \cdot -i)$, we get that (9.10) indeed defines a function in V_1, as required.

This Theorem 9.2 is, in the context of this chapter, too difficult to prove completely, but we shall show as an example the orthogonality claim that is being made. This is in fact quite straightforward. We observe first that it suffices to show $W_0 \perp V_0$. Let f be from E^*. Then we have that for any $k \in \mathbb{Z}^n$

$$\int_{\mathbb{R}^n} e^{-iy \cdot k} \hat{\psi}(y)\overline{\hat{\omega}_f(y)}\,dy$$

is the same as a fixed constant multiple of

$$\int_{\mathbb{R}^n} e^{i2y \cdot (f-k)} g(2y)\frac{|\hat{\psi}(y)|^2\,dy}{\sum_{\ell \in \mathbb{Z}^n}|\hat{\psi}(y+2\pi\ell)|^2} = \int_{\mathbb{T}^n} e^{i2y \cdot (f-k)} g(2y)\,dy,$$

the latter by standard 2π-periodisation of the integrand. This, however, vanishes because $f \neq 0$, so the inner product is zero, as required. Therefore we have established the orthogonality claim for $W_0 \perp V_0$, from which it follows by an inductive argument and from the condition (MRA1a) of the multiresolution analysis that the full orthogonality claim between spaces as asserted in the theorem statement is true.

9.4.2 Generalisations, review of further constructions

We now go into the more general concept of nonstationary MRA and observe that a considerable generalisation can be achieved by continuing to let

$$V_0 = \text{span}\left\{\psi(\cdot - j)\,\Big|\,j \in \mathbb{Z}^n\right\},$$

but introducing a new space

$$V_1 = \text{span}\left\{\psi_1(2 \cdot -j)\,\Big|\,j \in \mathbb{Z}^n\right\},$$

where ψ and ψ_1 are related not by scaling but only by the nonstationary refinement equation, or, in the Fourier domain, by the relation

(9.11) $$\hat{\psi} = a\left(\frac{1}{2}\cdot\right)\hat{\psi}_1$$

which they must obey with a 2π-periodic, square-integrable function $a(\cdot)$. In the special case when the nonstationary refinement equation holds for summable coefficients a_j of $a(\cdot)$, then $a(\cdot)$ is continuous, as we have noted before.

Now, we assume, additionally, that supp $\hat{\psi}$ = supp $\hat{\psi}_1$ = $\mathbb{R}^n$, which is true in all our examples and is true, in particular, when ψ and ψ_1 are compactly supported such as B-splines or box-splines. This is because the Fourier transforms of compactly supported functions are entire functions and therefore their (closed) support must be the whole space, unless they vanish identically. In the case of our radial basis functions, the fact that supp $\hat{\psi}$ = $\mathbb{R}^n$ for our example (9.2) can be verified directly by inspecting its Fourier transform and the Fourier transform of the radial basis function. It is a consequence of this that the above condition (MRA2) is immediately fulfilled in that case. Indeed, it would be sufficient that $\hat{\psi}$ and $\widehat{\psi_1}$ do not vanish at zero, that furthermore ψ and ψ_1 be integrable and square-integrable and that therefore both their Fourier transforms be continuous and bounded away from zero in a neighbourhood of the origin. Condition (MRA3) is always true in that case up to a subspace Y of dimension one or zero, i.e. the intersection of all V_j is Y (condition (MRA3) means that Y's dimension is zero) and the case $Y = \{0\}$ is automatically fulfilled if we are in the stationary setting. Therefore (MRA2) and (MRA3) need no further attention in the case considered here.

A general treatise on prewavelets is given in de Boor, DeVore and Ron (1993) that includes the proofs of the claims of the previous paragraph. The authors also show the sufficiency of supp $\hat{\psi}_j$ = $\mathbb{R}^n$ in order that the W_j are finitely generated shift-invariant spaces, i.e. generated by the multiinteger translates of just finitely many functions, and derive which are specific basis functions ω_e, $e \in E^*$, in V_{j+1} which generate the desired spaces W_j.

In Micchelli (1991), this subject is generalised by admitting scaling factors other than 2. In fact, scaling by integer matrices M with $|\det M| > 1$ is admitted where it is also required that $\lim_{j \to \infty} M^{-j} = 0$. This is usually achieved by requiring the sufficient condition that all of M's eigenvalues be larger than one in modulus. In this event, $|\det M| - 1$ prewavelets are required and their integer shifts together span each W_j. Except for some technical conditions which we do not detail here and which hold for our radial basis functions here, it is sufficient for the existence of the prewavelets that a square-integrable ψ which decays at least as $(1 + \|x\|)^{-n-\varepsilon}$ for $\varepsilon > 0$ and whose Fourier transform has no 2π-periodic zeros generates spaces V_j in the fashion described above, albeit by scaling $\psi_j := \psi(M^j \cdot)$, that satisfy the conditions of stationary with scaling by M MRA and a refinement equations, again with scaling by two componentwise, replaced by M. Hence, the V_1 for instance is now of the form

$$V_1 := \text{span}\left\{\psi_1(M \cdot - j) \,\middle|\, j \in \mathbb{Z}^n\right\},$$

where the standard choice for M which corresponds to our earlier choice of scaling by two is the unit matrix multiplied by two. The nesting property of the V_1 is now ensured by the refinement condition

$$(9.12) \qquad \psi(x) = \psi_0(x) = \sum_{j \in \mathbb{Z}^n} a_j \psi_1(Mx - j), \qquad x \in \mathbb{R}^n,$$

with the condition $a = \{a_i\}_{i \in \mathbb{Z}^n} \in \ell^1(\mathbb{Z}^n)$ as usual.

Utreras (1993) constructs prewavelets in the same way as in Theorem 9.2 and points to the important fact that no multiresolution analysis can be found for *shifts* of (4.4), i.e. when ϕ is replaced by $\phi(\sqrt{r^2 + c^2})$ for $c > 0$. More generally, he shows that the exponential (that is, too fast) decay of the Fourier transform $\hat{\phi}$ of the resulting basis function ϕ prohibits the existence of a multiresolution analysis generated by shifts and scales of ϕ or a linear combination of translates of ϕ. (It is the nestedness condition of the spaces in the sequence that fails.) This applies for example to the multiquadric radial basis function for odd n. This statement, however, only holds if the multiresolution analysis is required to be *stationary*, i.e. results from the dilates and shifts of just *one* function as in (9.3). On the other hand, several authors have studied the aforementioned *nonstationary* multiresolution analyses with radial basis functions, where the generating function of each V_j may be a different $L^2(\mathbb{R}^n)$ function ψ_j.

For instance, in the author's paper (1993b), univariate prewavelets from spaces spanned by (integer) translates of multiquadric and related functions are constructed. In order to get a square-integrable basis function first, derivatives of multiquadric functions are taken and convolved with B-splines. This is the same as taking divided differences of the radial basis function as in the beginning of this chapter, but it is more amenable to analysis because one can make use of the positivity of ϕ's distributional Fourier transform. After all, convolution of functions in the real domain means function multiplication in the Fourier domain (cf. Appendix). This fact is extensively used in the proofs of the results that follow. Let $\psi \in C(\mathbb{R})$ with

$$(9.13) \qquad |\psi(x)| = O(|x|^{-1-\varepsilon}), \qquad x \to \pm\infty,$$

$\hat{\psi}(t) > 0$, for all $t \in \mathbb{R}$, and $\hat{\psi}(0) = 1$ be given. The quantity ε is fixed and positive. We restrict ourselves here to the example $\psi(x) = \hat{k}\phi^{(2\lambda)}(x)$, where $2\lambda = \hat{n} + 1, \hat{k}$ is a suitable normalisation parameter, $\phi(x) = \tilde{\phi}(\sqrt{x^2 + c^2}), c \geq 0$, and $\tilde{\phi}$ is one of the functions $\tilde{\phi}(r) = r^{2\lambda - 1}, \lambda \in \mathbb{N}$, but the theory in Buhmann (1993b) is more general. Nevertheless, this covers the multiquadric example (for $\lambda = 1$). We consider the continuous convolutions – whose properties are summarised in the Appendix –

(9.14)
$$\begin{cases} C_j := B_j^{\text{c}} * \psi, & j \in \mathbb{Z}, \\ F_j := B_j^{\text{f}} * \psi, & j \in \mathbb{Z}. \end{cases}$$

The B-splines are the same as those in Chapter 2; the quantity $\hat{n}$ is their degree and B_j^{c} has knots in $\mathbb{Z}$ and support in $[j, j+\hat{n}+1]$, whereas B_j^{f} has knots in $\frac{1}{2}\mathbb{Z}$ and support in $\frac{1}{2}[j, j+\hat{n}+1]$. Thus, C_j and F_j are in $L^1(\mathbb{R}) \cap L^2(\mathbb{R})$, because the B-splines are and because $\psi \in L^1(\mathbb{R})$. We now define V_0 and V_1 as

(9.15)
$$\begin{cases} V_0 := \left\{ \displaystyle\sum_{j=-\infty}^{\infty} c_j C_j \,\middle|\, c = \{c_j\}_{j=-\infty}^{\infty} \in \ell^2(\mathbb{Z}) \right\}, \\[3mm] V_1 := \left\{ \displaystyle\sum_{j=-\infty}^{\infty} c_j F_j \,\middle|\, c = \{c_j\}_{j=-\infty}^{\infty} \in \ell^2(\mathbb{Z}) \right\}. \end{cases}$$

Then the $\omega_\ell := F_\ell - P F_\ell$, $\ell \in \mathbb{Z}$, is a prewavelet. Here $P : V_1 \to V_0$ is the same L^2-projection operator as above. It is furthermore true that the prewavelet decays at least as fast as

(9.16)
$$|\omega_\ell(x)| = O\big((1 + |x - \ell|)^{-3}\big), \qquad x \in \mathbb{R}.$$

So, although not of compact support or exponential decay, these prewavelets have quick algebraic decay (9.16) which is comparable with the properties of the quasi-interpolating basis functions ψ of the fourth chapter.

In summary, prewavelets with radial basis functions are also an admissible tool for approximation that has many useful properties which led to the development of wavelets and prewavelets. It is an alternative to the interpolation studied earlier in the book, depending on the particular application we wish to address.

10

Further Results and Open Problems

This book ends with a short chapter that has two goals. The first one is to catch up with some recent results on radial basis function approximation which have not been included in the main body of the book, so as to keep the book on a sufficiently introductory level and homogeneous as to the direction of the results. Nonetheless, there are many additional, new results which deserve to be mentioned in a monograph like this about radial basis functions. We take care of some of them by way of a short summary of results with a reference to a cited paper or book that is suitable for further study.

The second one is to mention several open problems that have, as far as the author is aware, not yet been solved at the time of writing this book and which are, partly, in close connection with the recent results in this chapter. (Since writing this book was a long term project, some very recent solutions may almost certainly have been overlooked by the author.) We hope to raise interest in the reader in those problems as subjects of general further, perhaps their own, research work.

10.1 Further results

This section enlists and briefly discusses a number of further results in the analysis of radial basis functions.

One of the very recent developments under much investigation now is the idea of the so-called radial basis function multilevel methods. The idea came up with the observation that radial basis functions with compact support have no remarkable convergence orders to speak of if they are used in the way classical radial basis functions are employed, i.e. with centres becoming dense in a domain, and therefore additional work for improvement is needed. We have already seen this to an extent in the sixth chapter where we briefly discussed their convergence properties. The first proposers of this idea were Narcowich, Schaback and Ward (1999), Floater and Iske (1996) and, later, Fasshauer (1999) who made particular use of this idea for the numerical solution of differential equations.

The idea is one of iterative refinement of an initial, quite crude approximation with radial basis functions with compact support by updating the residuals iteratively. For the first step, radial basis functions are employed for interpolation whose support sizes may even be such that *no* convergence at all could be expected if they were used without iteration, and only with the standard refinement of the spacing of the set of data points. Therefore the latter, standard approach is not the final approximation, but an iterative refinement of the residuals is applied, similar to the BFGP method of Chapter 7, by using at each level different radial basis functions and, especially, differently scaled ones.

In other words, an interpolant to the residual of the previous interpolation process is computed with different basis functions and/or scales and subsequently added to the previous residual. This process is repeated at each iteration. Unfortunately, there is, at present, not enough known about convergence orders of this idea of approximation but it works well in practice (Fasshauer, 1999, Chen *et al.*, 1999). The idea behind the approach is of course the notion of capturing various features of the approximand at different resolutions (frequencies) at different levels of the multilevel algorithm. In that, it resembles the well-known prewavelet approximations outlined in Chapter 9. One of the principal applications is the solution of (elliptic, in some applications nonlinear) partial differential equations by Galerkin methods, where radial basis functions with compact support are ideal for generating the test functions and the associated inner products for the so-called stiffness matrices, much like the piecewise polynomials of compact support in finite elements, but they have *a priori* the aforementioned, possibly bad, convergence behaviour.

Convergence analysis for radial basis functions of global support has been undertaken in much more general settings than indicated in our Chapter 5. For instance, Duchon-type radial basis functions which use inner products like those in Chapter 5 but apply fractional derivatives in very general settings have been studied in many contexts. They are often called thin-plate splines under tension. The analysis is in principle the same and uses the same analytic ideas, but requires some changes in the convergence orders for the scattered data interpolants of Chapter 5. The fractional derivatives result in nonintegral powers μ in condition (A1) on our radial basis function's Fourier transform. A suitable reference is the thesis by Bouhamidi (1992).

Very general results characterising approximation orders for gridded data and quasi-interpolation – in fact characterisations of the radial basis functions and other functions of global support for being able to provide such convergence orders – are given by Jia and Lei (1993). The results are general especially in the sense that they hold for L^p-spaces and L^p-convergence orders and not only for uniform convergence which is the most often considered choice. The necessary and sufficient conditions are expressed in terms of the so-called Strang and Fix conditions that were used implicitly many times in Chapter 4 and were described there. Especially for making the conditions necessary, the notion of controlled approximation order is required which does not feature in our book, whereas this paper is a suitable reference to it. On the whole, this is an excellent paper which nicely complements the work presented in this book in the fourth chapter. A closely related paper is the article by Halton and Light (1993).

If the multiquadric function is used with a fixed parameter c, then exponential convergence orders can be achieved, as we have noted in Chapter 5. Another paper in the same vein which established spectral convergence orders for multiquadric interpolation with fixed c and gridded data is the one by Buhmann and Dyn (1993). The spectral convergence orders depend only on the smoothness of the approximand, once a certain minimal smoothness is guaranteed, and increase linearly with its smoothness. It is measured in terms of the Fourier transform of the approximand.

A long and detailed article, especially about quasi-interpolation with radial basis functions on an integer grid, is the one by Dyn, Jackson, Levin and Ron (1992) which shows how quasi-interpolating basis functions and the uniform (Chebyshev) convergence orders are obtained from large, general classes of radial basis functions and how the connection with their Fourier transforms is made. It is shown in more general contexts than we did in Chapter 4 how a particular form of the Fourier transform and its singularity at zero is necessary and sufficient for the existence of suitable quasi-interpolants. It also details more explicitly how the quasi-interpolating basis functions that we called ψ are found from ϕ, our work by contrast focussing more on interpolation, where the Ls are prescribed by the cardinality conditions.

In the previous two chapters we have used only L^2-theory to explain our wavelets and least squares approximations. Binev and Jetter (1992) offer an approach to radial basis function approximation as a whole and in particular cardinal interpolation *only* by means of L^2-theory which facilitates many things in the theory through the exclusive use of Hilbert space and Fourier transform techniques. The results obtained contain not many surprises as compared with our results, but the exclusive use of this L^2-approach makes their contribution extremely elegant and provides an analysis of the L^2-stability of radial basis function bases which we have not considered.

In Chapter 5 we have used error estimates with the aid of the so-called power functions and native spaces where on each 'side' of the error estimate, the same radial basis function is used. An interesting generalisation which can lead to larger classes of approximands that may be included in the error estimate is that of Schaback (1996). He proposes the use of different radial basis functions for the approximation on one hand and for the error estimate and the power function involved therein on the other hand. If the radial basis functions used for the power function leads to a larger native space, then the class of approximands considered in the error estimate is enlarged. By these means, no better error estimates can be expected as the approximation spaces are not changed, the approximation order being a result of the choice of the space of approximants, but the spaces of the approximands which are admitted to the error estimate are substantially enlarged.

Two other papers about radial basis function approximation on infinitely many scattered data are the ones by Buhmann and Ron (1994) and Dyn and Ron (1995). In both cases, very particular methods are developed for converting the known approximation methods and results (especially quasi-interpolation) to the setting of scattered data. The second paper mentioned is more comprehensive, as it shows how to do this conversion in very general cases. The former treats, on the other hand, convergence orders not only in Chebyshev norm but for general L^p-norms. In both articles, interpolation is not specifically used but general approximation from the radial basis function spaces, and the approximants are not explicit.

Other extensions and variations of the 'gridded data convergence theory' include the work by Bejancu (1997) on finitely many gridded data in a unit cube where he shows that for all radial basis functions belonging to the class of (4.4) in the inside of this unit cube, the same convergence orders are obtained as with the full set of multiintegers scaled by a step size h. The uniform convergence is measured for this on any compact set strictly inside the unit cube. The main tool for this proof is to use the localness of the cardinal functions for the full grid which we know from Chapter 4 in order to show that the approximants on the finite cube and on the full scaled integer grid vary only by a magnitude that is bounded above by an $O(h^{2k})$ term.

We have already mentioned in Chapter 4 how convergence with radial basis functions can also occur when we restrict ourselves to the fixed integer grid and let certain parameters in the radial basis functions (e.g. parameters in multiquadrics and Gaussians) vary. Another recent contribution to this aspect of radial basis function interpolation is the article by Riemenschneider and Sivakumar (1999) which studies Gaussians and their convergence behaviour for varying parameters in the exponential.

Due to the many applications that suit radial basis functions for instance in geodesy, there is already a host of papers that specialise the radial basis function approximation and interpolation to spheres. Freeden and co-workers (1981, 1986, 1995) have made a very large impact on this aspect of approximation theory of radial basis functions. There are excellent and long review papers available from the work of this group (see the cited references) and we will therefore be brief in this section here. Of course, we no longer use the conventional Euclidean norm in connection with a univariate radial function when we approximate on the $(n - 1)$-sphere S^{n-1} within $\mathbb{R}^n$ but apply so-called geodesic distances. Therefore the standard notions of positive definite functions and conditional positive definiteness no longer apply, and one has to study new concepts of (conditionally) positive definite functions on the $(n - 1)$-sphere. These functions are often called zonal rather than radial basis functions.

This work started with Schoenberg (1942) who characterised positive definite functions on spheres as those ones whose expansions in series of Gegenbauer polynomials always have nonnegative coefficients. Xu and Cheney (1992) studied strict positive definiteness on spheres and gave necessary and sufficient conditions. This was further generalised by Ron and Sun in 1996. Recent papers by Jetter, Stöckker and Ward (1999) and Levesley *et al.* (1999) use native spaces, (semi-)inner products and reproducing kernels to derive approximation orders in very similar fashions to the work of Chapter 5.

A recent thesis by Hubbert (2002) generalises the approach of Chapter 5 to zonal basis functions, especially native spaces and the necessary extension theorems for obtaining error estimates.

Not only standard interpolation but also Hermite interpolation in very general settings was studied by Narcowich and Ward (1995). They not only addressed the Hermite interpolation problem but considered it also on manifolds rather than Euclidean spaces as we do in this book (see also our next section on open problems).

There is another aspect of radial basis function which is related to the questions of convergence we studied in this book, namely the question when general translates, that is with arbitrary knots, are dense in the set of continuous functions on a domain, say. An alternative question is for what sets Ξ of centres we get denseness in the space of continuous functions on a domain when the radial function ϕ is allowed to vary over a specified set of univariate continuous functions, $\phi \in \mathcal{A} \subset C(\mathbb{R}_+)$, say. These two and other related questions were studied especially by Pinkus, and we mention his 1999 paper from the bibliography. This question and paper are also related to applications of neural networks using radial basis function approximations, an application which has been alluded to many times in this book. A suitable recent reference is the paper by Evgeniou, Pontil and Poggio (2000).

10.2 Open problems

The first of the open problems which the reader is invited to pursue research work in is the question of the *saturation orders* of scattered data interpolation with radial basis functions for all dimensions, even from the apparently simple thin-plate spline class (4.4). At present, this is arguably the most urgent question in the approximation theory of radial basis functions, but the univariate thin-plate spline case has been settled by Powell (2001) as well as the case for $1 \le p \le 2$ by Johnson (2002).

As we have remarked at several places in the book, notably in Chapters 4 and 5, saturation orders are available for large classes of radial basis functions ϕ, including those defined in (4.4), so long as we work on *square grids of equally spaced* infinitely many data, but not enough is known for scattered sets of data sites Ξ, in spite of the fine results by Johnson which give *upper* bounds for approximation orders (1998a and b) and indeed saturation orders for special cases (2002). This question is especially pertinent if we formulate it for general L^p-approximations, $2 \le p \le \infty$, because it may well be that it can be answered once we have the complete answers by Johnson (2002) with restriction to $p \le 2$.

We known from Johnson's work that the gridded data results with the highest approximation orders are always unavailable in the general case of scattered data even when very smooth domain boundaries are required, to be contrasted with the highly plausible conjecture that the 'full' orders are obtained if we stay well away from domain boundaries or impose suitable, strong conditions on the approximand. On the other hand, what are the precise convergence orders obtained on the boundaries depending on their smoothness and that of the approximand for $p = \infty$, for example? Suitable boundary conditions for the univariate thin-plate spline case for getting the best possible convergence orders were given by Powell (2001).

These questions are, incidentally, particularly important in the context of using radial basis functions for the numerical solution of partial differential equations on domains.

Not only in the approximate numerical solution of partial differential equations is the study of approximations at the boundary of a domain important. It requires particular attention in the setting of the 'classical' radial basis functions studied in this work because of their global support and expected difficulties at the boundaries. For instance, the approximants may become unbounded outside the finite domain and make approximations at the boundary especially difficult.

A second important question to address is always the choice of various parameters within the radial basis functions, notably the notorious c parameter in the ubiquitous multiquadric radial basis function. Work has been done on these questions (Rippa, 1999) but it is mostly empirical. No doubt, the experimental work shows that there will be a data-dependence in the choice of c, at least on the (local) spacing of the data sites as in our section about parameter-dependent convergence of radial basis function interpolants. It is conjectured that there are also dependences of the smoothness (parameters) of the approximand, such as its second partial derivatives. This is underlined by the empirical results.

In the context of multiquadric interpolation and choice of parameters, there is still the question of existence and uniqueness of interpolants with different parameters $c = c_\xi$, the main problem being that the interpolation matrix is no longer a symmetric matrix and – presumably – the concepts of (conditional) positive definiteness or similar ones are no longer available. The same problems come up when radial basis functions of different compact support are used in the same approximant.

Of course within the context of choice of parameters, the question of optimal choices of centres arises too for interpolation with radial basis functions (see below and Chapter 8 for this point in the least squares setting). It is questionable from comparative work e.g. by Derrien whether grids form the optimal (regular) distribution of data sites, even when they are chosen to be function-value-independent. This can also be related to the condition numbers of interpolation matrices which are geometry-dependent, if independent of the number of centres as we have seen in Chapter 5. There are some practical experiments which indicate for example that hexagonal grids may be superior in two dimensions to square ones as far as the condition numbers of interpolation matrices are concerned (Derrien, private communication).

Incidentally, in practice it is not really the condition number of the direct interpolation (collocation) matrix which is relevant. Another quantity which is open for much research and analysis is the uniform (say) norm of the *interpolation operator*, even for uniform gridded data, and all the more for scattered data. Good estimates for those norms can be highly valuable for theory and implementations of radial basis function interpolation methods, because they are basis-independent in contrast with condition numbers which are not: on the contrary, we usually attempt to improve them by changing bases.

We have pointed out already at various places in the book that interpolation is not the only method of use with radial basis functions. Among others, we have mentioned quasi-interpolation and least squares approaches. Smoothing splines with additional parameters that can be fixed by so-called cross-validation can provide further options that have already partly been studied in the literature (e.g. Wahba, 1981, and the above empirical results about optimal c parameters use cross-validation for finding best cs). This will certainly be useful not only for the radial basis functions we mention in this book but for large classes of radial or indeed nonradial, multivariate functions that give rise to conditional positive definiteness.

Many applications that involve the numerical solution of partial differential equations which have already been outlined in the second chapter require evaluation of inner products of radial basis functions and their translates, *à la* Galerkin methods, for producing the entries of the stiffness matrices. It has so far not been looked at sufficiently how to do this efficiently unless the radial basis functions are of compact support. Ideally there could be recursive methods such as with polynomial splines to compute the inner products, i.e. integrals, but this will undoubtedly be a new and difficult area of research, again due to the unbounded support and the relative difficulty of the analytic expressions of the common radial basis functions, involving logarithms, square roots and the like. Most applications of radial basis functions for the numerical solution of partial differential equations are still open anyway, since there is much work to do for treating boundary conditions suitably.

Wavelets with radial basis functions have already been addressed in this book, not so wavelets with radial basis functions of compact support. Much as with spline wavelets which originate from B-splines this research could lead to highly relevant results and the creation of new, efficient wavelet methods in several dimensions.

Fast computations of radial basis function interpolants are already available, as we have seen in the central, seventh chapter of this book. However, not at all everything which could be done has been done in this vein, and an important gap in this work is the availability of fast practical methods in *high dimensions*. In two, three and four dimensions, algorithms are readily available, but it is really for large dimensions n when radial basis functions come into the centre of interest, due to the problematic spline and polynomial methods then for interpolation. We mentioned this in the third chapter. Therefore, there should be algorithms for moderate numbers of data points ξ (a few thousand, say) but large n – at present the bottleneck being the methods for fast evaluation in high dimension. This includes approximations on manifolds within high-dimensional spaces which should be specially considered within these high-dimensional applications.

Indeed, it is not at all uncommon to have approximations which happen, at face value, in very high-dimensional Euclidean spaces, but in fact the important phenomena (say, the qualitative behaviour of the solution of a partial differential equation or the asymptotic behaviour) happen in *lower-dimensional* manifolds or subspaces at least asymptotically. There are two things that can be helped in this context by the availability of radial basis functions in high dimensions: firstly they can help to identify those manifolds or subspaces (so this is a question of dimension reduction) and secondly they can themselves model the solution on those lower-dimensional structures. For this, it has to be kept in mind that, even after a dimension reduction, the structures may still be fairly high-dimensional.

It has already been mentioned in the chapter about least squares approximations that least squares approximations with free knots (here, with free centres) would be an interesting field of study and undoubtedly fruitful. This will be a nonlinear approximation method and probably give much better approximations than are available so far, because a data-dependent choice of centres should allow better fits for functions and data of various smoothness. Among some of the other fields mentioned in this section, this is one of the most warmly recommended new subjects for empirical and theoretical research.

Finally, neural networks lie between the sheer practical applications of radial basis function approximations and their theoretical study, because they themselves give rise to new analysis methods and approaches for the radial basis function spaces. They are particularly connected with the high-dimensional *ansatz* and their practical aspects demand smoothness as well as fast computation in high dimensions. Therefore the link between these two research areas should be fostered and enforced from the radial basis function side by particular attention to the needs of neural network approximations.

Appendix: Some Essentials
on Fourier Transforms

Since we require in this book so many facts about the Fourier transform in n-dimensional Euclidean space, we have collected in this appendix several essential facts. They are mostly taken from Stein and Weiss's excellent and famous book of 1971.

Definition. *Let f be an absolutely integrable function with n real variables. Then its Fourier transform, always denoted by $\hat{f}$, is the continuous function*

$$\hat{f}(x) = \int_{\mathbb{R}^n} e^{-ix \cdot t} f(t)\, dt, \qquad t \in \mathbb{R}^n.$$

This function $\hat{f}$ satisfies $\hat{f}(x) = o(1)$ for large argument, which is a consequence of the multidimensional Riemann–Lebesgue lemma.

We also observe the trivial fact that the Fourier transform of f is always uniformly bounded by the integral of $|f|$, that is by its L^1-norm. Moreover, the Fourier transform of f is uniformly continuous under that integrability condition.

Definition. *Let $\| \cdot \|_2$ denote the Euclidean norm defined by its square*

$$\|f\|_2^2 := \int_{\mathbb{R}^n} |f(t)|^2 \, dt.$$

Here, f is a function of n real variables. There is an inner product associated with this norm in the canonical way: for two functions f and g with finite Euclidean norm,

$$(f, g) = \int_{\mathbb{R}^n} f(t) \, \overline{g(t)} \, dt.$$

The space of all measurable $f \colon \mathbb{R}^n \to \mathbb{R}$ with finite Euclidean norm is called $L^2(\mathbb{R}^n)$.

In this definition, all measurable f that differ only on a set of measure zero are considered the same.

Theorem. *For two absolutely integrable functions f and g of n real variables, the convolution defined through*

$$f * g(x) := \int_{\mathbb{R}^n} f(x - t) g(t) \, dt, \qquad x \in \mathbb{R}^n,$$

satisfies

$$\widehat{f * g} = \hat{f} \cdot \hat{g}.$$

In particular, the convolution $f * g$ is itself absolutely integrable due to Young's inequality.

Theorem. *If f is absolutely integrable, so are $f(x - y)$, $x \in \mathbb{R}^n$, and $e^{ix \cdot y} f(x)$, $x \in \mathbb{R}^n$, for a fixed $y \in \mathbb{R}^n$, and their Fourier transforms are*

$$e^{-ix \cdot y} \hat{f}(x)$$

and

$$\hat{f}(x - y),$$

respectively. Moreover, the Fourier transform of $f(M \cdot)$, where M is a non-singular square $n \times n$ matrix, is $\hat{f}(M^{-T} \cdot)/|\det M|$. Here, the exponent '$-T$' means inverse and transpose.

Theorem. *If f and $x_1 f(x)$, $x \in \mathbb{R}^n$, are both absolutely integrable, where x_1 denotes the first component of x, then $\hat{f}$ is once continuously differentiable with respect to the first coordinate and the partial derivative is*

$$\frac{\partial \hat{f}(x)}{\partial x_1} = \widehat{(-it_1 f(t))}(x),$$

where t_1 is the first component of t.

Conversely we can also obtain the Fourier transform of the first partial derivative of f, if it exists and is absolutely integrable, as $ix_1 \hat{f}(x)$. Both of these statements can be generalised to higher order derivatives in a straightforward manner.

Theorem. *If f and its Fourier transform are both absolutely integrable, then the inversion formula*

$$f(x) = \frac{1}{(2\pi)^n} \int_{\mathbb{R}^n} e^{ix \cdot t} \hat{f}(t) \, dt, \qquad x \in \mathbb{R}^n,$$

holds. The process of inverting the Fourier transform is denoted by the symbol $\check{\hat{f}} = f$.

Furthermore, we observe that $\check{f} = \hat{f}(-\cdot)/(2\pi)^n$.

The bounded linear extension of the Fourier transform operator to the space of square-integrable functions $L^2(\mathbb{R}^n)$ is also called Fourier transform with the same notation, and it has the following properties. Note that the functions and their Fourier transforms are defined no longer pointwise but only almost everywhere, because the integrals no longer converge absolutely.

Theorem. *The Fourier transform is an isometry* $L^2(\mathbb{R}^n) \to L^2(\mathbb{R}^n)$ *in the sense that* $\|\hat{f}\|_2$ *agrees with* $\|f\|_2$ *up to a constant factor. (The factor is one if we normalise* $\hat{f}$ *by introducing an extra factor* $1/(2\pi)^{n/2}$.)

There also holds the famous Parseval–Plancherel identity:

Theorem. *For two n-variate and square-integrable functions* f *and* g *and their Fourier transforms* $\hat{f}$ *and* $\hat{g}$, *it is true that*

$$(\hat{f}, \hat{g}) = (2\pi)^n (f, g)$$

for the Euclidean inner product.

A closely related fact is that the $L^2(\mathbb{T}^n)$-norm of a Fourier series is the same as $(2\pi)^{n/2}$ times the $\ell^2(\mathbb{Z}^n)$-norm of its Fourier coefficients.

Commentary on the Bibliography

We finish this book by adding a few remarks about the current state of the literature on radial basis functions. To start with, we point out that there are many excellent reviews and research papers about radial basis function approximation, such as the ones by Dyn (1987, 1989), Hardy (1990) and Powell (1987, 1992a). The present book gives, unsurprisingly, the author's personal view and therefore depends on many of his own papers cited above. Moreover, there is another review with different emphasis recently published by the journal *Acta Numerica*, also by the same author but more focussed on surveying the very recent results and several applications. However, we wish to add to our fairly extensive list of references a few remarks that also concern the history of the development of radial basis functions.

Many ideas in and basic approaches to the analysis of radial basis functions date back to Atteia, and to Pierre-Jean Laurent and his (at the time) student in Grenoble Jean Duchon whose papers we have quoted in the Bibliography. The papers we have listed there are the standard references for anyone who is working with thin-plate splines and they initiated many of the ideas we have used in Chapter 5 of the book. They are also closely related to the work by Micchelli, and Madych and Nelson, about radial function interpolants, their unique existence and convergence properties. Duchon was certainly one of the pioneers on the important classes of thin-plate spline type radial basis functions (4.4) while the ideas on radial basis functions with compact support probably started with Askey whose paper (a technical report) we quote above as well. Duchon's approach was exclusively via the variational principles of Section 5.2.

With those papers having started the radial basis function idea in the 1970s as far as scattered data are concerned, much of what followed then in the 1980s and 1990s was based on the idea of studying data on grids for the analysis of convergence/approximation orders etc. Since the analysis is far more complete (and, in some sense, easier, as the reader will have seen by now) we have included the description of this analysis one chapter before the work of Duchon and its 'derivatives'. Much of that chapter goes back to the author's own work, but includes important contributions by Baxter (Theorem 4.19 describes his ideas for instance), Jackson (especially about quasi-interpolation as outlined in Section 4.2 and convergence), Powell (e.g. the proof of Theorem 4.7 is due to his ideas), joint work with Micchelli (1991, e.g. Theorem 4.10), and several others. The precursor of this work is in part the work by de Boor and collaborators, Dahmen

244   *Commentary on the bibliography*

and Micchelli and many others on *box-splines* because the box-splines (sometimes also called cube-splines, they are closely related to B-splines in multiple dimensions, but only defined on lattices) too were analysed on square grids first, and they gave beautiful applications to the Strang and Fix theory which we have quoted (but used only implicitly) in Chapter 4. In fact they were the reason why much of that theory has been developed. They benefited from the idea of square grids and the applications of Fourier analysis, the Poisson summation formula etc. in the same way as we have used in the analysis of radial basis functions too. They do have the simplification of dealing with basis functions (the box- or cube-splines) which have compact support. This difference is, from the analytical point of view, especially significant when proofs of approximation orders are established.

The multiquadric function (as well as the inverse multiquadric) has been especially fostered by Hardy who introduced their main applications in geophysics and gave them their name – often they are in fact called 'Hardy (inverse) multiquadrics' for this reason. Franke with his 1982 paper in *Mathematics of Computation* comparing several techniques of multivariate approximation brought the multiquadrics to the attention of many more mathematicians because of this very favourable review of their properties and convincing statements about their advantages in applications.

Few aspects of approximation theory can be discussed without mentioning the contributions of Schoenberg who envisaged many methods and theories suitable in multiple dimensions even before they were studied in detail generations later. We have mentioned in the Bibliography only his contributions to the nonsingularity of distance matrices (i.e., as they are sometimes called, linear basis functions) and complete monotonicity, but there are also contributions to the theory of approximation on equally spaced data with the use of Fourier transforms as in the Strang and Fix theory. He studied those especially for univariate, equally spaced data and the exponentially decaying Lagrange functions. The ideas about establishing existence and polynomial reproduction stem from his work. We recommend that the reader sees his *selecta* edited by Carl de Boor for this.

Micchelli, and Madych and Nelson, have extended the ideas of complete monotonicity and conditionally positive functions to apply to all those radial basis functions mentioned in Chapter 4 of this book, where Micchelli in his seminal 1986 paper concentrated on the nonsingularity of the radial basis function interpolation matrix and Madych and Nelson more on the convergence questions. Micchelli's work is mentioned in Chapters 2 and 5, while some of Madych and Nelson's contributions are quoted in Chapter 5. Later contributions of the 1990s are due to Schaback and Wu and others whose papers are also included in the bibliography. Finally, probably the sharpest upper bounds on the approximation orders are due to Michael Johnson who has developed into an expert in providing those 'inverse' results. Some of this work we have included in Chapters 4 and 5.

Many applications, both potential and topical, are mentioned especially in the Introduction and in Chapter 2, and we cite Hardy's papers as an excellent reference to applications, but in order to avail oneself of the methods for those practical applications, efficient computer implementations are needed as well. Some of the main contributors to the development of fast algorithms have been mentioned in Chapter 7, and we point here once more to the papers by Beatson, Goodsell, Light and Powell.

Using wavelets with all kinds of approximation spaces has been a topic of interest for more than 10 years now within approximation theory and now also within multivariate approximation theory. As far as the wavelets with radial basis functions are concerned, this is an area that is only beginning to develop now, and in Chapter 9 we have only reviewed some of the first moderate contributions. It is to be hoped, as was also pointed out in Chapter 10, that more will be known in the not too distant future. As to the general theory of wavelets, a reference to Daubechies' work for instance must not be missing from any typical bibliography on the subject.

The author of this book is grateful for all the fine contributions to the research into radial basis functions which have been provided by the many colleagues, many, but certainly not all, of whose papers are cited above, and which have made this monograph possible.

Bibliography

Abramowitz, M. and I. Stegun (1972) Handbook of Mathematical Functions, National Bureau of Standards, Washington, DC.

Arad, N., N. Dyn and D. Reisfeld (1994) 'Image warping by radial basis functions: applications to facial expressions', *Graphical Models and Image Processing* **56**, 161–172.

Askey, R. (1973) 'Radial characteristic functions', MRC Report 1262, University of Wisconsin–Madison.

Atteia, M. (1992) *Hilbertian Kernels and Spline Functions*, North-Holland, Amsterdam.

Ball, K. (1992) 'Eigenvalues of Euclidean distance matrices', *J. Approx. Th.* **68**, 74–82.

Ball, K., N. Sivakumar and J.D. Ward (1992) 'On the sensitivity of radial basis interpolation to minimal data separation distance', *Constructive Approx.* **8**, 401–426.

Barrodale, I. and C.A. Zala (1999) 'Mapping aerial photographs to orthomaps using thin plate splines', *Adv. Comp. Math.* **11**, 211–227.

Baxter, B.J.C. (1991) 'Conditionally positive functions and p-norm distance matrices', *Constructive Approx.* **7**, 427–440.

Baxter, B.J.C. (1992a) 'The interpolation theory of radial basis functions', PhD Thesis, University of Cambridge.

Baxter, B.J.C. (1992b) 'The asymptotic cardinal function of the multiquadric $\phi(r) = (r^2 + c^2)^{1/2}$ as $c \to \infty$', *Comp. Math. Appl.* **24**, No. 12, 1–6.

Baxter, B.J.C. and N. Sivakumar (1996) 'On shifted cardinal interpolation by Gaussians and multiquadrics', *J. Approx. Th.* **87**, 36–59.

Baxter, B.J.C., N. Sivakumar and J.D. Ward (1994) 'Regarding the p-norms of radial basis function interpolation matrices', *Constructive Approx.* **10**, 451–468.

Beatson, R.K., J.B. Cherrie and C.T. Mouat (1998) 'Fast fitting of radial basis functions: methods based on preconditioned GMRES iteration', *Adv. Comp. Math.* **11**, 253–270

Beatson, R.K, J.B. Cherrie and G.N. Newsam (2000) 'Fast evaluation of radial basis functions: methods for generalised multiquadrics in $\mathbb{R}^n$', Research Report, University of Canterbury (New Zealand).

Beatson, R.K. and N. Dyn (1992) 'Quasi-interpolation by generalized univariate multiquadrics', ms., 20pp.

Beatson, R.K. and N. Dyn (1996) 'Multiquadric B-splines', *J. Approx. Th.* **87**, 1–24.

Beatson, R.K., G. Goodsell and M.J.D. Powell (1996), 'On multigrid techniques for thin plate spline interpolation in two dimensions', in *The Mathematics of Numerical Analysis*, J. Renegar, M. Shub and S. Smale (eds.), Amer. Math. Soc. Providence, RI, 77–97.

Beatson, R.K. and L. Greengard (1997) 'A short course on fast multiple methods', in *Wavelets, Multilevel Methods and Elliptic PDEs*, J. Levesley, W. Light and M. Marletta (eds.), Oxford University Press, Oxford, 1–37.

Beatson, R.K. and W.A. Light (1992) 'Quasi-interpolation in the absence of polynomial reproduction', in *Numerical Methods of Approximation Theory*, D. Braess and L.L. Schumaker (eds.), Birkhäuser-Verlag, Basel, 21–39.

Beatson, R.K. and W.A. Light (1997) 'Fast evaluation of radial basis functions: methods for 2-dimensional polyharmonic splines', *IMA J. Numer. Anal.* **17**, 343–372.

Beatson, R.K. and G.N. Newsam (1992) 'Fast evaluation of radial basis functions: I', *Comp. Math. Appl.* **24**, No. 12, 7–19.

Beatson, R.K. and M.J.D. Powell (1992a) 'Univariate multiquadric approximation: quasi-interpolation to scattered data', *Constructive Approx.* **8**, 275–288.

Beatson, R.K. and M.J.D. Powell (1992b) 'Univariate interpolation on a regular finite grid by a multiquadric plus a linear polynomial', *IMA J. Numer. Anal.* **12**, 107–133.

Beatson, R.K. and M.J.D. Powell (1994) 'An iterative method for thin plate spline interpolation that employs approximations to Lagrange functions', in *Numerical Analysis 1993*, D.F. Griffiths and G.A. Watson (eds.), Longman, Harlow, Essex, 17–39.

Bejancu, A. (1999) 'Local accuracy for radial basis function interpolation on finite uniform grids', *J. Approx. Th.* **99**, 242–257.

Bennett, C. and B. Sharpley (1988) *Interpolation of Operators*, Academic Press, New York.

Bezhaev, A. Yu. and V.A. Vasilenko (2001) *Variational Theory of Spline*, Kluwer/Plenum, New York.

Binev, P. and K. Jetter (1992) 'Estimating the condition number for multivariate interpolation problems', in *Numerical Methods of Approximation Theory*, D. Braess and L.L. Schumaker (eds.), Birkhäuser-Verlag, Basel, 39–50.

Bishop, C. (1991) 'Improving the generalization properties of radial basis functions neural networks', *Neural Comp.* **3**, 579–588.

Bloom, T. (1981) 'Kergin-interpolation of entire functions on $\mathbb{C}^N$', *Duke Math. J.* **48**, 69–83

de Boor, C. (1990) 'Quasi-interpolants and approximation power of multivariate splines', in *Computation of Curves and Surfaces*, W. Dahmen, M. Gasca and C.A. Micchelli (eds.), Kluwer, Dordrecht, Netherlands, 313–345.

de Boor, C., R.A. DeVore and A. Ron (1992) 'Fourier analysis of the approximation power of principal shift-invariant spaces', *Constructive Approx.* **8**, 427–462.

de Boor, C., R.A. DeVore and A. Ron (1993) 'On the construction of multivariate (pre)wavelets', *Constructive Approx.* **9**, 123–166.

de Boor, C., K. Höllig and S. Riemenschneider (1993) *Box Splines*, Springer, New York.

de Boor, C. and A. Ron (1990) 'On multivariate polynomial interpolation', *Constructive Approx.* **6**, 287–302.

de Boor, C. and A. Ron (1992a) 'The least solution for the polynomial interpolation problem', *Math. Z.* **210**, 347–378.

de Boor, C. and A. Ron (1992b) 'Computational aspects of polynomial interpolation in several variables', *Math. Comp.* **58**, 705–727.

Böttcher, A. and B. Silbermann (1990) *Analysis of Toeplitz Operators*, Springer, Berlin.

Bouhamidi, A. (1992) 'Interpolation et approximation par des fonctions spline radiales', Thèse, Université de Nantes.

Braess, D. (1997) *Finite Elemente*, Springer, Berlin.

Bramble, J.H. and S.R. Hilbert (1971) 'Bounds for a class of linear functionals with applications to Hermite interpolation', *Numer. Math.* **16**, 362–369.

Brenner, S.C. and L.R. Scott (1994) *The Mathematical Theory of Finite Element Methods*, Springer, New York.

Broomhead, D. (1988) 'Radial basis functions, multi-variable functional interpolation and adaptive networks', Royal Signals and Radar Establishment, Memorandum 4148.

Buhmann, M.D. (1988) 'Convergence of univariate quasi-interpolation using multi-quadrics', *IMA J. Numer. Anal.* **8**, 365–383.

Buhmann, M.D. (1989) 'Multivariable interpolation using radial basis functions', PhD Thesis, University of Cambridge.

Buhmann, M.D. (1990a) 'Multivariate interpolation in odd-dimensional Euclidean spaces using multiquadrics', *Constructive Approx.* **6**, 21–34.

Buhmann, M.D. (1990b) 'Multivariate cardinal-interpolation with radial-basis functions', *Constructive Approx.* **6**, 225–255.

Buhmann, M.D. (1993a) 'On quasi-interpolation with radial basis functions', *J. Approx. Th.* **72**, 103–130.

Buhmann, M.D. (1993b) 'Discrete least squares approximations and prewavelets from radial function spaces', *Math. Proc. Cambridge Phil. Soc.* **114**, 533–558.

Buhmann, M.D. (1993c) 'New developments in the theory of radial basis function interpolation', in *Multivariate Approximation: from CAGD to Wavelets*, K. Jetter and F.I. Utreras (eds.), World Scientific, Singapore, 35–75.

Buhmann, M.D. (1995) 'Multiquadric pre-wavelets on non-equally spaced knots in one dimension', *Math. Comp.* **64**, 1611–1625.

Buhmann, M.D. (1998) 'Radial functions on compact support', *Proc. Edinburgh Math. Soc.* **41**, 33–46.

Buhmann, M.D. (2000) 'Radial basis functions', *Acta Numer.* **9**, 1–37.

Buhmann, M.D. (2001) 'A new class of radial basis functions with compact support', *Math. Comp.* **70**, 307–318.

Buhmann, M.D. and C.K. Chui (1993) 'A note on the local stability of translates of radial basis functions', *J. Approx. Th.* **74**, 36–40.

Buhmann, M.D., O. Davydov and T.N.T. Goodman (2001) 'Box-spline prewavelets with small support', *J. Approx. Th.* **112**, 16–27.

Buhmann, M.D., F. Derrien and A. LeMéhauté (1995) 'Spectral properties and knot removal for interpolation by pure radial sums', in *Mathematical Methods for Curves and Surfaces*, Morten Dæhlen, Tom Lyche and Larry L. Schumaker (eds.), Vanderbilt University Press, Nashville, Tenn. – London, 55–62.

Buhmann, M.D. and N. Dyn (1993) 'Spectral convergence of multiquadric interpolation', *Proc. Edinburgh Math. Soc.* **36**, 319–333.

Buhmann, M.D., N. Dyn and D. Levin (1995) 'On quasi-interpolation with radial basis functions on non-regular grids', *Constructive Approx.* **11**, 239–254.

Buhmann, M.D. and C.A. Micchelli (1991) 'Multiply monotone functions for cardinal interpolation', *Adv. Appl. Math.* **12**, 359–386.

Buhmann, M.D. and M.J.D. Powell (1990) 'Radial basis function interpolation on an infinite regular grid', in *Algorithms for Approximation II*, J.C. Mason and M.G. Cox (eds.), Chapman and Hall, London, 47–61.

Buhmann, M.D. and A. Ron (1994) 'Radial basis functions: l^p-approximation orders with scattered centres', in *Wavelets, Images, and Surface Fitting*, P.-J. Laurent, A. Le Méhauté, and L.L. Schumaker (eds.), A.K. Peters, Boston, Mass., 93–112.

Calderón, A., F. Spitzer and H. Widom (1959) 'Inversion of Toeplitz matrices', *Ill. J. Math.* **3**, 490–498.

Carlson, R.E. and T.A. Foley (1992) 'Interpolation of track data with radial basis methods', *Comp. Math. Appl.* **24**, 27–34.

Casdagli, M. (1989) 'Nonlinear prediction of chaotic time series', *Physica D*. **35**, 335–356.

zu, Castell, W. (2000) 'Recurrence relations for radial, positive definite functions', Technical Report, University of Erlangen.

Cavaretta, A.S., C.A. Micchelli and A. Sharma (1980) 'Multivariate interpolation and the Radon transform', *Math. Z.* **174**, 263–279.

Chen, C.S., C.A. Brebbia and H. Power (1999) 'Dual reciprocity method using compactly supported radial basis functions', *Comm. Numer. Math. Eng.* **15**, 137–150.

Cheney, W. and W.A. Light (1999) *A Course in Approximation Theory*, Brooks, Pacific Grove, Calif.

Chui, C.K. (1992) *An Introduction to Wavelets*, Academic Press, Boston, Mass.

Chui, C.K., K. Jetter and J.D. Ward (1992) 'Cardinal interpolation with differences of tempered functions', *Comp. Math. Appl.* **24**, No. 12, 35–48.

Chui, C.K. and J.Z. Wang (1991) 'A cardinal spline approach to wavelets' *Proc. Amer. Math. Soc.* **113**, 785–793.

Ciarlet, P.G. (1978) *The Finite Element Method for Elliptic Problems*, North-Holland, Amsterdam.

Dahmen, W. and C.A. Micchelli (1983a) 'Translates of multivariate splines', *Lin. Alg. Appl.* **52**, 217–234.

Dahmen, W. and C.A. Micchelli (1983b) 'Recent progress in multivariate splines', in *Approximation Theory IV*, C.K. Chui, L.L. Schumaker and J.D. Ward (eds.), Academic Press, New York, 27–121.

Daubechies, I. (1992) *Ten Lectures on Wavelets*, SIAM, Philadelphia.

Davis, P.J. (1963) *Interpolation and Approximation*, Dover, New York.

Duchon, J. (1976) 'Interpolation des fonctions de deux variables suivante le principe de la flexion des plaques minces', *Rev. Française Automat. Informat. Rech. Opér. Anal. Numér.* **10**, 5–12.

Duchon, J. (1978) 'Sur l'erreur d'interpolation des fonctions de plusieurs variables pas les D^m-splines', *Rev. Française Automat. Informat. Rech. Opér. Anal. Numér.* **12**, 325–334.

Duchon, J. (1979) 'Splines minimizing rotation-invariant semi-norms in Sobolev spaces', in *Constructive Theory of Functions of Several Variables*, W. Schempp and K. Zeller (eds.), Springer, Berlin–Heidelberg, 85–100.

Dyn, N. (1987) 'Interpolation of scattered data by radial functions', in *Topics in Multivariate Approximation*, C.K. Chui, L.L. Schumaker and F.I. Utreras (eds.), Academic Press, New York, 47–61.

Dyn, N. (1989) 'Interpolation and approximation by radial and related functions', in *Approximation Theory VI*, C.K. Chui, L.L. Schumaker, and J.D. Ward (eds.), Academic Press, New York, 211–234.

Dyn, N., I.R.H. Jackson, D. Levin and A. Ron (1992) 'On multivariate approximation by integer translates of a basis function', *Israel J. Math.* **78**, 95–130.

Dyn, N. and D. Levin (1981) 'Construction of surface spline interpolants of scattered data over finite domains', *Rev. Française Automat. Informat. Rech. Opér. Anal. Numér.* **16**, 201–209.

Dyn, N. and D. Levin (1983) 'Iterative solution of systems originating from integral equations and surface interpolation', *SIAM J. Numer. Anal.* **20**, 377–390.

Dyn, N., D. Levin and S. Rippa (1986) 'Numerical procedures for surface fitting of scattered data by radial functions', *SIAM J. Scient. Stat. Comp.* **7**, 639–659.

Dyn, N., W.A. Light and E.W. Cheney (1989) 'Interpolation by piecewise linear radial basis functions I', *J. Approx. Th.* **59**, 202–223.

Dyn, N., F.J. Narcowich and J.D. Ward (1997) 'A framework for approximation and interpolation on a Riemannian manifold', in *Approximation and Optimization*, M.D. Buhmann and A. Iserles (eds.), Cambridge University Press, Cambridge, 133–144.

Dyn, N. and A. Ron (1995) 'Radial basis function approximation: from gridded centers to scattered centers', *Proc. London Math. Soc.* **71**, 76–108.

Eckhorn, R. (1999) 'Neural mechanisms of scene segmentation: recordings from the visual cortex suggest basic circuits for linking field models', *IEEE Trans. Neural Net.* **10**, 1–16.

Evgeniou, T., M. Pontil and T. Poggio (2000) 'Regularization networks and support vector machines', *Adv. Comp. Math.* **12**, 1–50.

Farin, G. (1986) 'Triangular Bernstein–Bézier patches', *Comp.-Aided Geom. Design* **3**, 83–127.

Farwig, R. (1986) 'Rate of convergence of Shepard's global interpolation method', *Math. Comp.* **46**, 577–590.

Fasshauer, G. (1999) 'Solving differential equations with radial basis functions: multilevel methods and smoothing', *Adv. Comp. Math.* **11**, 139–159.

Faul, A.C. (2001) 'Iterative techniques for radial basis function interpolation', PhD Thesis, University of Cambridge.

Faul, A.C. and M.J.D. Powell (1999a) 'Proof of convergence of an iterative technique for thin plate spline interpolation in two dimensions', *Adv. Comp. Math.* **11**, 183–192.

Faul, A.C. and M.J.D. Powell (1999b) 'Krylov subspace methods for radial basis function interpolation', in *Numerical Analysis 1999*, D. F. Griffiths (ed.), Chapman and Hall, London, 115–141.

Floater, M. and A. Iske (1996) 'Multistep scattered data interpolation using compactly supported radial basis functions', *J. Comp. Appl. Math.* **73**, 65–78.

Fornberg, B. (1999) A practical guide to pseudospectral methods, Cambridge University Press, Cambridge.

Forster, O. (1984) *Analysis 3*, Vieweg, Braunschweig.

Franke, C. and R. Schaback (1998) 'Solving partial differential equations by collocation using radial basis functions', *Comp. Math. Appl.* **93**, 72–83.

Franke, R. (1982) 'Scattered data interpolation: tests of some methods', *Math. Comp.* **38**, 181–200.

Franke, R. and L.L. Schumaker (1987) 'A bibliography of multivariate approximation', in *Topics in Multivariate Approximation*, C.K. Chui, L.L. Schumaker and F.I. Utreras (eds.), Academic Press, New York, 275–335.

Freeden, W. (1981) 'On spherical spline interpolation and approximation', *Math. Meth. Appl. Sci.* **3**, 551–575.

Freeden, W. and P. Hermann (1986), 'Uniform approximation by spherical approximation', *Math. Z.* **193**, 265–275.

Freeden, W., M. Schreiner and R. Franke (1995) 'A survey on spherical spline approximation', Technical Report 95–157, University of Kaiserslautern.

Gasca, M. and T. Sauer (2000) 'Polynomial interpolation in several variables', *Adv. Comp. Math.* **12**, 377–410.

Gasper, G. (1975a) 'Positivity and special functions', in *The Theory and Applications of Special Functions*, R. Askey (ed.), Academic Press, New York, 375–434.

Gasper, G. (1975b) 'Positive integrals of Bessel functions', *SIAM J. Math. Anal.* **6**, 868–881.

Girosi, F. (1992) 'On some extensions of radial basis functions and their applications in artificial intelligence', *Comp. Math. Appl.* **24**, 61–80.

Golberg, M.A. and C.S. Chen (1996) 'A bibliography on radial basis function approximation', *Boundary Elements Comm.* **7**, 155–163.

Golomb, M. and H.F. Weinberger, H. F. (1959), 'Optimal approximation and error bounds', in *On Numerical Approximation*, R.E. Langer (ed.), University of Wisconsin Press, Madison, Wis., 117–190.

Golub, G.H and C.F. Van Loan (1989), *Matrix Computations*, 2nd ed., Johns Hopkins Press, Baltimore.

Goodman, T.N.T. (1983) 'Interpolation in minimum semi-norm, and multivariate B-splines', *J. Approx. Th.* **37**, 212–223.

Goodman, T.N.T. and A. Sharma (1984) 'Convergence of multivariate polynomials interpolating on a triangular array', *Trans. Amer. Math. Soc.* **285**, 141–157.

Gradsteyn, I.S. and I.M. Ryzhik (1980) *Table of Integrals, Series and Products*, Academic Press, San Diego.

Greengard, L. and V. Rokhlin (1987) 'A fast algorithm for particle simulations', *J. Comp. Phys.* **73**, 325–348.

Greengard, L. and V. Rokhlin (1997) 'A new version of the fast multipole method for the Laplace equation in three dimensions' *Acta Numer.* **6**, 229–269.

Guo, K., S. Hu and X. Sun (1993) 'Conditionally positive definite functions and Laplace–Stieltjes integrals', *J. Approx. Th.* **74**, 249–265.

Gutmann, H. M. (1999) 'A radial basis function method for global optimization', DAMTP Technical Report, University of Cambridge.

Gutmann, H. M. (2000) 'On the semi-norm of radial basis function interpolants', DAMTP Technical Report, University of Cambridge.

Hakopian, A. (1982) 'Multivariate divided differences and multivariate interpolation of Lagrange and Hermite type', *J. Approx. Th.* **34**, 286–305.

Halton, E.J. and W.A. Light (1993) 'On local and controlled approximation order', *J. Approx. Th.* **72**, 268–277.

Hardy, G.H., J.E. Littlewood and G. Pólya (1952), *Inequalities* 2nd ed., Cambridge University Press.

Hardy, R.L. (1971) 'Multiquadric equations of topography and other irregular surfaces', *J. Geophys. Res.* **76**, 1905–1915.

Hardy, R.L. (1990) 'Theory and applications of the multiquadric-biharmonic method', *Comp. Math. Appl.* **19**, 163–208.

Higham, N.J. (1996) *Accuracy and Stability of Numerical Algorithms*, SIAM, Philadelphia.

Hörmander, L. (1983) *Linear Partial Differential Operators*, Volume I, Springer, Berlin.

Hubbert, S. (2002) 'Radial basis function interpolation on the sphere', PhD Dissertation, Imperial College, London.

Jackson, I.R.H. (1988) 'Radial basis function methods for multivariable interpolation', PhD Dissertation, University of Cambridge.

Jackson, I.R.H. (1989) 'An order of convergence for some radial basis functions', *IMA J. Numer. Anal.* **9**, 567–587.

Jetter, K. (1992) 'Multivariate approximation: a view from cardinal interpolation', in *Approximation Theory VII*, E.W. Cheney, C.K. Chui and L.L. Schumaker (eds.), Academic Press, New York.

Jetter, K., S.D. Riemenschneider and Z. Shen (1994) 'Hermite interpolation on the lattice $\mathbb{Z}^{d}$', *SIAM J. Math. Anal.* **25**, 962–975.

Jetter, K. and J. Stöckler (1991) 'Algorithms for cardinal interpolation using box splines and radial basis functions', *Numer. Math.* **60**, 97–114.

Jetter, K., J. Stöckler and J.D. Ward (1999) 'Error estimates for scattered data interpolation on spheres', *Math. Comp.* **68**, 733–747.

Jia, R.Q. and J. Lei (1993) 'Approximation by multiinteger translates of functions having global support', *J. Approx. Th.* **72**, 2–23.

Johnson, M.J. (1997) 'An upper bound on the approximation power of principal shift-invariant spaces', *Constructive Approx.* **13**, 155–176.

Johnson, M.J. (1998a) 'A bound on the approximation order of surface splines', *Constructive Approx.* **14**, 429–438.

Johnson, M.J. (1998b) 'On the error in surface spline interpolation of a compactly supported function', manuscript, University of Kuwait.

Johnson, M.J. (2000) 'The L_2-approximation order of surface spline interpolation', *Math. Comp.* **70**, 719–737.

Johnson, M.J. (2002) 'The L_p-approximation order of surface spline interpolation for $1 \leq p \leq 2$', manuscript, University of Kuwait, to appear in *Constructive Approx.*

Jones, D.S. (1982) *The Theory of Generalised Functions*, Cambridge University Press, Cambridge.

Karlin, S. (1968) *Total Positivity*, Stanford University Press, Stanford, Calif.

Katznelson, Y. (1968) *An Introduction to Harmonic Analysis*, Dover, New York.

Kergin, P. (1980) 'A natural interpolation of C^k-functions', *J. Approx. Th.* **29**, 278–293.

Kremper, A., T. Schanze and R. Eckhorn (2002) 'Classification of cortical signals with a generalized correlation classifier based on radial basis functions', *J. Neurosci. Meth.* **116**, 179–187.

Laurent, P.-J. (1972) *Approximation et optimisation*, Hermann, Paris.

Laurent, P.-J. (1991) 'Inf-convolution splines', *Constructive Approx.* **7**, 469–484.

Levesley, J. (1994) 'Local stability of translates of polyharmonic splines in even space dimension', *Numer. Funct. Anal. Opt.* **15**, 327–333.

Levesley, J., W.A. Light, D. Ragozin and X. Sun (1999) 'A simple approach to the variational theory for interpolation on spheres', in *New Developments in Approximation Theory*, M.D. Buhmann, M. Felten, D. Mache and M.W. Müller, (eds.), Birkhäuser-Verlag, Basel, 117–143.

Light, W.A. and E.W. Cheney (1992a) 'Quasi-interpolation with translates of a function having non-compact support', *Constructive Approx.* **8**, 35–48.

Light, W.A. and E.W. Cheney (1992b) 'Interpolation by periodic radial basis functions', *J. Math. Anal. Appl.* **168**, 111–130.

Light, W.A. and H. Wayne (1998) 'Some remarks on power functions and error estimates for radial basis function approximation', *J. Approx. Th.* **92**, 245–266.

Lorentz, G.G., M. v. Golitschek and Y. Makovoz (1996) *Constructive Approximation, Advanced Problems*, Springer, Berlin.

Madych, W.R. (1990) 'Polyharmonic splines, multiscale analysis, and entire functions', in *Multivariate Approximation and Interpolation*, K. Jetter and W. Haußmann (eds.), Birkhäuser-Verlag, Basel, 205–215.

Madych, W.R. and S.A. Nelson (1990a), 'Polyharmonic cardinal splines', *J. Approx. Th.* **60**, 141–156.

Madych, W.R. and S.A. Nelson (1990b) 'Multivariate interpolation and conditionally positive definite functions II', *Math. Comp.* **54**, 211–230.

Madych, W.R. and S.A. Nelson (1992) 'Bounds on multivariate polynomials and exponential error estimates for multiquadric interpolation', *J. Approx. Th.* **70**, 94–114.

Mairhuber, J.C. (1956) 'On Haar's theorem concerning Chebyshev approximation problems having unique solutions', *Proc. Amer. Math. Soc.* **7**, 609–615.

Mattner, L. (1997) 'Strict definiteness of integrals via complete monotonicity of derivatives', *Trans. Amer. Math. Soc.* **349**, 3321–3342.

Matveev, O.V. (1997) 'On a method for interpolating functions on chaotic nets', *Math. Notes* **62**, 339–349.

McLain, D.H. (1974) 'Computer construction of surfaces through arbitrary points', *Information Processing* **74**, Proceedings of the IFIP Congress 74, Stockholm, 717–721.

Meyer, Y. (1990) *Ondelettes et Opérateurs I*, Hermann, Paris.

Micchelli, C.A. (1980) 'A constructive approach to Kergin interpolation in $\mathbb{R}^k$, multivariate B-splines and Lagrange-interpolation', *Rocky Mountain J. Math.* **10**, 485–497.

Micchelli, C.A. (1986) 'Interpolation of scattered data: distance matrices and conditionally positive definite functions', *Constructive Approx.* **1**, 11–22.

Micchelli, C.A. (1991) 'Using the refinement equation for the construction of pre-wavelets IV: cube splines and elliptic splines united', IBM report RC 17195, October 1991.

Micchelli, C.A. and P. Milman (1980) 'A formula for Kergin-interpolation in $\mathbb{R}^k$', *J. Approx. Th.* **29**, 294–296.

Micchelli, C.A., C. Rabut and F. Utreras (1991) 'Using the refinement equation for the construction of pre-wavelets: elliptic splines', *Numer. Alg.* **1**, 331–351.

Misiewicz, J.K. and D.St P. Richards (1994) 'Positivity of integrals of Bessel functions', *SIAM J. Math. Anal.* **25**, 596–601.

Narcowich, F.J., R. Schaback and J.D. Ward (1999), Multilevel interpolation and approximation, *Appl. Comp. Harm. Anal.* **7**, 243–261.

Narcowich, F.J, Sivakumar and J.D. Ward (1994) 'On condition numbers associated with radial-function interpolation', *J. Math. Anal. Appl.* **186**, 457–485.

Narcowich, F.J. and J.D. Ward (1991) 'Norms of inverses and condition numbers of matrices associated with scattered data', *J. Approx. Th.* **64**, 69–94.

Narcowich, F.J. and J.D. Ward (1992) 'Norm estimates for the inverses of a general class of scattered-data radial-function interpolation matrices', *J. Approx. Th.* **69**, 84–109.

Narcowich, F.J. and J.D. Ward (1995) 'Generalized Hermite interpolation by positive definite kernels on a Riemannian manifold', *J. Math. Anal. Appl.* **190**, 165–193.

Olver, F. (1974) *Asymptotics and Special Functions*, Academic Press, New York.

Pinkus, A. (1995) 'Some density problems in multivariate approximation', in *Approximation Theory: Proceedings of the International Dortmund Meeting IDOMAT 95*, M.W. Müller, M. Felten and D.H. Mache (eds.), Akademie Verlag, 277–284.

Pinkus, A. (1996) 'TDI Subspaces of $C(\mathbb{R}^d)$ and some density problem from neural networks', *J. Approx. Th.* **85**, 269–287.

Pinkus, A. (1999) 'Approximation theory of the MLP model in neural networks', *Acta Numer.* **8**, 143–196.

Pollandt, R. (1997) 'Solving nonlinear equations of mechanics with the boundary element method and radial basis functions', *Int. J. Numer. Meth. Eng.* **40**, 61–73.

Powell, M.J.D. (1981) *Approximation Theory and Methods*, Cambridge University Press, Cambridge.

Powell, M.J.D. (1987) 'Radial basis functions for multivariable interpolation: a review', in *Algorithms for Approximation*, J.C. Mason and M.G. Cox (eds.), Oxford University Press, Oxford, 143–167.

Powell, M.J.D. (1990) 'Univariate multiquadric approximation: reproduction of linear polynomials', in *Multivariate Approximation and Interpolation*, K. Jetter and W. Haußmann (eds.), Birkhäuser-Verlag, Basel, 227–240.

Powell, M.J.D. (1991) 'Univariate multiquadric interpolation: some recent results', in *Curves and Surfaces*, P.J. Laurent, A. LeMéhauté, and L.L. Schumaker (eds.), Academic Press, New York, 371–382.

Powell, M.J.D. (1992a) 'The theory of radial basis function approximation in 1990', in *Advances in Numerical Analysis II: Wavelets, Subdivision, and Radial Functions*, W.A. Light (ed.), Oxford University Press, Oxford, 105–210.

Powell, M.J.D. (1992b) 'Tabulation of thin plate splines on a very fine two-dimensional grid', in *Numerical Methods of Approximation Theory*, D. Braess and L.L. Schumaker (eds.), Birkhäuser-Verlag, Basel, 221–244.

Powell, M.J.D. (1993) 'Truncated Laurent expansions for the fast evaluation of thin plate splines', *Numer. Alg.* **5**, 99–120.

Powell, M.J.D. (1994a) 'Some algorithms for thin plate spline interpolation to functions of two variables', in *Advances in Computational Mathematics*, H.P. Dikshit and C.A. Micchelli (eds.), World Scientific (Singapore), 303–319.

Powell, M.J.D. (1994b) 'The uniform convergence of thin-plate spline interpolation in two dimensions', *Numer. Math.* **67**, 107–128.

Powell, M.J.D. (1996) 'A review of algorithms for thin plate spline interpolation', in *Advanced Topics in Multivariate Approximation*, F. Fontanella, K. Jetter and P.J. Laurent (eds.), World Scientific (Singapore), 303–322.

Powell, M.J.D. (1997) 'A new iterative algorithm for thin plate spline interpolation in two dimensions', *Ann. Numer. Math.* **4**, 519–527.

Powell, M.J.D. (2001) Talk at the 4th conference on Algorithms for Approximation, Huddersfield.

Powell, M.J.D. and M.A. Sabin (1977) 'Piecewise quadratic approximations on triangles', *ACM Trans. Math. Software* **3**, 316–325.

Quak, E., N. Sivakumar and J.D. Ward (1991) 'Least squares approximation by radial functions', *SIAM J. Numer. Anal.* **24**, 1043–1066.

Rabut, C. (1990) 'B-splines polyharmoniques cardinales: interpolation, quasi-interpolation, filtrage', Thèse d'Etat, Université de Toulouse.

Reimer, M. (1990) *Constructive Theory of Multivariate Functions with an Application to Tomography*, BI Wissenschaftsverlag, Mannheim.

Riemenschneider, S.D. and Sivakumar (1999) 'Gaussian radial basis functions: cardinal interpolation of ℓ^p and power growth data', *Adv. Comp. Math.* **11**, 229–251.

Rippa, S. (1999) 'An algorithm for selecting a good parameter c in radial basis function interpolation', *Adv. Comp. Math.* **11**, 193–210.

Ron, A. (1992) 'The L_2-approximation orders of principal shift-invariant spaces generated by a radial basis function', in *Numerical Methods of Approximation Theory*, D. Braess and L.L. Schumaker (eds.), Birkhäuser-Verlag, Basel, 245–268.

Ron, A. and X. Sun (1996) 'Strictly positive definite functions on spheres in Euclidean spaces', *Math. Comp.* **65**, 1513–1530.

Rudin, W. (1991) *Functional Analysis*, McGraw–Hill, New York.

Saitoh, S. (1988) Theory of reproducing kernels and its applications, Longman, Harlow, Essex.

Sauer, T. (1995) 'Computational aspects of multivariate polynomial interpolation', *Adv. Comp. Math.* **3**, 219–237.

Sauer, T. and Y. Xu (1995) 'On multivariate Lagrange interpolation', *Math. Comp.* **64**, 1147–1170.

Schaback, R. (1993) 'Comparison of radial basis function interpolants', in *Multivariate Approximation: from CAGD to Wavelets*, K. Jetter and F.I. Utreras (eds.), World Scientific, Singapore, 293–305.

Schaback, R. (1994) 'Lower bounds for norms of inverses of interpolation matrices for radial basis functions', *J. Approx. Th.* **79**, 287–306.

Schaback, R. (1995a) 'Error estimates and condition numbers for radial basis function interpolation', *Adv. Comp. Math.* **3**, 251–264.

Schaback, R. (1995b) 'Multivariate interpolation and approximation by translates of a basis function', in *Approximation Theory VIII, Volume 1: Approximation and Interpolation*, Charles K. Chui and L.L. Schumaker (eds.), World Scientific, Singapore, 491–514.

Schaback, R. (1996) 'Approximation by radial basis functions with finitely many centers', *Constructive Approx.* **12**, 331–340.

Schaback, R. (1997) 'Optimal recovery in translation-invariant spaces of functions', *Ann. Numer. Math.* **4**, 547–555.

Schaback, R. (1999) 'Native spaces for radial basis functions I', in *New Developments in Approximation Theory*, M.D. Buhmann, M. Felten, D. Mache and M.W. Müller (eds.), Birkhäuser-Verlag, Basel, 255–282.

Schaback, R. and H. Wendland (1998) 'Inverse and saturation theorems for radial basis function interpolation', Technical Report, University of Göttingen.

Schaback, R. and Z. Wu (1995) 'Operators on radial basis functions', *J. Comp. Appl. Math.* **73**, 257–270.

Schoenberg, I.J. (1938) 'Metric spaces and completely monotone functions', *Ann. Math.* **39**, 811–841.

Schoenberg, I.J. (1942) 'Positive definite functions on spheres', *Duke Math. J.* **9**, 96–108.

Schoenberg, I.J. (1988) *Selected Papers*, Birkhäuser-Verlag, Boston, Mass. – Basel.

Schumaker, L.L. (1981) *Spline Functions: Basic Theory*, Wiley, New York.

Sibson, R. and G. Stone (1991) 'Computation of thin plate splines', *SIAM J. Sci. Stat. Comp.* **12**, 1304–1313.

Sivakumar, N. and Ward, J.D. (1993) 'On the least squares fit by radial function to multidimensional scattered data', *Numer. Math.* **65** (1993), 219–243.

Sloan, I.H. (1995) 'Polynomial interpolation and hyperinterpolation over general regions', *J. Approx. Th.* **83**, 238–254.

Sonar, T. (1996) 'Optimal recovery using thin-plate splines in finite volume methods for the numerical solution of hyperbolic conservation laws', *IMA J. Numer. Anal.* **16**, 549–581.

Stein, E.M. (1970) *Singular Integrals and Differentiability Properties of Functions*, Princeton University Press, Princeton, NJ.

Stein, E.M. and G. Weiss (1971) *Introduction to Fourier Analysis on Euclidean Spaces*, Princeton University Press, Princeton, NJ.

Steward, J. (1976) 'Positive definite functions and generalizations, a historical survey', *Rocky Mountains Math. J.* **6**, 409–434.

Strang, G. and J. Fix (1973) *An Analysis of the Finite Element Method*, Prentice–Hall, Englewood Cliffs, NJ.

Sun, X. (1992) 'Norm estimates for inverses of Euclidean distance matrices', *J. Approx. Th.* **70**, 339–347.

Sun, X. (1993a) 'Solvability of multivariate interpolation by radial or related functions', *J. Approx. Th.* **72**, 252–267.

Sun, X. (1993b) 'Conditionally positive definite functions and their application to multivariate interpolations', *J. Approx. Th.* **74**, 159–180.

Utreras, F.I. (1993) 'Multiresolution and pre-wavelets using radial basis functions', in *Multivariate Approximation: from CAGD to Wavelets*, K. Jetter and F.I. Utreras (eds.), World Scientific, 321–333.

Wahba, G. (1981) 'Spline interpolation and smoothing on the sphere', *SIAM J. Sci. Stat. Comp.* **2**, 5–16.

Wendland, H. (1995) 'Piecewise polynomial, positive definite and compactly supported radial functions of minimal degree', *Adv. Comp. Math.* **4**, 389–396.

Wendland, H. (1997) 'Sobolev-type error estimates for interpolation by radial basis functions', in *Surface Fitting and Multiresolution Methods*, A. LeMéhauté and L.L. Schumaker (eds.), Vanderbilt University Press, Nashville, Tenn., 337–344.

Wendland, H. (1998) 'Error estimates for interpolation by radial basis functions of minimal degree, *J. Approx. Th.* **93**, 258–272.

Werner, H. and R. Schaback (1979) *Praktische Mathematik II*, 2nd ed., Springer, Berlin.

Widder, D. V. (1946) *The Laplace Transform*, Princeton University Press, Princeton, NJ.

Wiener, N. (1933) *The Fourier Transform and Certain of Its Applications*, Cambridge University Press, Cambridge.

Williamson, R.E. (1956) 'Multiply monotone functions and their Laplace transforms', *Duke Math. J.* **23**, 189–207.

Wu, Z. (1992) 'Hermite–Birkhoff interpolation of scattered data by radial basis functions', *Approx. Th. Appl.* **8**, 1–10.

Wu, Z. (1995a) 'Characterization of positive definite radial functions', in *Mathematical Methods for Curves and Surfaces*, T. Lyche, M. Daehlen and L.L. Schumaker (eds.), Vanderbilt University Press, Nashville, Tenn., 573–578.

Wu, Z. (1995b) 'Multivariate compactly supported positive definite radial functions', *Adv. Comp. Math.* **4**, 283–292.

Wu, Z. and R. Schaback (1993) 'Local error estimates for radial basis function interpolation of scattered data', *IMA J. Numer. Anal.* **13**, 13–27.

Wu, Z. and R. Schaback (1996) 'Operators on radial functions', *J. Comp. Appl. Math.* **73**, 257–270.

Xu, Y. and E.W. Cheney (1992) 'Interpolation by periodic radial functions', *Comp. Math. Appl.* **24**, No. 12, 201–215.

Yosida, S. (1968) *Functional Analysis*, Springer, New York.

Index